New Directions in Bioprocess Modeling and Control

Maximizing Process Analytical Technology Benefits

Second Edition

New Directions in Bioprocess Modeling and Control

Maximizing Process Analytical Technology Benefits

Second Edition

Gregory K. McMillan
Christopher Stuart
Rehman Fazeem
Zachary Sample
Timothy Schieffer

Notice

The information presented in this publication is for the general education of the reader. Because neither the author nor the publisher has any control over the use of the information by the reader, both the author and the publisher disclaim any and all liability of any kind arising out of such use. The reader is expected to exercise sound professional judgment in using any of the information presented in a particular application.

Additionally, neither the author nor the publisher has investigated or considered the effect of any patents on the ability of the reader to use any of the information in a particular application. The reader is responsible for reviewing any possible patents that may affect any particular use of the information presented.

Any references to commercial products in the work are cited as examples only. Neither the author nor the publisher endorses any referenced commercial product. Any trademarks or tradenames referenced belong to the respective owner of the mark or name. Neither the author nor the publisher makes any representation regarding the availability of any referenced commercial product at any time. The manufacturer's instructions on the use of any commercial product must be followed at all times, even if in conflict with the information in this publication.

Printed in the United States of America.
Version: 1.0

ISBN-13: 978-1-64331-104-3 (print)
ISBN-13: 978-1-64331-105-0 (ePub)
ISBN-13: 978-1-64331-106-7 (Kindle)

ISA
67 T. W. Alexander Drive
P.O. Box 12277
Research Triangle Park, NC 27709

Library of Congress Cataloging-in-Publication Data in process

Dedication

This book is dedicated to everyone working in the process industry who is seeking to advance process performance by gaining a greater understanding of first-principle process relationships and the opportunities to improve automation system dynamics and strategies. The authors are especially appreciative of practitioners who are willing to invest the time to use the digital twin and first-principle process simulations with automation dynamics to develop a deeper knowledge and cultivate innovation to get the most out of the largely untapped capability of today's measurements and control systems to achieve consistent and reliable process control improvements.

Contents

Acknowledgments

I am extremely fortunate to have been in a uniquely positive creative environment for my entire 50-year career. I have worked with exceptionally talented professionals who were industry leaders in particular areas of expertise, many at the forefront of using modeling for better understanding and development of process control opportunities. I was also given the freedom to explore, discover, detail, implement, and document advancements. The subsequent publication of what I had learned was encouraged, which is increasingly rare for users compared to producers of automation systems. This culture in my tenure in Monsanto's Engineering Technology (ET) department was nurtured by its director and creator, the late Dr. James Fair, professor emeritus at the University of Texas. Steady-state modeling was the focus for process and mechanical design improvement and dynamic modeling for process control improvement for gaining deeper knowledge and fast, flexible testing. As far as the sources of my accomplishments, I estimate 50% are from models, 20% from conversations with experts, and 30% from field experience.

Upon retirement from Solutia, the spinoff of Monsanto's chemical business, I was fortunate to continue to explore the synergy of modeling and control working for Mark Nixon, director of DeltaV Research and Development (R&D), one of the great minds in developing new control system capabilities. I was able to explore and document the benefits of automation tools using batch profile slopes, wireless devices, and analyzers for more intelligent and productive batch and continuous control.

Working part-time in the development and demonstration of dynamic models for key unit operations and control system enhancements, particularly for the bioprocesses, has been a great finale to my career thanks to Nobin William, vice president of Simulation R&D at Emerson Process Systems and Solutions, who provides extraordinary support and knowledge on dynamic simulation. This is complemented by a positive work environment with creative energy and talent particularly exemplified by the coauthors and vice presidents Dave Sextro and Mart Berutti.

I greatly appreciate the model prototyping by Dean Cook, with advanced capability as seen in Appendix M, and the bioreactor model implementation with control systems in a digital twin by Timothy Schieffer, Ricardo Dunia, and Sean Sullivan at Emerson Process Systems and Solutions. I also appreciate Texas A&M University using the prototype model (noted in Chapters 6 and 9) plus new future models in a digital twin.

I also thank the cofounder of the ISA Mentor Program, Hunter Vegas, and the program resources Brian Hrankowsky, Luis Navas, and Hector Torres and longtime friend Owen Campney for excellent technical reviews. Finally, I thank my spouse Carol, who has supported all my endeavors for more than 33 years.

Gregory K. McMillan

Preface

The breakthroughs since the last edition of this book can potentially revolutionize advances in bioprocess performance. The modeling breakthroughs include new simple equations that can be easily initialized and continuously improved to model the effect of the concentrations of all components, temperature, and pH on cell growth rate and product formation rate. The automation breakthroughs include an enhanced proportional-integral-derivative (PID) controller that enables the use of at-line and off-line analyzers for closed-loop control growth rate and production rate. The enhanced PID controller is simple to implement, easy to tune, and amazingly resilient to measurement failures and increases in the analysis time. To support these breakthroughs are blocks to model automation system dynamics, measure system performance, compute batch slopes, predict endpoints, and rapidly estimate PID tuning settings. To provide a better focus on the individuality of technologies, equations, figures, and tables use the section numbers.

About the Authors

Gregory K. McMillan is a retired senior fellow from Solutia and an ISA Fellow. He is presently working part-time as a senior principal software engineer in Process Simulation Development at Emerson Process Systems and Solutions. McMillan was an adjunct professor in the Washington University Saint Louis Chemical Engineering department from 2001 to 2004. He received the ISA "Kermit Fischer Environmental" Award for pH control in 1991, received the *Control* magazine "Engineer of the Year" Award for the Process Industry in 1994, was inducted into the *Control Magazine* "Process Automation Hall of Fame" in 2001, was honored by *InTech Magazine* in 2003 as one of the most influential innovators in automation, and received the ISA Life Achievement Award in 2010. He received the ISA Mentoring Excellence Award in 2020. He is the author of numerous ISA books on process control, his most recent being *Advances in Reactor Measurement and Control* and *Good Tuning: A Pocket Guide, 4th Edition*. McMillan has been the monthly "Control Talk" columnist for *Control* magazine since 2002. McMillan is the founder and co-leader with Hunter Vegas of the ISA Mentor Program for industry practitioners and started the ISA 5.9 Standards and Practices committee on PID Algorithms and Performance. He has a bachelor of science degree in engineering physics from Kansas University and a master of science degree in control theory from Missouri University of Science and Technology all with a process emphasis.

Christopher Stuart is a senior software engineer for Emerson's Process Simulation Development group. He holds a bachelor of science degree in chemical engineering with a biochemical emphasis from Missouri University of Science and Technology. Stuart's research interests are in efficient algorithms for thermodynamic calculations and dynamic simulations.

Rehman Fazeem is an advanced chemical engineer at Honeywell's UOP, formally known as Universal Oil Products, in the UOP Process Technology (UPT) division. His areas of expertise are process design, simulation, equipment design, process control, automation, chemical reaction technology, and catalyst deactivation. Fazeem has more than 10 years of experience across process industries such as oil and gas, refining, petrochemicals, chemicals, paper and pulp, and pharmaceuticals. Before Honeywell, he worked at Emerson as lead project engineer in the Process Simulation group under the Process Systems and Solutions (PSS) division. He has a master of science degree in chemical engineering from Missouri University of Science and Technology.

Zachary Sample has more than seven years of experience delivering cross-industry simulation solutions in various roles ranging from technical implementation and engineering to project and sales management. In these roles, he has had a passion for helping process companies better leverage simulation to improve automation and operational performance. Sample has a bachelor of science degree in chemical engineering from Missouri University of Science and Technology and a master of science degree in chemical engineering from North Carolina State University.

Timothy Schieffer is a lead project engineer in Emerson's Process Systems and Solutions with more than 5 years of experience building first-principle dynamic simulations. He has helped implement the bioreactor model with the latest advances running at several hundred times real time in a digital twin. Schieffer has a bachelor of science degree in chemical engineering from Missouri University of Science and Technology.

1 Opportunities

1.1 Introduction

The cell is the ultimate reactor. It is the essential and, in some cases, the only possible route to the production of complex compounds [1].

Bioreactors (fermenters) are the key units of operation in biopharmaceutical, brewing, biochemical, biofuel, and activated sludge processes. Each bioreactor relies on the performance of billions of these individual reactors or cells. Process control attempts to influence the sophisticated metabolic reactions inside the cell by controlling the environment immediately outside of the cell [2].

Process control helps regulate and make repeatable the individual sophisticated internal reactions of billions of cells by controlling their extracellular environment.

To control a process variable, we must measure it directly or infer it from other measurements. The key process variables of a cell's environment that are measured are compositions and conditions, such as temperature, pH, dissolved oxygen (DO) and carbon dioxide, and substrate. In process control, the compositions and conditions in the broth or media and vent gas are termed *process outputs*, and the quantity, composition, and condition of feeds or charges into the bioreactor or of coolant into the coils or jacket are termed *process inputs*. Seed cultures, nutrients, substrates, air, and oxygen are process inputs. Understanding process inputs versus process outputs and the transfer of variability is important. Process inputs are not only the manipulated flows of streams entering the process but are also the pH, temperature, composition, and physical properties of the streams, which may not be constantly creating disturbances. Process outputs are created in the process as a result of kinetics and the material, energy, and charge balances. Many process inputs and process outputs are unfortunately not measured. In basic feedback control, process outputs that are measured or inferred are control system inputs maintained at set points by manipulating control system outputs that become process inputs. The goal of process control is to transfer variability from important process outputs to the process inputs designed to be manipulated. How well the process outputs are controlled is determined by process design and control system design, nonidealities, and dynamics, such as loop dead time, measurement resolution, repeatability, noise, the tuning of the controllers, and the resolution of the control valves and variable speed drives [3].

Section 1.2 of this chapter discusses the major sources of variability, the definition of process inputs and outputs, the availability of measurements, the differences between experimental and first-principle models, the quality and quantity of data, and the different modes of batch operation. Section 1.3 discusses the different levels of control and the effect each has on the selection of model inputs and on the setup of basic feedback loops for different types of cell cultures. Section 1.4 introduces the important topics of online yield and capacity performance indicators. Section 1.5 outlines the use of model predictive control for optimization. Section 1.6 summarizes the reasons behind, the drivers of, and the tools used for the Process Analytical Technology (PAT) initiative.

Models offer benefits before they are put online. It is the authors' experience that significant improvements result from the process knowledge and insight that are gained when building the experimental and first-principle models for process monitoring and control. The benefits often come in ways not directly attributable to the associated technologies, such as changes in batch set points, end points, and phases.

Significant improvements result from the process knowledge and insight that are gained when building the experimental and first-principle models for process monitoring and control.

Doing modeling in the process development and early commercialization stages is advantageous because it increases process efficiency and provides ongoing opportunities for improving process control. When benchtop and pilot plant systems use the same industrial control systems and configuration expertise that are employed in manufacturing, applications of modeling and control can be developed as an integral part of the process definition and ported for industrial production via the control definition. The advanced technologies discussed in this book—model predictive control

(Chapter 4), the digital twin (Chapter 5), first-principle models (Chapter 6), analytical technologies (Chapter 7), and data analytics (Chapter 8)—are important tools for maximizing the benefits from process modeling and control. The synergistic discovery of knowledge is consistent with the intent behind the Process Analyzer and Process Control Tools sections of the Food and Drug Administration (FDA) *Guidance for Industry: PAT—A Framework for Innovative Pharmaceutical Development, Manufacturing, and Quality Assurance* document, as the following excerpts show.

The synergistic knowledge discovered through the integration of analyzers, models, and controls is the essence of the PAT opportunity.

In the Process Analyzer section of the FDA's guidance document, we read:

For certain applications, sensor-based measurements can provide a useful process signature that may be related to the underlying process steps or transformations. Based on the level of process understanding, these signatures may also be useful for process monitoring, control, and end point determination when these patterns or signatures relate to product and process quality.

In the Process Control Tools section of the guidance document, we read:

Strategies should accommodate the attributes of input materials, the ability, and reliability of process analyzers to measure critical attributes, and the achievement of process end points to ensure consistent quality of the output materials and the final product.

Design and optimization of drug formulations and manufacturing processes within the PAT framework can include the following steps (the sequence of steps can vary):

- *Identify and measure critical material and process attributes relating to product quality*
- *Design a process measurement system that allows real-time or near real-time (e.g., on-, in-, or at-line) monitoring of all critical attributes*
- *Design process controls that provide adjustments to ensure all critical attributes are controlled*
- *Develop mathematical relationships between product quality attributes and measurements of critical material and process attributes*

Within the PAT framework, a process end point is not a fixed time; rather it is the achievement of the desired material attribute…Quality decisions should be based on process understanding and the prediction and control of relevant process/product attributes.

The discovery and implementation of more optimal batch profiles, end points, and cycle times are encouraged by the PAT guidance document. The proper integration of technologies can speed up rather than slow down the time to market. Ideally, the optimization process is coincident with the commercialization process. Waiting until a pharmaceutical is out of patent protection increases the risk of not having the lowest cost position. The return on investment (ROI) gained by optimizing the production process for a new drug can be enough to justify the use of analyzers, models, and more intelligent control. The goal of this book is to provide the basis for taking advantage of this opportunity.

Making the optimization process coincident with the commercialization process can lead to a low-cost position, an important factor for a competitive advantage when a drug loses patent protection.

Learning Objectives

- Be able to track down the major sources of variability
- Recognize how instruments aggravate or mitigate variability
- Understand how to reduce variability through different levels of control
- Know the basic setup of loops for different types of bioreactors
- Appreciate the implications of missing online and infrequent lab measurements
- Become familiar with the requirements and capabilities of various models
- Understand how the level of control affects the selection of model inputs
- Discover optimization opportunities
- Be aware of the business drivers for the Process Analytical Technology (PAT) initiative

1.2 Analysis of Variability

In the pulp and paper industry, poor feedback control increases the variability of important process outputs. This phenomenon is rare for bioreactors because of the large volume, the slowness of the kinetics, and the lack of interactions. However, the performance the bioreactors achieve still depends on how well the controllers are set up and tuned to deal with the sources of variability. Most feedback control implementation problems involving bioreactors can be traced to nonrepresentative measurements or mechanical design limitations in equipment, agitation, piping, and injection.

The bulk velocity in a bioreactor for animal cell cultures is extremely low and ranges from 0.01 to 0.1 ft/s (dimensionless scale of agitation ranges from 0.1 to 1.0, and power input per unit volume ranges from 10 to 100 W/m^3). For bioreactors with fungi and bacterial cultures, the degree of agitation is comparable to chemical reactors with gas dispersion: bulk velocity is 0.8 to 1.0 (ft/s) or larger (scale of agitation ranges from 8 to 10 and power-per-unit volume is 1000 W/m^3 or more) [4, 5, 6]. For ethanol, the scale of agitation ranges from 1 to 2 in the bioreactor and 2 to 4 in the mix tanks.

Low fluid velocities result in larger mixing time delays, less broth or media uniformity and gas dispersion, and lower mass transfer rates. The response time of a clean electrode and thermowell is several times larger in an animal cell culture. A more important consideration is the increased propensity for coatings caused by low fluid velocity. A 1 mm coating on a pH measurement electrode can cause its response time to increase from 10 s to 7 min [7]. Chapter 2 on process dynamics discusses how these time delays and time lags add up to a total loop dead time that determines the loop's ultimate performance.

A coating on the reference electrode junction can cause a drift of 0.5 pH or more during the batch. Because the peak in the growth rate of the biomass with pH is relatively sharp, the pH drift can affect batch performance.

Even if electrodes could be withdrawn, cleaned, and calibrated during a batch, the result would be disruptive because the time required for equilibration of the reference junction in the broth or media is significant [7]. Thus, even if the optimum pH does not change during the batch, there is an opportunity to counteract a drift in measurement by slowly trimming the pH set point. In continuous processes, it is quite common for operations to home in on a more optimum set point to compensate for sensor offset. However, in batch processes, the practice is to use fixed set points.

The low fluid velocities in animal cell bioreactors increase the total loop dead time and the need for optimizing the pH set point.

Load cells or mass inventories are frequently used as a standard way of adding some materials. Many believe that load cells and mass inventories are accurate standard technology, but the reality is that they are often the source of significant variability. To use load cells, you must be sure that all piping and other connections to the vessel offer no vertical forces on the vessel (this includes operators who lean on the vessel when looking through the tank site glass). This is often difficult to achieve in reactors because many steam sterilization and air/nutrient feed lines are connected to them. Also, changing ambient temperature conditions causes metal fittings and/or the support structure to expand and contract and temperatures in the jacket cooling medium to change, which then changes the medium's density, weight, and so on. These changes can potentially change a load cell measurement unless the load cell is carefully engineered not to be affected or the change is adjusted for by correcting the model [8].

Poor control loop performance in bioreactors can often be traced back to nonrepresentative measurements or mechanical design limitations.

Totalized Coriolis mass flowmeters are used instead of load cells to achieve higher accuracy in charge measurements. These meters also provide an extremely precise density measurement for verifying the composition of charges. In the case of ethanol production, the density measurement on the feed to the fermenter provides an inference of the percentage of starch and sugar solids. Adding both a differential pressure measurement across the meter for laminar flow and a neural network produces a viscosity measurement that is an inference of starch-to-sugar conversion from the liquefaction process. Accurate mass flow measurements also facilitate an accurate mass flow ratio of feed to recycle flow at various points in the ethanol process.

A higher level of control, termed *optimization*, occurs when cell growth and product formation rates, which are indicative of intracellular reaction rates, are inferred from oxygen uptake rates (OURs) and carbon dioxide production rates (CPRs). Mass spectrometers can be used to provide online indications of the OUR and CPR by way of measuring the oxygen and carbon dioxide in the off-gas [1, 10]. Alternatively, the air or oxygen feed can be momentarily shut off and the rate of change in the DO used to infer the OUR. For ethanol production, the simple addition of a pressure measurement at the bottom of the fermenter enables us to infer the CPR from the broth's or media's loss in weight. These rates potentially can be made more consistent and maximized at various points in the batch by adjusting the set point for key process outputs, such as substrate concentration, dissolved gas concentration, and pH [8, 9]. Increasingly, substrate concentration is being measured and controlled to determine and optimize cell growth and product formation rates.

Most of the variability in the product or effluent can be tracked down to unknown variability in the process inputs. Measurable changes in seed cells are reduced by not transferring the seed tank until appropriate criteria are met. The process inputs that vary the most and with the greatest consequences are the unknown differences in the internal composition and metabolic pathways of cells. It is important to remember that a cell is the most sophisticated reactor known, and genetic engineering is still in its early stages. Fortunately, a digital twin that includes the actual control system configuration and operator interface with models of the process and automation system dynamics added as described in Chapter 5 can achieve its objectives using relatively simple expressions for the kinetics and mass transfer rates, with the parameters fit to experimental data and combined with equations for the material, charge, and energy balances.

We describe these procedures in Chapter 6 on first-principle models.

The causes of variability in the seed cells are analogous to the causes of variability in people: the current and historical environment of the culture and genetics (heredity) [8]. In the case of environment, we must consider the entire environment that the cell culture has been exposed to because the cell culture began as a few frozen cells in a liquid nitrogen storage container. These cells must be carefully thawed in a controlled fashion, and then go through several growth stages in different progressively larger equipment sets, typically lasting many days. During these stages, the cells must be given all the appropriate nutrients and be controlled at the appropriate temperature, pH, and the like—and all of this is performed in a sterile environment.

These "early" steps are typically not as well controlled as the fermenter or bioreactor itself because they entail more manual operations. Therefore, the opportunity for excursions, operator mistakes (i.e., for variability), is significant [8].

There is a low, but finite, probability that a cell will mutate while dividing to become two cells. With each succeeding generation of cells, the probability increases that one or more of the cells has mutated. Note that by the time a culture is making product in a fermenter, perhaps 30 or more generations have occurred since the start of a few frozen cells. A mutated cell theoretically might produce the same or more product than a standard cell, but in most cases it produces less (or perhaps none).

Companies try to develop industrial processes to limit the number of cell generations because this minimizes concerns regarding cell mutations. The FDA sometimes checks on this during its reviews. Progressive mutation is one of the reasons bioprocesses are not typically run continuously (i.e., for extremely long periods of time) to maximize the probability that the culture will remain in a relatively pure state [8].

This is not to say that other bioreactor inputs cannot cause significant variability. The use of complex raw materials as nutrient sources, such as soybean meal or fish meal, is one of the most obvious significant sources of variability. This is the case when a process shifts from one ocean fish harvest or soybean harvest to another. However, these kinds of changes cause sudden shifts in the productivity of all bioreactors making a product. The changes in nutrients align with any new lot of raw material being used rather than varying in a random lot-to-lot fashion, which might be associated with instrument noise or drift [8].

Process Inputs and Outputs

Figure 1-2 shows many of the process inputs and outputs that are possible for a bioreactor. Process inputs are typically quantities of reagents, substrates, nutrients, gases, enzymes, cells, or surfactants added to the vessel and cooling water circulated through

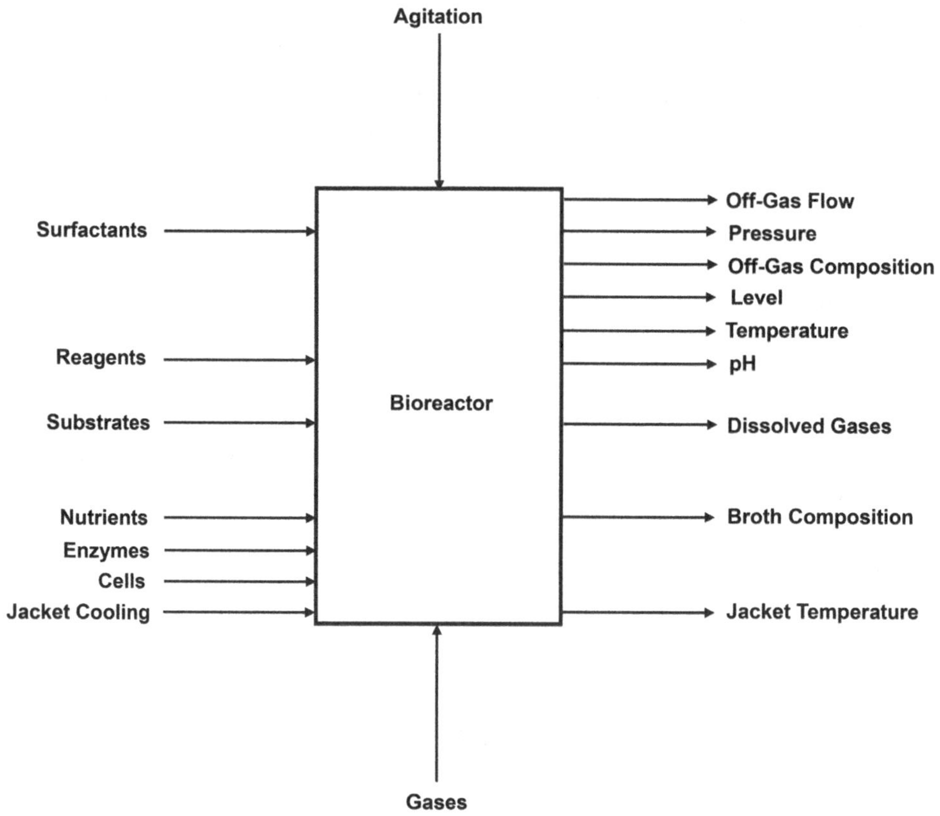

Figure 1-2. Bioreactor process inputs and outputs.

the jacket. The flows of these process inputs are set by a final element, such as a pump or fan speed. Another type of process input is agitation set by motor speed.

There are also unknown and often undesirable inputs such as impurities and contaminants. The process inputs determine the process outputs, such as broth or media and jacket temperature, pressure, pH, dissolved gases, broth or media composition, and off-gas composition and flow [1, 10].

Field and Laboratory Measurements

Even though the process outputs are indications of bioreactor performance, only the simpler measurements, such as temperature, pressure, pH, and differential pressure, are typically installed on bioreactors. This is particularly true in the brewing and biofuel industries. The next most common measurement is DO for aerobic processes. Dissolved carbon dioxide electrodes have significantly improved since their appearance in the 1980s and are increasingly being used. In particular, their use is increasing for animal cell cultures when the pH is controlled by adding carbon dioxide, for the aeration and stripping of carbon dioxide in the broth or media by an air sparge, or for the sweeping of carbon dioxide from the head space by an inert gas [1]. Companies seeking higher levels of control add "at-line" measurements of both components, such as glucose probes for substrate concentration and optical density instruments for biomass concentration.

Pharmaceutical manufacturers take advantage of sophisticated types of off-gas analyses online. These analyses are generated by sending gas sample streams through field mass spectrometers that have a much better resolution, repeatability, reliability, and response time than other analyzers. Figures in Chapter 7 illustrate inline analyzers with sensors inserted into the process and at-line analyzers with sensors automatically measuring samples automatically extracted that should be used in labs, pilot plants, and production units. Microprocessors and software have made possible diagnostics, computation capability, and user interfaces that greatly improve the mass spec's functionality and maintainability [1, 10].

Bioreactors do not have enough field measurements of process outputs particularly via in-line or at-line analyzers.

The most important process output, broth or media composition, is measured in the laboratory. When this measurement is done in a laboratory, the time delay between when the sample was taken and when the analysis result is available to the control system is large and inconsistent. So too is the time interval between data points. The infrequency and variability in the timing of laboratory analysis makes it difficult to correlate process outputs with process inputs and to construct experimental models, such as partial least squares (PLS) and artificial neural network (ANN) models. This infrequency and timing variability also make it nearly impossible to use the results for basic control directly. Inferential measurements (virtual sensors), which are discussed here and in Chapter 5, can provide rapid as well as more reliable measurements with less noise in broth or media composition, such as sugar and biomass concentration.

The laboratory measurement data of the most important process output, broth or media composition, is so sparse and delayed that conventional PLS and ANN modeling is difficult, making it unlikely that the lab data can be used directly for basic process control.

Experimental Models

Process models can be used to prevent abnormal situations, to detect bad batches, to experiment (exploring what-if scenarios), to prototype control systems, to check out configurations, to train personnel in operations, to predict process performance (batch end point, yield, and/or cycle time), and to optimize in real time.

Experimental models correlate a series of process outputs to process inputs. The two major types of experimental models used in industry (model-predictive control—MPC, and data analytics) are discussed in detail in Chapters 4 and 8, respectively. Some important fundamental conceptual differences distinguish these two models.

ANN models predict a process output as a nonlinear function of process inputs with adjustable time delays. PLS models in data analytics software predict a process output from a linear combination of a reduced set of independent variables called *latent variables* or *principal components,* which are themselves a linear combination of process inputs with adjustable time delays that are critical for synchronization of process inputs with process outputs in continuous processes. The principal component analysis (PCA) eliminates correlations between process inputs. Both ANN and PLS are primarily steady-state methods reformulated for batch processes. In batch processes, the insertion of a time delay on each input for dynamic compensation is not needed for prediction of batch endpoints. The piecewise linear fit of a PLS model is also able to deal with the nonlinearities of batch profiles. Consequently, data analytics potentially has more readily usable capability in batch processes.

The PLS models in data analytics are better able to predict batch end points than other experimental models because of a piecewise linear fit to nonlinear batch profiles, elimination of correlations between inputs, and no requirement for insertion of time delays on PLS model inputs.

Normally, the ANN and PLS process inputs and outputs are not incremental and do not have a time constant or process gain [11]. MPC provides a future trajectory (time response) for the change in a process output based on the linear combination of the effects of past changes in process inputs. The trajectory includes the effect of a steady-state gain and time constant for self-regulating (exponential) responses or an integrating gain for integrating (ramping) responses, besides a time delay. Model identification software can be used to develop online dynamic estimators (ODE) with future trajectories for use in MPC [11].

ANN models predict nonlinear behavior, PLS and PCA models eliminate correlations between inputs, and MPC and ODE models provide the exponential and ramping time response for deviations about operating points.

To be capable of identifying a good quality model, the important inputs must have known changes that are at least five times larger than the noise associated with the input. Also, at least five changes in each input are needed in both directions at different process conditions, production rates, and batch times. In other areas of process control, sufficient quantity and quality of data are ensured by conducting step or bump tests and pseudo random binary sequences (PRBSs), in the case of MPC, or by designing experiments, in the case of PLS or ANN. It is rare that historical data can be completely relied on to generate these experimental models because there is usually not enough excitation of measured process inputs. Experimental models require extensive measured changes in process outputs for measured changes in process inputs [3, 11].

To build good experimental models, there must be at least five changes in each important input in both directions at different operating points that are each at least five times larger than the noise band.

Because each bioreactor batch may be worth hundreds of thousands to millions of dollars (for high value-added products), opportunities for introducing perturbations are severely restricted. Because each batch takes days to weeks to complete, even if perturbations could be made within the specification limits set by the plant's procedures for validating and managing change procedures, it might take months to years before enough data points are available to develop an MPC, PLS, or ANN model.

It takes months to years to get enough batches with identifiable changes in process inputs that are large enough to develop an MPC, PLS, or ANN model.

Processes that are subject to federal Current Good Manufacturing Practices (CGMPs), such as those for medicines, are required to define all critical process parameters (parameters that could affect product quality) and the corresponding "proven acceptable ranges." For fermentations and bioreactors that make pharmaceuticals, parameters such as pH and temperature are typical critical process parameters. In these cases, it would not be permitted to deliberately perturb the pH or temperature outside the "proven acceptable range" (which is a fairly narrow range of values for many fermentations) just for the sake of obtaining model information. The reason is that such process pH and temperature perturbations would be defined as process deviations that could cause the production lot to be rejected for the purposes of making product for market [2, 8].

In the early applications of PCA to batch processes, a *golden batch* for multivariate statistical process control (MSPC) was defined to be the median of the batch profiles for a fixed set of inputs. This definition is misleading because it is not representative of the best possible theoretical or practical batch, and it assumes fixed process inputs that may not be achieved in plant operation. The best batch profile must be found by theory and experimentation for any process input in order to meet the control objectives.

Chapter 8, "Data Analytics," discusses the relative merits of using multiway and model-based PCA to generate reference batch profiles.

The "golden batch" is not the median of the batch profiles but is the best batch profile achievable based on theory and experimentation for a variable set of inputs.

There is a multiplicative effect for biological process kinetics that creates restrictions on experimental methods to analyze or predict cell growth or product formation. While the incentive is greater for high-value biologic products, there are challenges with models of biological processes due to multiplicative effects (neural networks and data analytic models assume additive effects). Almost every first-principle model has a specific growth rate and product formation rate that is the result of a multiplication of factors each between 0 and 1 to detail the effect of temperature, pH, DO, glucose, amino acid (e.g., glutamine), and inhibitors (e.g., lactic acid). Thus, each factor changes the effect of every other factor. One way to understand this is by realizing that if the temperature is too high, cells are not going to grow and may in fact die. It does not matter if there is enough oxygen or glucose. Similarly, if there is not enough oxygen, it does not matter if all the other conditions are fine. One way to address this problem is to make all factors as close to 1 and as constant as possible except for the one of interest. Data analytics can be used to identify the limitation and/or inhibition first-principle model parameter for one condition, such as the effect of glucose concentration via the Michaelis-Menten equation if all other factors are constant and nearly 1.

First-principle dynamic models in a digital twin can identify the golden batch and deal with the multiplicative effect plus the issues of not having sufficient data due to long batch times, concerns about creating a bad batch, and lack of knowledge of the acceptable range of process parameters.

The severe restrictions on experimental methods due to the multiplicative effect of kinetics, long batch times, extreme cost of bad batches, and lack of knowledge of acceptable operating ranges can be addressed with digital twin.

Dynamic First-Principle Models

An emerging opportunity is the use of dynamic first-principle models in digital twins. These models have accurate mass and energy balances and kinetics, which include the activities of the nutrients and cells to address many of the issues created by the lack of sufficient test and analysis data [11]. These models can provide real-time inferential measurements of broth or media composition without the noise and delay associated with analytical results. The models can be run faster than real time for rapid testing and for designing experiments. The results have many uses, including discovering the best batch profile for the current set of inputs, developing PCA logic for batch performance monitoring, and identifying MPC, PLS, and ANN models. The models can fill in the remaining profile of "in progress" batches for MSPC and predict key performance indicators (KPIs), such as batch end points, yield, and cycle times.

Dynamic first-principle models in a digital twin can be used to develop KPI, "golden batch" profiles, batch performance metrics, and MPC, PLS, and ANN models.

Inferential measurements from even the best experimental and first-principle models do not eliminate the need for composition measurements. Field analyzers provide more frequent data with better timing than laboratory analysis. This can reduce the model development and verification time from years to weeks. Initially, these measurements are needed to identify experimental and first-principle model parameters. Chapters 4 and 5 present techniques that show how these models can be used to provide a future (predicted) composition. The inferential measurement improves the reliability and reduces a process composition's noise and dead time for feedback control and optimization [11]. However, drift and unknowns are inevitable, and feedback correction is essential. A delayed and lagged composition value is synchronized with a field or laboratory measurement, and a portion of the difference is used for feedback correction of the future value [11]. A field analyzer or a probe, as simple as a dissolved carbon dioxide electrode or turbidity, can provide a full correction within a batch phase. Due to the extreme length and value of batches, the correction within a batch by online inferential measurements is a great contrast to laboratory measurement, where multiple batches are generally needed to get enough samples to approach a full correction.

Experimental and first-principle models do not diminish but instead accentuate the value of field analyzers, and vice versa.

Manual, Sequential-Batch, Fed-Batch, and Continuous Operation Modes

Besides determining the availability and accuracy of measurements of the process inputs and outputs, the mode of operation also determines the selection, implementation, and performance of modeling and control methods. The main modes of operation are manual, sequential-batch, fed-batch, and continuous. These modes affect how much variability is introduced into the process inputs, which in turn has an impact on process outputs and batch profiles.

In the manual mode, the operator starts and stops the addition of the process inputs. Typically, the concentration and duration of the amount added (batch charge) is specified, but the actual timing is inconsistent because of the differences in operator knowledge, skills, attentiveness, and goals. The batch record for pharmaceuticals enforces the timing as much as humanly possible, but this may not be the case for biofuel production, where manual enzyme and nutrient addition and yeast pitching are practiced.

Manual operation introduces variability into process inputs.

In the sequential-batch mode, a batch control system automatically prepares and sequences the addition of process inputs. Frequently, the charge is a totalized flow or a weight from a tank on load cells. However, the concentration of constituent raw materials and the prepared charge are seldom measured online. In biopharmaceutical processes, a lab analysis is performed to confirm the composition of raw materials.

However, lab procedures may not exist for measuring all the properties that affect a cell metabolism. In particular, not all the properties of corn are measured that affect the enzyme charges and batch cycle time for ethanol and the nutritional content and flowability of the by-product of distiller's dried grains and solubles (DDGS) sold as animal and fish food. The variability of starch and nutrient content, which depends on field and weather conditions, has a major impact on the performance of an ethanol plant. For the brewing industry, the quality of the water, such as alkalinity, affects the pH and taste of the product [12].

Sequential-batch operation reduces the variability of the mass and timing of process inputs but not necessarily the variability in the concentrations of charges.

Trace quantities of nutrients or contaminants may not be recognized as having an important effect on biological processes. The internal structure of the cells is not measured and is seldom even modeled. Consequently, the variability in the intended and undesired components of process inputs is not completely addressed by sequential-batch operation. Online analyzers and laboratory analysis typically focus on extracellular composition.

For processes that are subject to government CGMPs, such as in pharmaceutical manufacture, the raw material lots that the processes use must be assayed before the processes start to verify "fitness for use." In some cases, a guaranteed "certificate of analysis" from the supplier is sufficient. In general, scientists who are developing a manufacturing process must determine what parameters can affect product quality, which includes raw material concentrations and purities. After those parameters are identified, the process must be tested for these parameters before it is implemented [8].

In the brewing, wine, and biofuel industries, awareness is growing that variability in the composition and quality of raw materials (e.g., hops, grapes, and corn) caused by location, weather, and water should be measured [12].

The largest suspect source of variability in well-automated batch and continuous operations are changes in the composition of inoculums and recycle streams, and to a lesser degree, reagents and substrates for the pharmaceutical industry and raw materials and water for other industries.

In fed-batch operation, the flows of some of the process inputs are manipulated by a process controller. In these cases, variability in a process output is transferred to a process input. Common examples of fed-batch operation are the use of a pH controller to manipulate a reagent flow and a DO controller to manipulate an air or oxygen flow.

Installing a glucose probe or using a first-principle model to inferentially measure glucose opens up opportunities for a concentration controller to manipulate substrate feed. Fed-batch operation is sometimes called *semi-continuous* because control loops continuously manipulate process inputs. However, fed-batch is still a batch operation because the discharge flow is zero until the end of the batch.

For continuous bioreactors, there is a continuous discharge flow from the bioreactor and continuous flows of process inputs. When the bioreactor is running in the production mode, there may also be some recycle of effluent to recover raw materials or to do further conversion. There may be a start-up sequence and some of the same issues associated with batch sequences. Once the bioreactor reaches operating conditions, the focus shifts to throughput rate, load disturbance rejection, and process degradation as a result of contamination and recycle. Although continuous bioreactors can significantly increase capacity, practical issues have limited their use in industry primarily to treating activated sludge and producing biodiesel fuels. A few industrial biopharmaceutical processes are essentially continuous, although they are better known as *perfusion* processes. These are characterized by continuous feeds and continuous withdrawal of broth or media (containing product). They run for several weeks or months per batch. However, they still eventually must be stopped and cycled to start a new batch because of mutation and other concerns [8].

1.3 Transfer of Variability

The variability of key process outputs, typically the composition and quality of products, is important in any plant. Theoretically, reducing product variability allows some plants to run at higher capacity or better efficiency by operating closer to constraints. However, there is a natural tendency not to run very close to product quality constraints in pharmaceutical batch processes because a single excursion across such constraints could cause an entire lot to be deemed unacceptable for market, thereby wasting weeks of time and effort.

Another key factor in reducing variability is the number of runs needed in plant trials to evaluate suggestions for process modifications. For example, a typical bioprocess plant's coefficient of variation (COV) that is 100% times the ratio of the standard deviation to the mean is about 10%, which is somewhat higher than the track record for small molecule chemical processes for pharmaceuticals. Most proposed plant modifications (nutrient changes, culture strain changes, etc.) have a suggested improvement of about 5%. The number of production fermentation runs needed to statistically verify that a 5% improvement is occurring, in an environment where plant COV is 10%, will be high. If plant variability could be improved such that COV is 5%, then significantly fewer expensive and time-consuming production fermentations would be needed to determine whether the suggested improvement was real or not. Typically, nearly all proposed improvements are initially done at pilot scale. However, many changes do not scale well to production-size equipment (a common problem for bioprocesses), which means that production plant trials are also needed [8]. Note that improvements do not automatically translate into dollars. Often a set point or mode of operation must be changed to move the operating point closer to the optimum to realize the benefits [13].

For manual operation, operators' adjustments are a major contributor to product variability. To reduce this source of variability, batch automation software provides sequences that fix the timing and quantity of each process action. The remaining product variability can be attributed to changes in the cells and feeds and to the progression of the batch itself.

Adding good feedback control, such as a well-tuned proportional-integral-derivative (PID) controller, controls a process output, such as pH or DO at a set point, by manipulating process inputs, such as reagent and gas flows. Changes that would have occurred in pH or DO now appear as changes in reagent and gas flow. Note that it is not possible to fix both the process outputs and process inputs, and that attempts to fix all process inputs by discrete actions will not compensate for changes in the quantity, quality, and the composition of process inputs. Feedback control offers plants the ability to compensate for unknowns by transferring variability from process outputs (controlled variables) to process inputs (manipulated variables). When the changes (disturbances) are measured, feedforward control can be added to feedback control to preemptively change a process input. This is achieved by modeling the relationship between the measured disturbance and the manipulated process input [3].

Modeling and process control open new ways to minimize variability in a process's efficiency and capacity. Trying to predict and fix all the optimum steps in a profile of process inputs is infeasible. Instead, process control offers plants the opportunity to transfer variability in process outputs of direct or indirect value to process inputs and to provide a continuous profile that compensates for unknowns.

Discrete Process Actions

Often when manufacturing is faced with variability in process outputs, it chooses to focus completely on fixing process inputs. Operations and process design departments often request discrete process actions or steps to add specific amounts of reagents, gases, and substrate or they set cooling rates at specific times in the batch to anticipate batch requirements. "If ... then ..." is the predominant mode of thinking.

The predominant solution chosen by plants to reduce variability in batch operation is to set process inputs (perform discrete process actions) on a predictable time schedule.

For continuous operation, there is theoretically a steady state, but even here unknown disturbances, such as influent composition and recycle, cause the process outputs to perpetually move. For batch operations, predicting input requirements is even more problematic in that there is no steady state, as evidenced by the batch profiles. The reagent demand, OUR, CPR, and substrate consumption have a profile that is linked to the biomass and product profiles. The actual amount of a process input that is needed is a moving target in a batch, and feedback correction (i.e., closed-loop control) is needed to compensate for unknowns in concentrations and biological mechanisms. Preprogrammed process actions (i.e., open-loop control) may work well in laboratories and some pilot plants, and they can help a control loop get started (by initializing the controller outputs). However, preprogrammed process actions should not interfere with the proper operation of feedback controllers [14, 15]. Most of the conflicts between discrete actions and loops can be traced to a lack of fundamental understanding of the "what, how, when, and why" of basic feedback control for batch operations that have a moving target. For example, batch sequences may position a control valve in the remote output (ROUT) mode where the PID controller is effectively in manual at various points in the batch to achieve a preprogrammed feed, coolant, or reagent profile. However, it would be better in this case to change the set point in a remote cascade (RCAS mode where the PID controller is effectively in automatic) and to allow the PID controller to correct for unknowns and disturbances through feedback control. This requires that configuration specialists understand the dynamics and controller tuning for batch profiles. To provide this understanding, Chapters 2 and 3, on dynamics and tuning (respectively), discuss in detail the implications of the integrating response of batch processes compared to the self-regulating response of continuous processes.

Process actions can be useful to get a control loop started, but they should not retain too much control and interfere with the operation of basic feedback control.

Feedback Control

It is important that the feedback control be allowed to do its job. Not only is the immediate performance better, but over the long haul the performance will be more sustainable. The definition of feedback control means that the control system is by nature self-correcting to some extent and thus able to deal with changing conditions. However, its effectiveness depends on the tuning of the controllers. Signal characterization and the scheduling of tuning settings can help compensate for the nonlinearities and changes in automation system and batch dynamics. The feedback control task is usually separated into two levels. The basic or lower level control may use PID controllers. To account for interactions and operating constraints and facilitate optimization, MPC may be used for the advanced or upper level [3, 11]. In each level of feedback control, manipulated variables are adjusted to keep controlled variables at a set point or target.

Feedback control corrects for unknowns and unidentified disturbances and eliminates the many knowledge and timing limitations of preset, sequential, and manual adjustments. Feedback correction greatly reduces the variability in key process variables that are selected to be the controlled variables.

Feedback control transfers variability in process outputs to process inputs to compensate for disturbances or differences in processing conditions [3]. For bioreactors, it is desirable to maximize this transfer so the variability in process outputs is minimized. For example, consider the following three situations: no feedback control, basic control (PID), and advanced control (MPC). Figure 1-3a shows a preprogrammed schedule of steps in the process inputs (reagent flow and oxygen-enriched airflow). The process outputs, pH substrate concentration, and DO vary with batch demand and

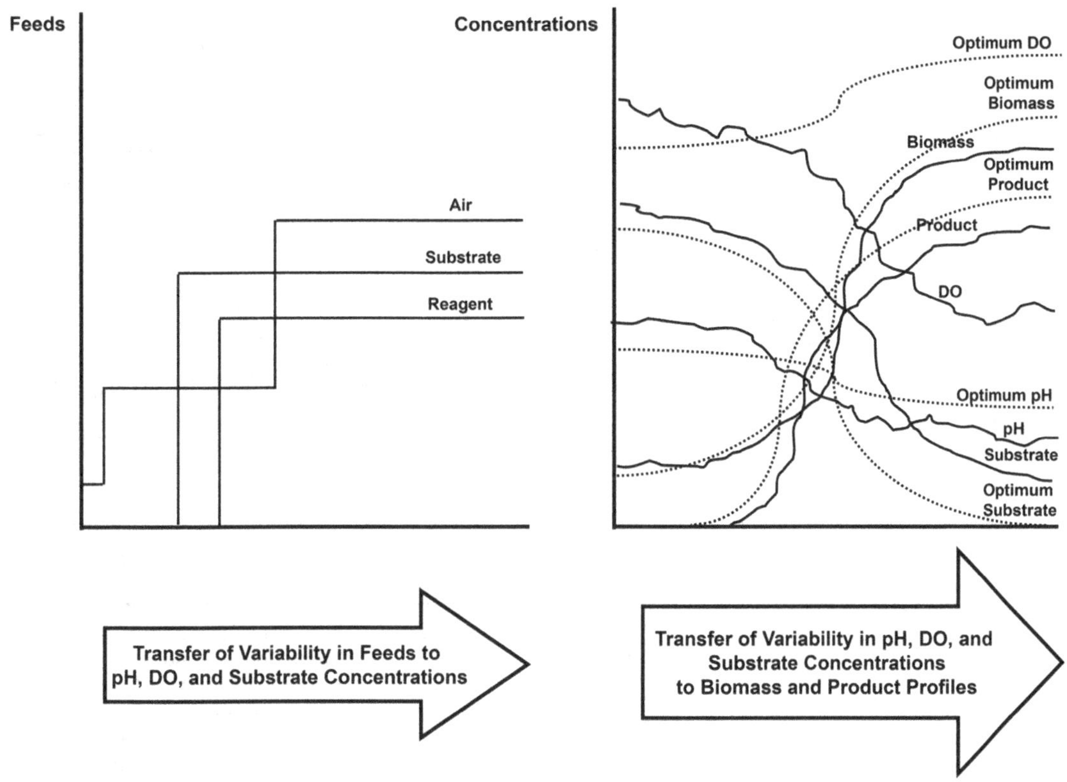

Figure 1-3a. Bioreactor inputs and outputs for no feedback control.

with changes in the processing and concentrations of these and other process inputs. Also, the average values of process outputs are offset from the optimum because of bias errors in the measurements and unknowns in the kinetics and the process inputs. In Figure 1-3b, PID loops manipulate the reagent and oxygen-enriched airflow to draw straight lines on pH and DO. Notice that there is still an offset from the optimum values of pH and DO and that the other process outputs, biomass and product concentration, still have variability and an offset compared to the best batch profile (golden batch) for the current set of inputs. In Figure 1-3c, an MPC whose set points are based on the golden batch profile manipulates the substrate and DO set points to provide a more uniform and optimum profile in the biomass and product concentrations [9]. Using an MPC has been demonstrated in two cases involving DO control to do a better job than a PID controller in dealing with noise and dead time. Chapter 4, "Model Predictive Control," details how the development of the digital twin and an innovative translation of controlled variables have addressed most of the practical problems that have hindered biomass and product profile optimization to date. Figure 1-3c also shows how the pH may be trimmed to account for pH drift or for the pH requirements during different phases in the fermenter, such as the saccharification, growth, and product formation phases for ethanol production.

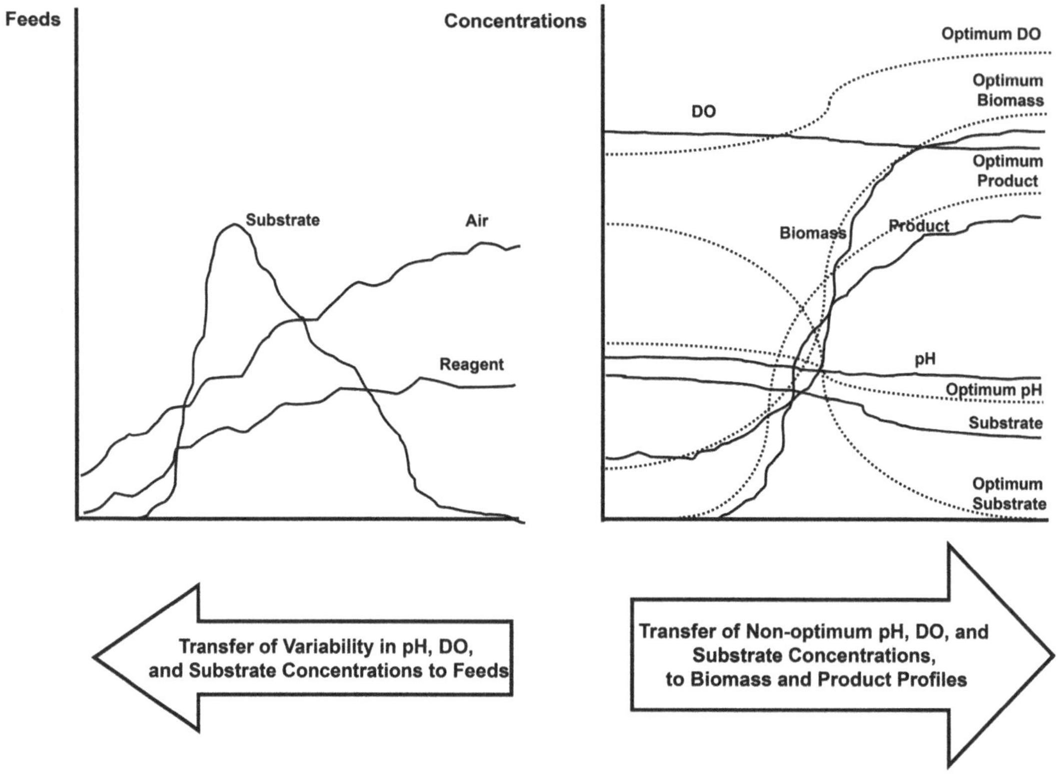

Figure 1-3b. Bioreactor inputs and outputs for basic PID control.

In summary, good batch manager software is essential. When it is combined with good measurements and final elements, manufacturers can more effectively track down the sources of variability and implement varying levels of process control to transfer variability from process outputs to process inputs. Transferring variability is the key to identifying abnormal batches and achieving more optimum and consistent operating points, profiles, and end points.

Selecting MSPC, PLS, and ANN Inputs

The transfer of variability profoundly effects which inputs and outputs are appropriate for MSPC batch analysis and for PLS and ANN models, yet this is a topic that is not usually discussed in literature or recognized by the user. For pH control, the input for MSPC, PLS, and ANN would change from pH to reagent flow once the loop is put in the automatic mode (i.e., the loop is closed).

> ***Feedback control transfers variability from process outputs to process inputs and profoundly affects which inputs are chosen for MSPC batch analysis and PLS and ANN models. As loops are closed, the***

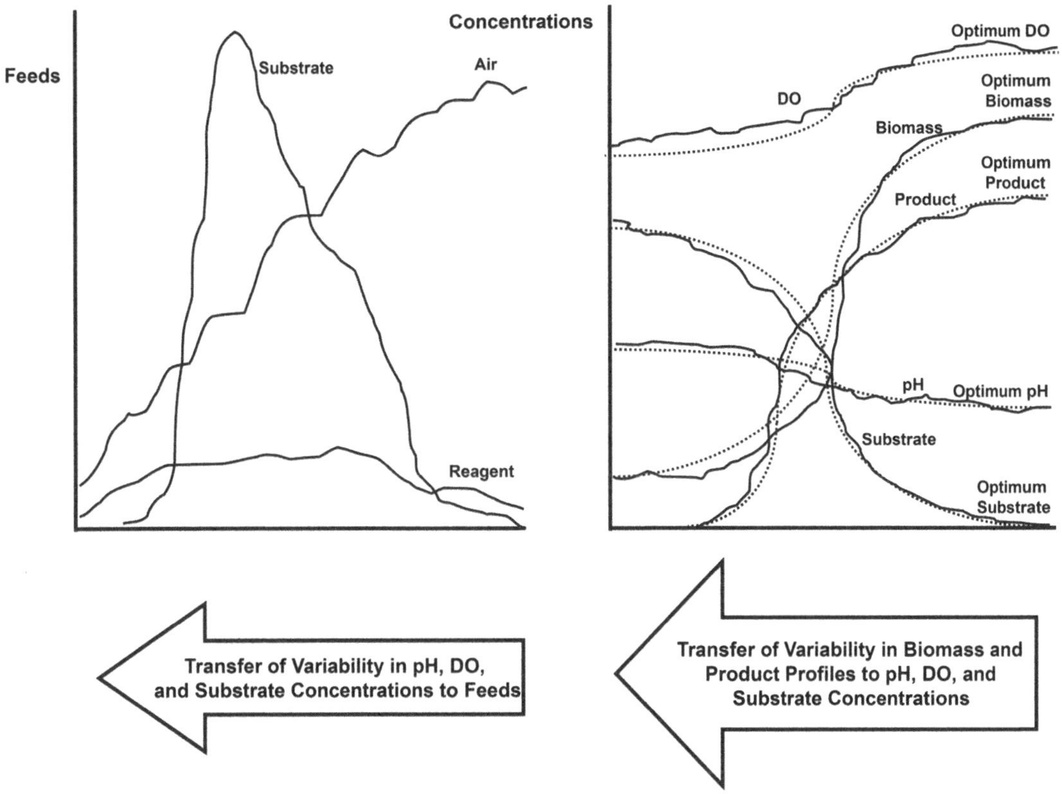

Figure 1-3c. Bioreactor inputs and outputs for advanced MPC control.

MSPC, PLS, and ANN inputs move from controlled variables to manipulated variables.

Setting Up Feedback Loops

How the loops for a bioreactor are set up will depend on the product and the degree of sophistication needed to meet market requirements.

However, it is useful to consider how the different types of cell cultures affect the selection of the process outputs to be controlled and the process inputs to be manipulated. For example, Figure 1-3d shows the setup of the basic feedback control loops for a fungi cell culture used for antibiotic production. The temperature controller manipulates the jacket temperature, the pH controller manipulates ammonia reagent flow, the DO controller output is split to manipulate airflow and agitation based on the load, and finally the glucose concentration controller manipulates the glucose feed rate. If these loops have fixed set points, the MSPC, PLS, and ANN inputs would be the PID controller outputs (i.e., manipulated jacket temperature, ammonia flow, airflow, pressure, agitation, and glucose flow). If the set points of these control loops were

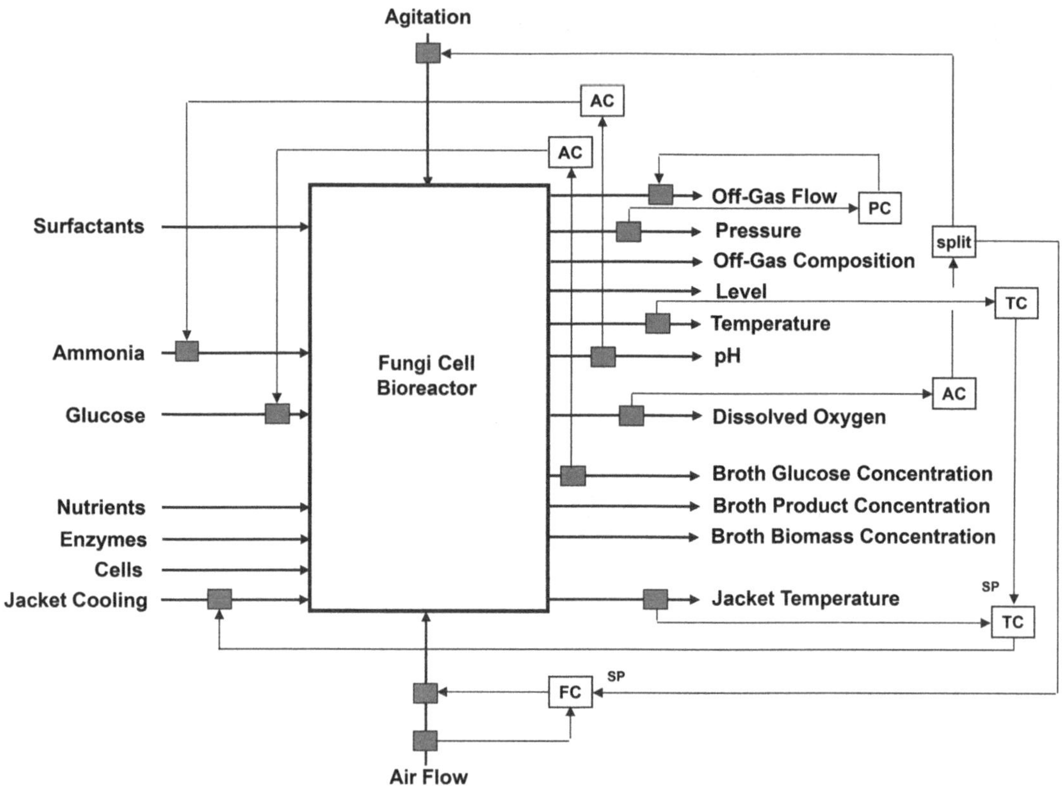

Figure 1-3d. Basic feedback loops for a fungi cell culture for antibiotic production.

continually manipulated for more optimal operation by an MPC, then the MSPC, PLS, and ANN inputs would be the MPC outputs (i.e., manipulated temperature, pH, DO, and glucose concentration set points). The seed culture charge concentration and timing and the off-gas concentrations of components, such as oxygen and carbon dioxide, are important process inputs and outputs for analysis and modeling. They can also be used to infer the OUR and CPR. The MSPC, PLS, and ANN outputs nearly always include the biomass concentration and/or growth rate and the product formation rate and/or concentration because these are essential to a bioreactor's operation.

An increase in head pressure increases the equilibrium concentration for DO via Henry's law and therefore increases the driving force for mass transfer. However, pressure may not be manipulated for DO because it also increases the concentration of dissolved carbon dioxide, which is usually not measured.

Figure 1-3e shows the setup of the basic feedback control loops for a simultaneous saccharification and fermentation of a yeast cell culture used to produce ethanol from corn. The loops are similar to Figure 1-3d except that the reagents are different and there is no DO control.

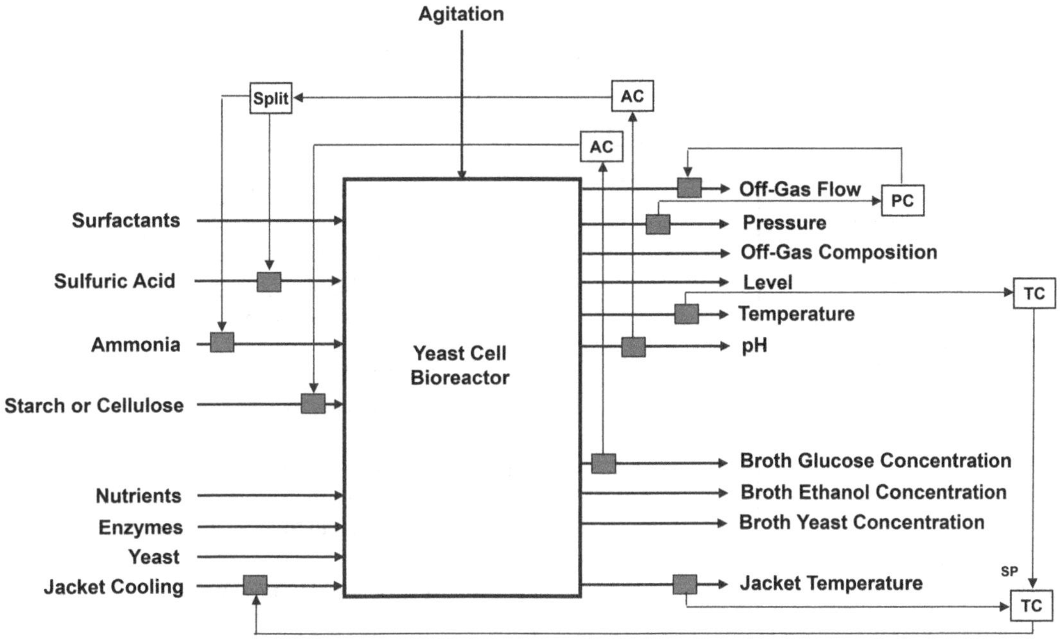

Figure 1-3e. Basic feedback loops for a yeast cell culture for ethanol production.

The substrate is a starch or cellulose that is converted into glucose (saccharification) by an enzyme and then fermented to ethanol by yeast. In this case, the glucose concentration controller manipulates the liquefied starch or cellulose feed. This feed pH is controlled by adding ammonia and sulfuric acid. The nitrogen, phosphorous, minerals, and trace elements in the feed are balanced by adding various chlorides, phosphates, sulfates, and yeast extract. In the bioreactor, the yeast, ethanol, and enzyme concentrations are important. Once the loops are closed, the MSPC, PLS, and ANN inputs become jacket temperature, ammonia and sulfuric reagents, and starch or cellulose feed. The amounts and timing of the enzyme and yeast charge (process inputs) and the off-gas carbon dioxide concentration (process output) would be important inputs for analysis/modeling.

Figure 1-3f shows the setup of the basic feedback control loops for a bacterial cell culture used in biopharmaceutical production. The setup of the loops is similar to Figure 1-3d except that the airflow may be enriched by oxygen. Acid does not need to be added for bacterial cell culture fermentations unless, perhaps, as a contingency to counter the mistaken addition of too much base. Bacterial cultures generate two sources of acid naturally: CO_2 production (carbonic acid) and acetate formation (acetic acid).

Figure 1-3g shows the setup of the basic feedback control loops for an animal cell culture used in biopharmaceutical production. The setup of loops is similar to Figure 1-3f except that the reagents are carbon dioxide and sodium carbonate or sodium

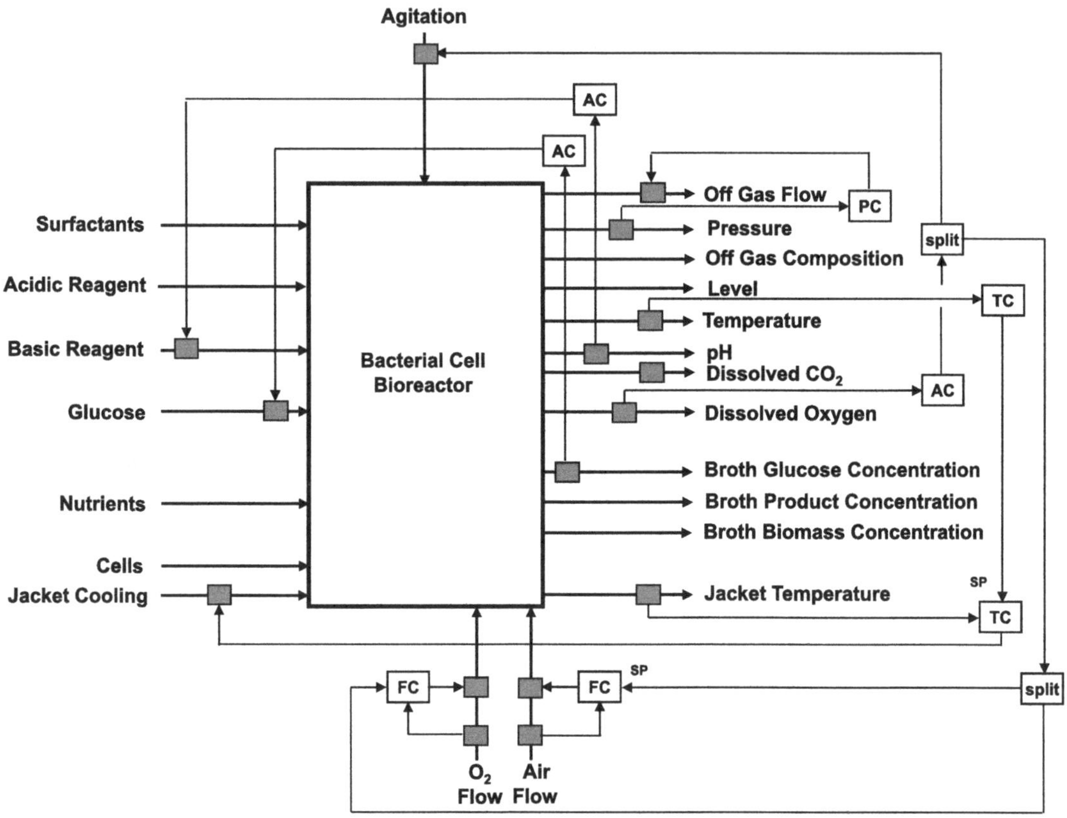

Figure 1-3f. Basic feedback loops for a bacterial cell culture for pharmaceutical production.

hydroxide, there are two substrates (i.e., glucose and glutamine), and actual serum and/or various amino acids, growth factors, vitamins, and minerals are added to duplicate a blood serum as closely as possible. Also, vessel pressure and agitation are not manipulated. An increase in vessel pressure increases the dissolved carbon dioxide concentration, and an increase in agitation causes cell lysis (breakage of the animal cells), which are more fragile because they do not have cell walls. Besides the usual biomass, substrate, and product concentrations, it may also be important to measure by-products, such as lactic acid, which indicate cell health. An enhanced PID to control batch profile slopes can deal with large and variable at-line analysis and lab analysis sample times without the need for retuning [19].

1.4 Online Indication of Performance

A KPI of a bioreactor's process efficiency is yield. It can be computed from the product mass at the end of the batch, which is the product end point concentration (mg/L). Though yield is commonly stated just in terms of mg/L of product, the fundamental definition of *yield* takes into account the batch volume and substrate charged. In this book, the yield is expressed in terms of a particular substrate and is computed as the

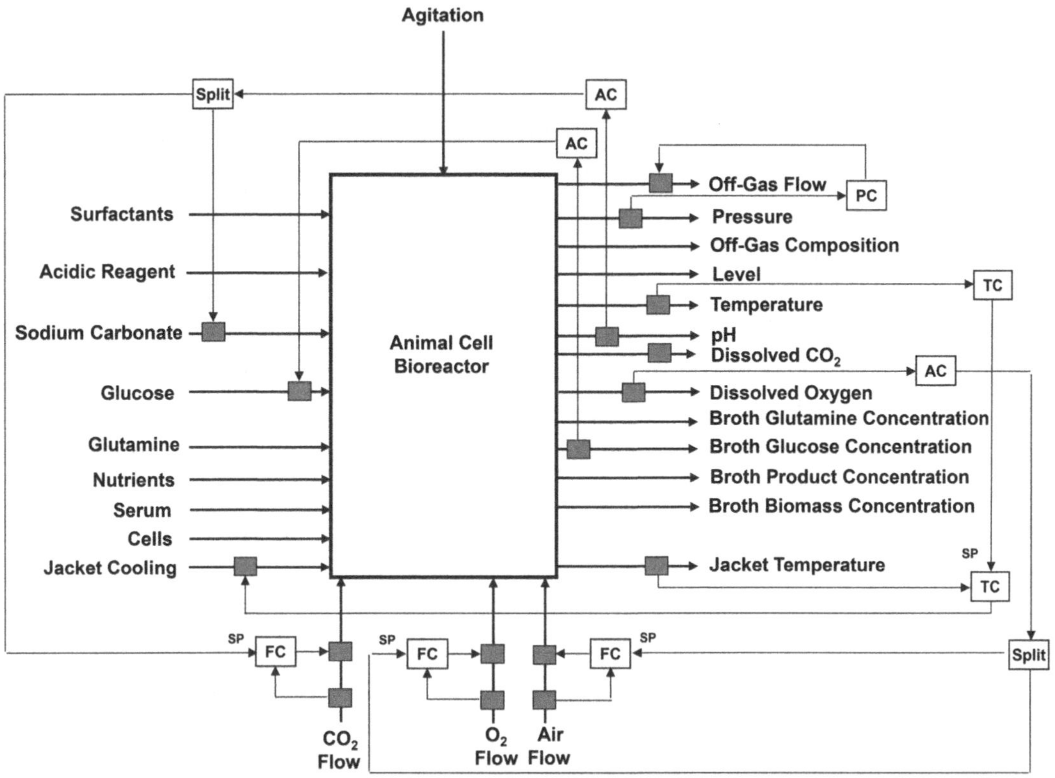

Figure 1-3g. Basic feedback loops for an animal cell culture for pharmaceutical production.

kg/L of product multiplied by the broth or media volume (L) and divided by the total mass of substrate charged (kg). This dimensionless number for yield can be compared to the yield coefficients used in the kinetic expressions. If the actual yield is divided by the yield coefficient and multiplied by 100%, then the relative yield becomes a percentage of maximum practical yield. This substrate yield is important because of the large quantities of substrate used. For animal cell cultures, the serum can be the largest material cost.

A KPI of a bioreactor's process capacity is production rate. It can be computed as the product mass at the end of the batch divided by the batch cycle time (kg product/h). The time spent holding the batch or waiting on other unit operations is important for plant capacity. However, an increase in the bioreactor capacity eventually shows up in additional plant capacity as these bottlenecks are identified and eliminated. Greater bioreactor capacity means more time for maintenance and process improvements. If the plant is not sold out, it also means more time that the plant can be idle. Higher product concentrations can also translate into reduced purification and waste treatment costs.

Greater bioreactor production capacity can result in improved maintenance and processes in addition to lower downstream processing costs.

Expressing bioreactor yield and production rate as a percentage of the maximum practical value demonstrated in industrial benchtop and pilot plant systems provides a benchmark and better scaling for recognizing changes in bioprocess efficiency and capacity. The prediction of yield and production rate is filtered to reduce noise and displayed on trend charts in the operations control room. Online trending of properly processed KPIs, such as reagent use, cell growth rate, death rate, and product formation rate, at key points in the batch improves the performance of operations, control system support, maintenance, and process support by increasing the plant's awareness of where it has been and where it is going. In Figure 1-4a, a simple linear program (LP) optimizer can identify an MPC's optimum operating point, which is the vertex intersection of operating limits for the manipulated variables (i.e., DO and glucose) and the controlled variables (i.e., biomass and product concentration). The optimum vertex depends on the value of yield and capacity with changes in market demand, operating costs, and process limits. If the vertex moves, a high-fidelity first-principle model and real-time optimization are needed to identify the new vertex.

Online trending of KPIs improves the technical support, operation, and maintenance of the bioreactor and its control systems.

Digital twins with dynamic first-principle models can predict KPIs and process variables as the batch progresses [20]. They do this by running faster than real time to create profiles into the future of yield, capacity, biomass, and product. Additionally, these digital twins running real time can be automatically adapted by a combination of model predictive control and neural networks to provide fast and reliable inferential

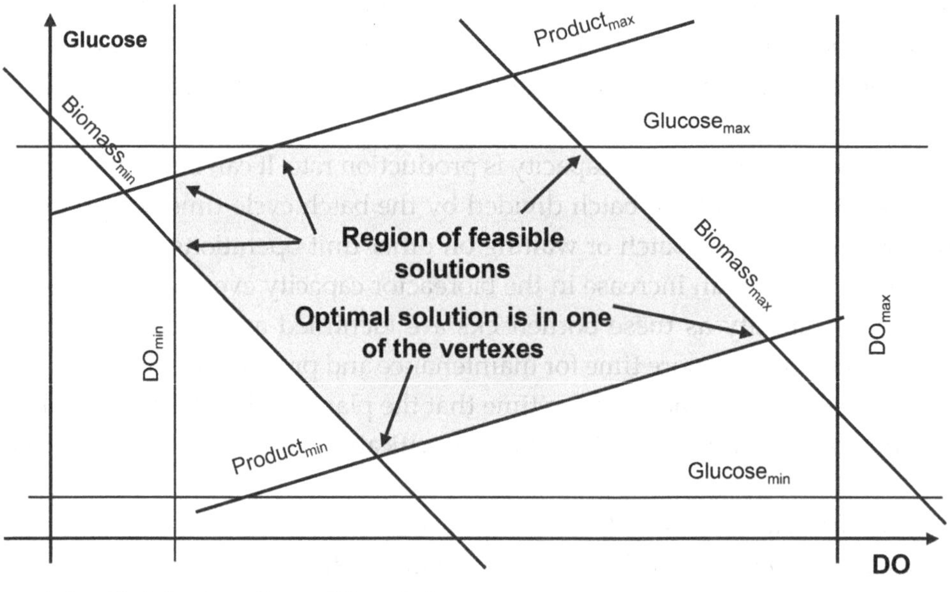

Figure 1-4a. Variation in optimum DO and glucose.

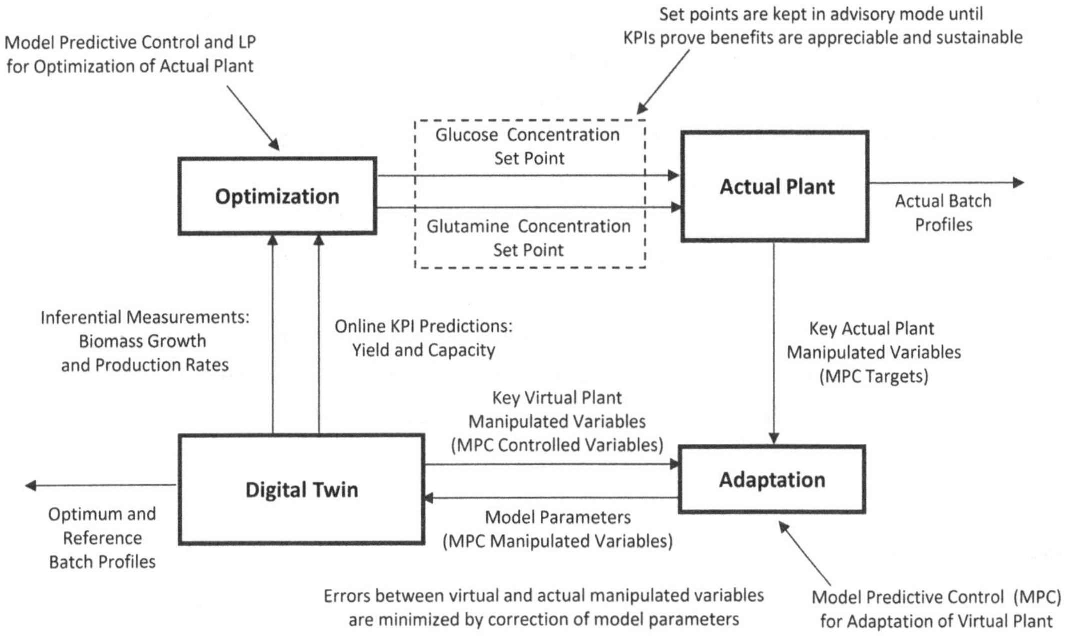

Figure 1-4b. Online KPI and inferential measurements from an adapted virtual plant (digital twin).

measurements that become the controlled and optimization variables. Figure 1-4b shows this general setup, which we discuss in Chapter 5, "Digital Twin."

An adapted digital twin can provide online predictions of yield and capacity as well as inferential measurements of growth and production rates for optimization.

The digital twin can also act as a knowledge base for exploring the opportunities for making additional measurements [20]. The digital twin can quantify and demonstrate the value of additional Coriolis meters for mass flow, density, and viscosity measurements as well as additional conventional and analytical measurements. For example, in ethanol production a direct density and inferential viscosity measurement of the laminar fermenter feed from a Coriolis meter and differential pressure sensor may provide important information about starch and dextrin content. This can be useful for optimizing the enzyme charge. Similarly, a pressure sensor on the bottom of the fermenter may provide an accurate inferential measurement of the loss-in-weight of the carbon dioxide evolution rate, which is useful for adapting the digital twin's kinetics.

1.5 Optimizing Performance

Inferential measurements of biomass, product, and substrate concentrations can be used as the controlled variables for model predictive control. The set points or targets for these concentrations are the golden batch profiles used to achieve production objectives. Alternately, inferential biomass growth rate and product formation rates can be used as the controlled variables to help remove the effects of bias errors. They can also be used to prevent the MPC from attempting to force growth for formation rates that are too fast or too slow while the MPC is attempting to match a profile. In this case, the set points or targets are the slopes of the golden batch profiles. In both cases, the penalties on error for the controlled variables for the MPC and an LP optimizer can set the relative importance of the maximization of yield versus capacity.

The use of biomass growth and product formation rates as MPC-controlled variables eliminates bias errors and avoids the risk of forcing rates that are too fast or slow.

1.6 PAT

Origins of PAT

In the 1950s, a new citric acid production facility had 50,000 L fermenters. Batches ran for a month, and the substrate was molasses from sugarcane. Forty years later the same plant produced a higher-quality citric acid using 150,000 L fermenters. Batches were completed in less than a week, and expensive cane sugar substrate was replaced by fructose from cornstarch.

The FDA has long held that quality cannot be tested into products. Quality must be built into a process. Traditionally, quality has been verified after the manufacture of a product. Process steps that are successfully operated in a repeatable manner are accepted as being validated. Generally, a process step is considered validated after three consecutive successful batches.

This methodology provides a high probability that a high-quality product will be attained, but the uncertainty of managing process changes through regulatory submissions has discouraged innovation. The reason behind the PAT initiative is to allow pharmaceutical manufacturers to continuously improve processes along with the rest of the chemical process industry and at the same time assure a high-quality product.

FDA Initiative

In an effort to increase the safety, efficiency, and affordability of medicines, the FDA has proposed a new framework for regulating pharmaceutical development, manufacturing, and quality assurance. The primary focus of the initiative is to reduce variability by gaining a better understanding of a process than can be obtained by the traditional approach. Note that PAT is a "guidance." PAT is not contained in the Code of Federal Regulations, and companies are not required by law to use PAT.

A draft report explaining the new regulatory framework was published in 2002 and titled [16] *Pharmaceutical CGMPs for the 21st Century—A Risk-Based Approach.*

Along with the final version of this report, industry guidance was published in September 2004. It was titled [17] *Guidance for Industry: PAT—A Framework for Innovative Pharmaceutical Development, Manufacturing, and Quality Assurance.*

Although PAT stands for *process analytical technology,* it has become an acronym for designing quality into a process. PAT is formally defined in section IV of the *Guidance for Industry*:

> *The Agency considers PAT to be a system for designing, analyzing, and controlling manufacturing through timely measurements (i.e., during processing) of critical quality and performance attributes of raw and in-process materials and processes, with the goal of ensuring final product quality. It is important to note that the term analytical in PAT is viewed broadly to include chemical, physical, microbiological, mathematical, and risk analysis conducted in an integrated manner. The goal of PAT is to enhance understanding and control the manufacturing process, which is consistent with our current drug quality system: quality cannot be tested into products; it should be built-in or should be by design.*

The focus of this new regulatory environment is on discovering process variation and controlling that variation when it might contribute to patient risk. The process variation is discovered by identifying and measuring critical quality attributes in a timely fashion. In this way, processes can be developed and controlled in such a way that the quality of product is guaranteed.

Being able to analyze the sources of process variability and control that variability provides engineers with a basis for better transitions from lab to production, promoting innovation including equipment and process control improvements.

Business Drivers

According to Pisano [18], a biopharmaceutical manufacturer that can develop a process rapidly and efficiently reaps the benefits of accelerated time-to-market; rapid ramp-up of yield, quality, and capacity from initial to normal long-term levels; and a stronger proprietary position. Although Pisano's comment was made in the context of the then-existing regulatory environment, the goal of the PAT framework is to help manufacturers reap just these benefits. The *Guidance for Industry* states:

> *Gains in quality, safety, and/or efficiency...are likely to come from:*
>
> - *Reducing production cycle times by using on-, in-, and/or at-line measurements and controls*
> - *Preventing rejects, scrap, and re-processing*

- *Real-time release*
- *Increasing automation to improve operator safety and reduce human errors*
- *Improving energy and material use and increasing capacity*

Facilitating continuous processing to improve efficiency and manage variability. For example, use of dedicated, small-scale equipment (to eliminate certain scale-up issues).

PAT Tools

Guidance for Industry PAT — A Framework for Innovative Pharmaceutical Development, Manufacturing, and Quality Assurance (Pharmaceutical CGMPs September 2004) states:

> *There are many tools available that enable process understanding for scientific, risk-managed pharmaceutical development, manufacture, and quality assurance. These tools, when used within a system, can provide effective and efficient means for acquiring information to facilitate process understanding, continuous improvement, and development of risk-mitigation strategies. In the PAT framework, these tools can be categorized according to the following:*
>
> - *Multivariate tools for design, data acquisition and analysis*
> - *Process analyzers*
> - *Process control tools*
> - *Continuous improvement and knowledge management tools*

PAT opens up opportunities for analyzers and for the integrated use of tools for analysis, control, and modeling to improve plant performance.

Process control as well as continuous improvement and knowledge management are the two PAT tool categories that are most applicable to the subject of this book: the study of bioreactor performance.

One type of process control is the optimization of KPIs. Examples of KPI discussed in this book are the predicted product yield and the predicted production rate of bioreactors. Basic feedback control loops are improved, and then dynamic first-principle models are integrated with model predictive control in a digital twin to ensure product quality while maximizing these KPIs.

1.7 Insights

Many challenges exist for biological processes as to the complexity and multiplicative nature of cellular response, regulation of change, separation of knowledge bases, and protection of proprietary knowledge. As a result, it is tempting to simply sequence flows that affect DO, pH, and cell growth. While it may seem like the right thing to do, based on traditional lab experimentation, it is unrealistic to expect that the timing and magnitude of flows can deal with the many unknowns. Although flow sequencing can be repeatable, the most important response of the controlled variables is not, and the batch behavior is confusing. By simply using PID control, repeatability of the controlled variables is achieved. Differences in batches can be seen in the manipulated flows that are automatically set to deal with change. It is ironic that in a misguided attempt to minimize change, flows are sequenced, which makes the batch much more vulnerable to change.

By using PID control, the source and nature of differences in batches can also be more readily diagnosed. The manipulated flows offer great knowledge as key parts of inferential measurements for cell growth rate and product formation rate as detailed in Chapters 6, 7, and 9.

The essential aspects of PID control are well within the realm of understanding of all professionals in research, development, design, support, and maintenance. The use of the digital twin and dynamic first-principle models can extend the knowledge to include the process. Flexible, fast, virtual experimentation can be done with the actual system configuration, historian, graphics, and alarms. The use of the digital twin in the lab, pilot plant, and production unit promotes research and development (R&D) achievements to be rapidly and readily carried over to industrial production. The time from the first successful lab experiment to the saleable product is greatly reduced. The sooner the saleable product is ready, the longer the new product will be on the market, which for high-value biologics translates to a huge gain in profits, especially considering patent time limitations.

Advances in at-line and in-line analyzers greatly enhance the fidelity and capability of models and the control system to maximize cell health and productivity. Advances in the computation of batch profile slopes by an innovative future value block (Chapter 9) have opened the door to optimization of profiles. MPC can be added to work with PID control. MPC is particularly suited to the optimization of profile slopes because of its patience in making changes, resilience to noise, minimization of interactions, knowledge of dynamics, and the sophistication of optimization.

Modeling and control provides a wealth of opportunities for all types of professionals in research and development and in the plant to make the most of their skills making innovation a reality.

1.8 Best Practices

1. Use a digital twin in the lab, pilot plant, and full-scale production plant for knowledge discovery, implementation, and operational consistency.

2. Develop a first-principle model using the generalized, easily set kinetics described in the completely new Chapter 6 to predict cell response without the need for proprietary knowledge.

3. Use in-line sensors (e.g., dissolved oxygen (DO) and carbon dioxide, dielectric spectroscopy, pH, and turbidity) with the best repeatability.

4. Use at-line analyzers (e.g., BioProfile FLEX, mass spectrometers, and near-infrared spectroscopy) with the best repeatability to provide feedback measurements of actual concentrations of all components, cell size, and cell health.

5. Use a PID controller to control DO, glucose, glutamine, pH, pressure, and temperature.

6. Use model predictive control (MPC) to optimize cell growth rate and product formation rate.

7. Use an innovative future value block to provide the slopes of batch profiles for cells and product, and to predict the end points with a good signal-to-noise ratio.

8. Use the same digital twin developed in research and development (R&D) for plant design, process control improvement, operator training, diagnostics, and performance monitoring by the automation engineer working with R&D, design, operations, and maintenance.

References

1. Gregory K. McMillan, *Advances in Reactor Measurement and Control* (Research Triangle Park, NC: ISA [International Society of Automation], 2015).

2. Joseph S. Alford, "Bioprocess Control: Advances and Challenges" (presentation at CPC-7 published in *Computers and Chemical Engineering* (2007)).

3. Gregory K. McMillan and P. Hunter Vegas, *Process/Industrial Instruments and Controls Handbook*, 6th ed. (New York: McGraw-Hill Education, 2019).

4. Shuichi Aiba, Arthur E. Humphrey, and Nancy F. Millis, *Biochemical Engineering*, 2nd ed. (Cambridge, MA: Academic Press, 1973).

5. "CE Refresher—Liquid Agitation," a 12-part series of articles from *Chemical Engineering* (1975–1976). Reprint, Dayton, OH: Chemineer (1985).

6. Alvin W. Nienow, Christian Langheinreich, Neil C. Stevenson, Nicholas A. Emery, Timothy M. Clayton, and Nigel K. H. Slater, "Homogenisation and Oxygen Transfer Rates in Large Agitated and Sparged Animal Cell Bioreactors: Some Implications for Growth and Production," *Cytotechnology* 22 (1996): 87–94.

7. Gregory K. McMillan and Robert Cameron, *Advanced pH Measurement and Control*, 3rd ed. (Research Triangle Park, NC: ISA [International Society of Automation], 2005).

8. Joseph S. Alford, e-mail correspondence, 2006.

9. Wilson Grant, Gregory K. McMillan, and Michael Boudreau, "PAT Benefits from the Application of Modeling and Advanced Process Control of Bioreactors," *Emerson Global Users Exchange*, Nashville (October 2005).

10. Joseph S. Alford, "Bioprocess Mass Spectrometry: A PAT Application," *Journal of Process Analytical Technology* (May–June 2006).

11. Gregory K. McMillan, "Process Modeling" and "Advanced Process Control," in *A Guide to the Automation Body of Knowledge*, 3rd ed., ed. Nicholas P. Sands and Ian Verhappen (Research Triangle Park, NC: ISA [International Society of Automation], 2018).

12. Dave Anderson, "Watch the Water," *InTech* 52 (November–December 2005).

13. Gregory K. McMillan and Stan Weiner, "The Bad Hall of Fame" (Control Talk *Column*), *Control* (March 2006).

14. Gregory K. McMillan and Stan Weiner, "Life Is a Batch" (Control Talk *Column*), *Control* (June 2005).

15. Gregory K. McMillan and Stan Weiner, "Grounded in Reality" (Control Talk *Column*), *Control* (December 2005).

16. *Pharmaceutical CGMPs for the 21st Century–A Risk-Based Approach*, final report, Department of Health and Human Services, US Food and Drug Administration (September 2004).

17. *Guidance for Industry: PAT—A Framework for Innovative Pharmaceutical Development, Manufacturing, and Quality Assurance*, Department of Health and Human Services, US Food and Drug Administration (September 2004).

18. Gary P. Pisano, *The Development Factory: Unlocking the Potential of Process Innovation* (Boston: Harvard Business School Press, 1997).

19. Gregory K. McMillan, "Wireless: Overcoming Challenges in PID Control & Analyzer Applications," *InTech* 57 (July–August 2010).

20. Gregory K. McMillan and Martin E. Berutti, "Virtual Plant Virtuosity," *Control* (August 2017).

2 Dynamics

2.1 Introduction

The dynamics of a process response can be modeled by three types of parameters: dead time, time constant, and open-loop gain. The open-loop gain, more commonly known as the *process gain*, can be for a self-regulating process or an integrating process. The dynamics are identified from the plant's operation, and they determine both the process's ultimate performance and the actual performance of the control system. The process dynamics are used to compute proportional-integral-derivative (PID) controller settings. They could also be employed directly to predict the trajectory of process outputs from past process inputs in future applications of model predictive control (MPC) for bioreactors. Although MPC technology has been used almost exclusively on continuous processes, recent batch applications are promising, and it is anticipated that bioreactor applications are possible.

Section 2.2 of this chapter discusses the largest sources of dead time in the automation and process systems, the impact of total dead time on the controller's capability and tuning, and the procedure for estimating the performance based on the tuning settings. Section 2.3 details the characteristics of a self-regulating process response, the way small time constants effectively become dead time, and the reasons the location of the largest time constant affects control. Section 2.4 delves into the integrating process response of the more important process variables as well as the implications of this response for process testing and controller tuning,

Learning Objectives

- Understand how dead time sets the ultimate performance limit and maximum possible aggressiveness of the controller tuning
- Learn how the design of equipment and automation systems affects dead time
- Become familiar with ways to estimate loop performance from tuning settings
- Be aware of how limit cycles originate from the final element resolution
- Learn the three basic parameters used to define the process response
- Recognize the differences between self-regulating and integrating responses
- Be able to convert self-regulating response parameters into an integrating parameter

2.2 Performance Limits

Impact of Loop Dead Time

Control-loop dead time is the period after a change in the controller output (manipulated variable) in which there is no recognizable change in the process variable (controlled variable). For feedback control, the controller output must change a process input to cause a change in a process output that is measured and seen by the controller.

The performance of the control loop is ultimately limited by the loop dead time, which is the time it takes to make and see a change. However, the actual performance is determined by the controller tuning. The dead time sets how aggressively a controller can be tuned without becoming unstable. In practice, there is always a trade-off between performance and robustness (i.e., the controller's relative ability to be stable despite changes in process dynamics). In the process industry, controllers are generally tuned with settings that are far from the performance limit allowed by process dynamics [7]. This is particularly true for the primary loops on bioreactors because the process variable's rate of change—as determined by a large process time constant or a small integrating process gain—is slow compared to the size of the process dead time.

The ultimate limit to controller performance is determined by the loop dead time, but the actual limit is usually set by the controller tuning.

The control loop must respond to changes in its set point and loads (disturbances). The disturbances can originate from changes to any of the process inputs, as discussed in Chapter 1, such as flows, charges, and intracellular metabolism and composition. Set point changes to primary loops, such as dissolved oxygen (DO), pressure, pH, and temperature, are rather minimal, although the optimum pH and temperature may change from the growth to production phases. However, the set points to the secondary loops (changes in gas or reagent flow, agitation, and coolant demand) are continual and cover a wide range. This is so they can deal with the one to three order-of-magnitude changes in biomass during the first half of the batch [1].

As MPC or real-time optimization (RTO) applications are developed, set point changes to the primary loops become more frequent. Fortunately, the bioreactor response to disturbances is slow because the volume is large compared to flow rates and the growth, product formation, and mutation rates are slow.

Sliding stem control valves are rarely used within the sterile envelope of the bioreactor. Here, the sterile envelope is defined (envisioned) as including the bioreactor and the piping network inside the sterile filters. The few control valves that are used here are easy to clean and sterilize [1]. One type available is a clean-in-place (CIP), sanitize-in-place (SIP), and self-draining, sliding-stem angle control valve with a packless design, metal diaphragm, and a low-friction roller-bearing mechanism [2]. A wider spectrum of valves is used outside the sterile envelope. A sterile filter is often used downstream of the valve to ensure that whatever is going into the fermenter is sterile.

The control valve's resolution and deadband should be determined per ANSI/ISA-75.25.01-2000 (R2010), *Test Procedure for Control Valve Response Measurement from Step Inputs,* and ANSI/ISA-TR75.25.02-2000 (R2010), *Control Valve Response Measurement from Step Inputs.* Although it is difficult to generalize, sliding stem valves that have a packless or low-friction design with digital positioners have a resolution limit of 0.05% to 0.1% and a deadband of 0.1% to 0.2%. Rotary control valves tend to have a larger resolution limit and deadband. Control valves without positioners have an order-of-magnitude or larger resolution limit and deadband [3, 4, 5].

A common divisive problem is on-off valves posing as throttling valves. The on-off valves look attractive because of their low leakage and high capacity. A common mistake for a variable speed drive (VSD) is to use the standard low-resolution input card and to add deadband and rate limiting in the inverter setup in a misguided concern for motor speed changes and overload. A generously sized totally enclosed fan cooled (TEFC) NEMA Class B motor with Class F insulation and a pulse-width modulated inverter with built-in cascade control of speed to torque should be used to maximize rangeability and responsiveness.

For the small actuators used on valves for bioreactor control, the prestroke dead time and stroking lag times associated with pressure changes in the actuator are negligible. The largest source of dead time in a valve response is the resolution limit and deadband whenever the rate of change is slow in the controller output. The dead time can be estimated as the resolution limit or deadband for changes in the same direction and opposite direction, respectively, divided by the rate of change of the controller output [3, 4, 5]. However, as the total size of the change approaches the resolution limit, the dead time gets much larger, particularly in pneumatic positioners [5]. Some digital positioners can be tuned to minimize this increase in dead time. The larger and faster changes in the controller's output from less sluggish tuning settings can minimize this dead time.

The control valve dead time from resolution and deadband can be minimized by using a digital positioner without integral action and higher gain.

The start of a change in flow as a result of a change in speed is almost instantaneous unless a deadband has been configured in the VSD to prevent a response to noise. The rate of change of speed might be limited in the VSD for motor load, but the rate limiting is usually fast enough so that it is not a consideration. In smaller-scale benchtop and pilot plant–size bioreactors, there is significant use of peristaltic pumps for liquid additions such as base and substrate. This is because it is easy to take the tubing out of the pump roller bar mechanisms and clean and sterilize the tubing associated with these feeds [1].

Impact of Controller Tuning Settings

In a single-loop configuration, the output of the process controller (e.g., DO, pH, or substrate) goes directly to a VSD or a control valve. In a cascade loop, the output of the primary process controller becomes the set point of a secondary controller. The most common secondary controller is the flow or speed controller. These secondary loops compensate for load changes and disturbances before they affect the primary process controller. However, the secondary controller response must be five times faster than the primary controller response, or interaction will develop between the two controllers. The secondary controller must be tuned for a fast response. The performance of a cascade control system that has a slow secondary loop may be worse than if the primary controller output directly manipulates valve position. This is the case for gas pressure control.

In general, for variable speed pumps, the speed controllers should be left in the variable frequency drive (VFD) rather than being brought into the control room so that the speed response is fast enough for a fast primary controller. There may also be a cascade of speed to torque control in the VFD that may be compromised if the speed control is moved. VSDs are generally, but not necessarily, VFDs.

The secondary controller must be tuned to be five times faster than the primary controller, or cascade control can do more harm than good.

In many distributed control systems (DCSs) and programmable logic controller (PLC) systems, the tuning settings for a PID controller are gain (K_c), reset time (T_i), and rate time (T_d). The gain setting is dimensionless, and the time settings are in seconds per repeat for reset and in seconds for rate. For reset time, the setting may be listed in terms of seconds with the seconds per repeat implied. In some controllers, proportional band in percent ($100\%/K_c$) is used instead of gain, and reset action in repeats per minute ($1/(T_i/60)$) is used instead of a reset time in seconds per repeat. Because these settings are inversely related to the previously cited settings, it is critical that users understand what type of tuning settings are used as well as their units and the effect of PID form. If a Series Form is not used, the user must make sure the rate time is appreciably less than the reset time to prevent oscillations. The Series Form inherently prevents the effective rate time from exceeding 1/4 the effective reset time through interaction factors in time domain. The effect of PID form and structure on performance is discussed in Chapter 3, "Basic Control."

The controller action is the opposite of the process action for negative feedback control if any reverse action of a final element is accounted for by a signal reversal. This reversal can be achieved in field devices, but it is preferable that it be done for maintainability and visibility in the blocks in the configuration, such as the *splitter* and the *analog output* (AO) blocks. The process action is the direction of the process variable change for a change in the manipulated variable (e.g., flow or speed). Note that direct and reverse action may be defined differently in different systems and that an "increase-to-close" (fail open) valve may not have a signal reversal built into the configuration or valve positioner requiring the configured control action to be the same as the process action.

Signal reversals typically occur when the controller output sets multiple flows (i.e., multiple manipulated variables) by either split-ranged or simultaneous action. Often for temperature control, the cooling valve is reverse-acting (fail open) and the heating valve is direct-acting (fail closed), and for gas pressure control, the vent valve is reverse-acting (fail open) and the nitrogen valve is direct-acting (fail closed). The signal reversal is normally done in *splitter* and AO blocks for split-ranged and simultaneous action, respectively. The output action can also be reversed in the controller block, but this is not the preferred practice unless there is no *splitter* and AO block.

Most industrial fermentations only use a base addition because the metabolic products are acidic in nature. However, in some applications a pH controller is split-ranged between acid and base flows where the signal is reversed for the acid flow. For example, the acid flow decreases, and when this flow reaches zero, the base flow increases as the controller output increases, which increases the pH (direct process action). In this case, the pH controller has reverse control action, and the "splitter" block reverses the signal for the acid flow in its output array specification.

Although a DO controller typically manipulates one process input (a single flow or agitation speed), in some applications a DO controller is set up to simultaneously manipulate air and oxygen flows to provide an oxygen-enriched gas flow. For example, the airflow decreases and the oxygen flow increases as the controller output increases. In this case, the DO controller has reverse action, and the analog output block for the airflow has a signal reversal (i.e., increase to close) specified.

If the conversion between gain and proportional band and between reset action and reset time or controller action is wrong, nothing else matters.

Equation 2-2a for integrated error is derived, as shown in Appendix D, from the response of a Standard Form of the proportional-integral (PI) controller to a load upset where the open-loop error (E_o) is the magnitude of the error if the controller is in manual. The integrated error (E_i) is the total area between the process variable and the set point for the controller in automatic [3, 4]. The equation shows that the integrated error is inversely related to controller gain and directly proportional to reset time.

$$E_I = \frac{1}{K_o \bullet K_c} \bullet T_i \bullet E_o \tag{2-2a}$$

where

E_i = integrated error (percent seconds)
E_o = open-loop error (percent)
K_c = controller gain
K_o = open-loop self-regulating process gain (PV percent/CO percent) (dimensionless) (e.g., product of valve or VSD gain, process gain, and measurement gain as detailed in Equation 2-4b)
T_i = controller reset time (seconds)

If we take the ratio of the integrated errors at a constant open-loop gain for new and original tuning settings and subtract this ratio from 1, the open-loop error cancels out. We end up with the expression within the parentheses in Equation 2-2b, which gives an estimate of the fractional improvement in integrated error for load changes achieved by more aggressive tuning. If we multiply this fraction improvement by the control error, the result is the possible improvement in the standard deviation as a result of better tuning. The equation assumes that the loop is stable. If the loop is oscillating, it might be better to have less aggressive tuning, but this case is rare except at the start of the batch when demand is extremely low. Equation 2-2b does not address the portion of the standard deviation that is caused by limit cycles or noise because these are set by the devices, equipment, and process. Though Equation 2-2a was derived for load changes, it also gives a relative indication of the performance improvement for set point changes. A PID structure of integral action on error (E) and proportional derivative action on the process variable (PV), denoted as "I on E, PD on PV" in Chapter 3, can be used to eliminate overshoot for set point changes from tuning designed to minimize variability from disturbances. When implementing cascade control, Equation 2-2a should first be used to improve the tuning of the secondary loop before cascade is applied to the primary loop.

$$\Delta S_c \cong \max imum\left[\left(1.0 - \frac{K_{co}}{K_{cn}} \bullet \frac{T_{in}}{T_{io}}\right), 0.0\right] \bullet \Delta E_c \tag{2-2b}$$

where

ΔE_c = control error caused by load or set point changes (percent)
K_{cn} = new controller gain
K_{co} = old controller gain
ΔS_c = improvement in standard deviation from new controller tuning (percent)
T_{in} = new controller reset time (seconds)
T_{io} = old controller reset time (seconds)

The control error (ΔE_c) caused by load or set point changes can be estimated from the mean absolute error (MAE_c) over a test period. The current test time is the module's execution time added to the last value of the test time multiplied by a fractional "forgetting factor" determined by the supervision time, as detailed in Equation 2-2c. Similarly, the integrated absolute error is the absolute error added to a last value of the integrated absolute error forgotten to a degree again determined by the supervision time, as detailed in Equation 2-2d. The mean absolute error is just the integrated absolute error divided by the test time, as shown in Equation 2- 2e. Finally, the control error (ΔE_c) for use in Equation 2-2b is computed, per Equation 2-2f, as the absolute difference between the mean absolute error and the standard deviation of the capability of the process (S_{cap}) set by noise. This can be estimated from the moving rate average of successive differences of the deviation from set point, as illustrated by Equations 2-2g and 2-2h. The supervision time is typically either the duration of an operator shift or a batch phase. If the configuration is not set up so the set point tracks the process variable when the PID control algorithm is in the manual or remote output modes, then the error should be zeroed out until the controller is in service (automatic, cascade, or remote cascade modes).

$$T_n = T_e + (1 - 1/T_s) \bullet T_{n-1} \tag{2-2c}$$

$$IAE_n \cong |CV_n - SP_n| + (1 - 1/T_s) \bullet IAE_{n-1} \tag{2-2d}$$

$$MAE_c \cong IAE_c / T_n \tag{2-2e}$$

$$\Delta E_c \cong |MAE_c - S_{cap}| \tag{2-2f}$$

$$MR_n = r \bullet [|CV_n - SP_n| - |CV_{n-1} - SP_{n-1}|] + (1 - r) \bullet MR_{n-1} \tag{2-2g}$$

$$S_{cap} \cong MR_n / 1.128 \tag{2-2h}$$

where

CV_n = controlled variable for current scan (percent)
CV_{n-1} = controlled variable for last scan (percent)
IAE_n = integrated absolute error for test time (percent seconds)
MAE_n = mean absolute error for test time (percent)
MR_n = moving rate average at scan n (percent)
r = weighting factor (0.02)
SP_n = set point for current scan (percent)
SP_{n-1} = set point for last scan (percent)
S_{cap} = standard deviation of the capability of process as determined by noise (percent)
T_e = module execution time (seconds)
T_s = process supervision time (seconds)
T_n = current test time (seconds)

An adaptive controller has been developed that identifies the process model and the impact of new controller tuning settings for Equation 2-2b. This makes it possible to estimate the possible improvement in the standard deviation through better tuning [5, 6, 7]. The subsequent Equations 2-2c to 2-2h provide the statistical analysis to confirm the actual improvement achieved after better tuning.

The improvement in control-loop performance can be estimated from the change in tuning settings and the mean absolute error.

The controller tuning's dependence on process dynamics, including dead time, is detailed in Chapter 3. It is important to remember that although the most aggressive tuning settings and the ultimate performance depends on the relative size of the dead time, the actual performance still depends on the tuning settings used. In other words, if a controller is detuned, it will perform the same as a controller with more dead time.

Sources of Dead Time

The purpose of a control system is to recognize and deal with change. Figure 2-2 shows the path from a change in the controller output to a change in the process variable seen by a DO, pH, and substrate controller that is manipulating gas, reagent, or substrate flow by means of a VSD. A dead time or a time lag is associated with each block in the path. When DO is controlled by manipulating the agitation speed, there are no piping or injection time delays associated with manipulating a flow. The total loop dead time can be estimated as the sum of the time delays and time lags of each block [3, 4, 5]. It can be visualized as the total delay required for a signal to make one cycle around the loop. The starting point of the change can be anywhere in the loop.

The controller has a dead time that is, on average, 1/2 of the module execution time, which is typically 1 to 5 s. Excluding resolution limits, the VSD's dead time is negligible unless a deadband has been added for noise rejection.

The injection delay is the time it takes for the change in flow to make it into the broth or media. It includes any transportation delay in the piping from the point of the mixing of two streams, such as might occur in oxygen enrichment for DO control. If there is a dip tube, a transportation delay exists in the tube if the stream composition changes from mixing of streams or backfilling of the dip tube with broth or media when the pump is off.

The transportation delay is equal to the piping or dip tube volume divided by the flow. Because reagent flows are so small, using a dip tube creates a huge reagent delivery delay for pH control [6]. Fortunately, the reagent is normally dumped on the surface. However, if the agitation is not noticeably breaking the surface or there is more

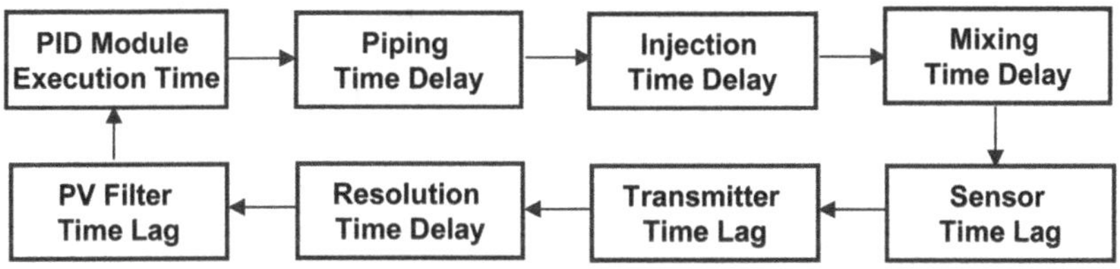

Figure 2-2. Sources of loop dead times for DO, pH, and substrate control with fast, precise VSD.

radial (swirling) than axial (vertical) movement of the broth or media, there is still a considerable delay before the reagent is pulled down from the surface [6]. If there is more than one reagent, each reagent should have its own piping and dip tube. Separate reagent piping prevents the transportation delay detailed in the first expression of Equation 2-2i. This delay is the time for the propagation of a change in the reagent concentration from the point of mixing of reagents to the point of injection.

Once the change in flow has entered the broth or media, it takes time for it to be sufficiently dispersed and transported to the sensor. The location of the sensor is a key factor. If the sensor is too close to the injection, short-circuiting occurs, producing an inconsistent reading. If the sensor is too close to the wall or a baffle, the lower fluid velocity translates into a slower dispersion and a slower inherent electrode response, which is greatly aggravated by an increased propensity for fouling [6]. In the extreme case of stagnation, the sensor does not see a broth or media that is representative of the mixture. If the sensor was located externally, there would also be a transportation delay associated with the piping from the bioreactor to the sensor.

Slow fluid velocities increase the fouling and time lag of electrodes.

The process dead time caused by liquid mixing (ignoring transportation, injection, stagnation, and sensor delays) for a well-mixed vessel with axial agitation can be estimated as the turnover time. This is approximately the broth or media volume divided by the volumetric injection flow and the agitator pumping rate [6]. Some bioreactors are not well mixed, so the actual mixing dead time is larger than the turnover time. This is especially the case for animal cell cultures, where agitation is minimal because the cells are easily damaged due to the lack of a cell membrane. The largest dead time for a pH or substrate sensor inserted into a bioreactor is usually the sum of the injection transportation delay and the mixing time delay. This process dead time can be approximated by Equation 2-2i for well-mixed bacterial, fungi, and yeast cultures and is rivaled only by the dead time from the final element's resolution limit.

If the agitation is not breaking the surface, there is a considerable time delay before the reagent is sucked down and dispersed according to the turnover time.

$$\theta_p = V_i / F_i + [V_b / (F_i + F_a + F_r)] \bullet K_z \tag{2-2i}$$

where

F_a = volumetric agitator pumping rate (meters cubed per second)
F_i = volumetric injection flow (meters cubed per second)

F_r = volumetric recirculation flow including the eductor effect (meters cubed per second)
K_z = geometric factor (0.5 for baffled vertical well-mixed vessel)
V_i = injection (piping and dip tube) volume for concentration change (meters cubed)
V_b = back mixed volume for concentration change (meters cubed)
θ_b = process dead time for concentration change (meters cubed)

The largest source of pH loop dead time is mixing and injection time delays.

The process dead time from mixing is an order of magnitude smaller for DO than pH control. The degree and rate of dispersion for the DO response is the result of both the bubble velocity and the pattern and mass transfer coefficient set by the agitation speed, sparger design, and injection flow and broth or media viscosity. The dead time of a clean DO sensor is about four times larger than the dead time of a clean pH sensor, but the net result is that the loop dead time is still much less for DO than for pH.

As a general rule, the loop dead time for DO control is considerably less than the dead time for pH and substrate control.

For a change to be readily distinguished from noise, the excursion in the process variable must exceed the resolution limit and noise band of the measurement. This dead time is on average about 1/2 the resolution limit or noise band, whichever is significantly larger, divided by the rate of change of the process variable. The resolution limit for an analog-to-digital (A/D) converter with one signed bit is about 0.05% and 0.003% of span, for a 12-bit and 16-bit A/D, respectively. If the process variable is changing at 0.0001% per second (which is not uncommon for the composition response of the liquid phase), then the dead time can be estimated as 250 s and 15 s for a 12-bit and 16-bit A/D, respectively. The actual dead time can be almost zero, or twice as large as these numbers, because it depends on whether the measurement's starting point is at the opposite or near end of the resolution limit before the change. The dead time from the DO resolution limit is negligible because the DO's rate of change is so fast.

The dead time from the resolution limit of the measurement is variable and significant for slowly changing process variables and 12-bit A/D converters.

In practice, the total loop dead time should be estimated based on the time between an appreciable change in the controller output or set point and the first noticeable change in the measurement. If there are no set point changes with the controller in automatic or output changes with the controller in manual, then you can roughly estimate the minimum dead time as the time delay from the start of the flow set by the batch sequence to the start of the measurement's response.

The time between the start of flow and the start of the measurement's response provides an estimate of the minimum loop dead time.

The observed dead time increases as the rate of change in the controller output decreases. This is principally because of the increase in dead time from a slower ramp rate through the final element resolution limit and through the measurement resolution limit and noise band. The sensor may also respond more slowly because of the smaller concentration gradient at the tip of the electrode, particularly if there is any coating. A small set point change has the same effect because it corresponds to a small change in the controller output. The observed dead time for the gradual changes in load caused by the changes in the slope of the growth or product formation rates after the lag phase is much larger than the dead time observed for set point changes. This is because the change in the controller output over the loop's response time is miniscule. The control error over the loop's response time, following from Equation 2-2d, is incredibly small. Even significant fractional improvement in performance has a minimal effect, as Equation 2-2b suggests. This explains why even with far less than optimal tuning, DO and pH loops still draw a straight line. The principal advantage gained from improved tuning of primary controllers reveals itself as these controllers work to catch up with the increase in demand associated with the five to ten doubling of cells during the exponential growth phase. The substrate concentration rapidly drops, and the substrate feeding is started. Before the exponential growth phase is well established, there is very little oxygen, reagent, and substrate demand, so the primary controllers do not see or have to deal with significant load changes, and secondary set point changes are minimal [1]. DO controllers tend to cycle at the beginning of the batch because meters and final elements are sized and controllers are tuned for peak oxygen demand.

The improvement gained from improved tuning of primary controllers is most observable after the exponential growth phase is well established.

The steepness of the slope rather than the change in slope of the biomass profile during the exponential growth phase determines how much work the PID controller has to do to meet the DO and substrate demands. In other words, the magnitude of the controller output's rate of change (i.e., the size of shift in controller output within a batch phase) is a better indicator of load than a change in the controller's output rate of change. The tuning settings may change at different points in the batch, but the job for the PID controller is greatest at the controller output's maximum rate of change.

Sluggish tuning of the secondary loops for DO control can have significant consequences at any time in the batch process. This is because it can cause slow and nearly sustained cycling of the DO controller, which is a result of interaction with the flow or speed loops.

Limit Cycles

A resolution limit anywhere in the control loop causes a limit cycle. The period (T_o) of the limit cycle depends on the loop gains and the controller tuning settings, according to Equation 2-2j for balanced self-regulating processes, developed by Hägglund [8].

The limit cycle may be hidden by data compression, attenuated by signal filtering, and disguised by noise:

$$T_o = Max\{4 \bullet T_i, 4 \bullet T_i \bullet [(1/K_o \bullet K_c) - 1.0]\} \tag{2-2j}$$

The amplitude of the limit cycle depends on the open-loop gain (product of the valve or VSD gain, process gain, and measurement gain) and the valve or VSD signal resolution per Equation 2-2k [9].

$$A_o = S_v \bullet K_o \tag{2-2k}$$

where

A_o = limit cycle amplitude (percent)
K_c = controller gain (percent/percent)
K_o = open-loop gain (percent/percent)
S_v = signal resolution of valve or VSD (percent)
T_i = integral mode reset time (seconds)
T_o = limit cycle period (seconds)

Sustained cycling is not necessarily a bad thing in that it might increase mass transfer rates and stimulate biological processes. However, if the peak in the plot of the biomass growth rate or product formation rate versus the process variable is relatively sharp, the variability can result in suboptimal batch profiles.

2.3 Self-Regulating Processes

The process response is fast and self-regulating for the secondary loops in the bioreactor cascade control systems. Flow, speed, and coolant temperature reach a new steady state relatively quickly in response to a manual change in the secondary controller's output.

Secondary loops generally have a fast self-regulating process response.

Figure 2-3 shows the open-loop self-regulating response of the process to a manual change in the controller output. It is termed *open-loop* because the controller is in manual, and it is termed *self-regulating* because the process decelerates to a new steady state. The process response in Figure 2-3 can be described by a steady-state open-loop gain, a loop dead time, and a single time constant. This approximation is called a *first-order* model.

The loop dead time (τ_d) is the time from the start of the change in the controller output to the time at which an excursion of the process variable goes beyond its noise band. The time from the start of the change of the process variable to about 63% of its final value is the open-loop time constant (τ_o). The final change in the process variable (in percent) divided by the change in the manual controller output (in percent) is the steady-state, open-loop gain (K_p).

Loop dead time delays the controller's ability to recognize and deal with change. If there was no noise, perfect control would be possible if the dead time were zero, in

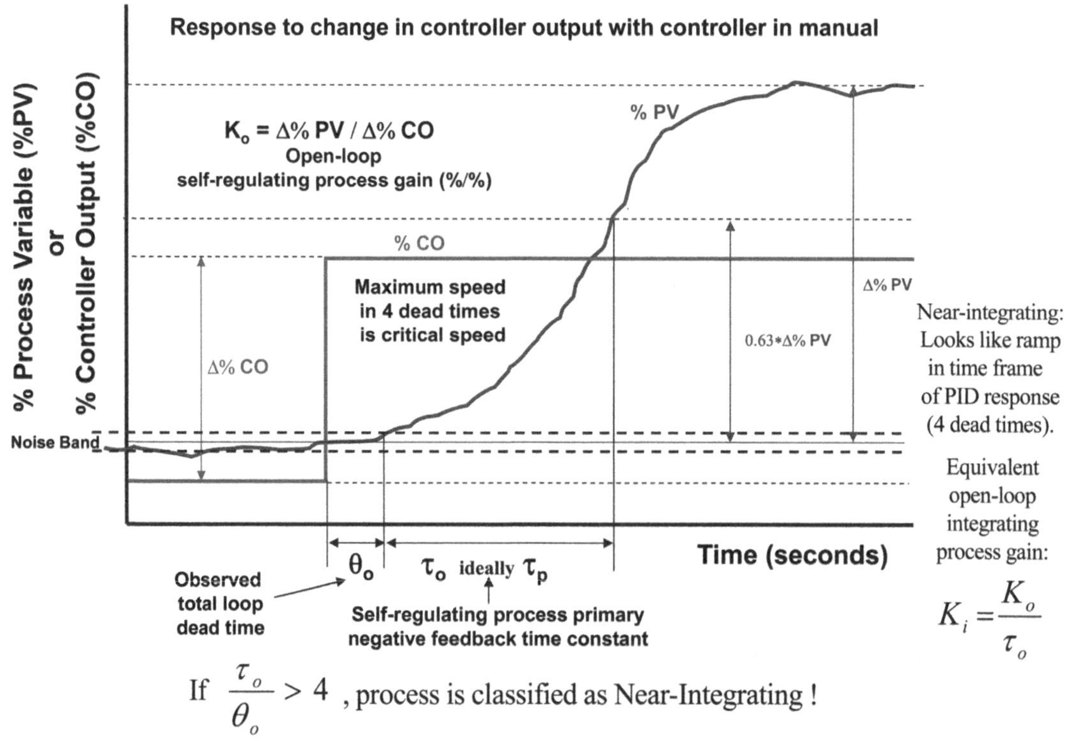

Figure 2-3. Self-regulating process response.

that the controller could immediately see and compensate for a change. The controller would be stable for even the most aggressive tuning settings.

If the loop dead time were zero, perfect control would be theoretically possible.

For the control of liquid flow and agitation by manipulating a VSD with a negligible deadband, the process dead time is small, and most of the loop dead time is set by the module execution time and resolution of the VSD. However, the improvement in control of the secondary loop from PI module execution times of less than 1 s is overshadowed by other considerations, such as proper tuning of the PI controller. The module execution times of primary controllers should be larger than 1 s to maximize the signal-to-noise ratio. This is because the real DO, pH, or substrate change over 1 s is within the noise band. Simply stated, the true change of the process variable within the module execution time should be five times larger than the noise.

Larger module execution times are also desirable because they reduce controller loading, but the execution time must be less than twice the oscillation period to avoid severe aliasing. The primary module execution time should also be less than 1/5 the process dead time and time constant, whichever is smallest. The input and output cards and measurement devices are normally set up for oversampling to provide five scans for the fastest module execution time.

The module execution times of primary loops should be longer than 1 s to maximize the signal-to-noise ratio and minimize the loading.

A potentially large source of loop dead time for controlling coolant temperature is the thermal time lags for heat transfer between media and piping transportation delays. These lags and delays depend heavily on the coolant system design. Another significant source of dead time is the thermowell lag time, which varies from 6 to 60 s depending on the fluid velocity, fit of the sensing element within the well, and degree of coating [3]. For slowly changing process variables, the dead time that results from the final element's resolution is significant. However, it can be minimized by using sliding stem valves with digital positioners or a special high-resolution VSD [4, 5]. The resolution of a standard VSD is about 0.1 Hz in 60 Hz, which is about 0.17%. The resolution of rotary valves varies from about 0.25% to 5.0% depending on the degree of shaft windup and the type of positioner, packing, seating surfaces, and feedback mechanism. The resolution of sliding stem valves that have low friction packing and digital positioners typically ranges from 0.1% to 0.5% [4].

The heat transfer and transportation delays set during the design of coolant systems are the largest sources of dead time for coolant temperature control.

The open-loop time constant can be approximated as the largest time constant in the control loop wherever it is located. The open-loop time constant slows down the process variable's excursion rate and gives the controller a chance to catch up with a load change. If a large time constant in the process downstream of the point where disturbances enter is larger than the loop dead time, it will improve control by slowing down these disturbances and by effectively filtering out process noise. Stated another way, a small dead-time-to-time-constant ratio improves the potential for good control within the loop. What is achieved in practice depends on the tuning settings, as shown in Equation 2-2b. Often, a large time constant is problematic because open-loop tests that wait until a steady state is reached take too long. The time to reach 98% of the final response is approximately four time constants plus the dead time, as shown in Equation 2-3. For a slow process and to deal with the nonlinear response of sensors and valves, an 86% response time that is approximately two time constants plus dead time is more effective in judging performance. Often the reason a loop is difficult is that the loop is too slow even if the slowness is in the process time constant, which offers the potential for better control.

$$T_{98} = 4 \bullet \tau_o + \theta_o \tag{2-3}$$

where

T_{98} = time to reach 98% of final value (98% response time) (seconds)
τ_o = open-loop time constant (primary time constant) (seconds)
θ_o = total loop dead time (seconds)

Although a large process time constant might offer improved control of a properly tuned single-loop controller, it can cause problems if it is in the secondary loop of a cascade control system. If the time constant in a secondary loop is not significantly less than the time constant in the primary loop, a fundamental rule is potentially violated: the secondary loop must be much faster than the primary loop. Some negative consequences can be avoided by using tuning methods that make the secondary loop faster. The discussion on tuning in Chapter 3 addresses this concern in detail and provides a solution.

Although a large process time constant can improve the potential for better loop performance, it creates tuning and cascade control issues.

A large time constant caused by a sensor lag time, damping adjustment in the transmitter, or filter on the process variable in the control system also smooths out noise. However, it attenuates the control system's view of the true changes in the process.

A large time constant in the measurement path creates an illusion of better control [3, 4, 5]. It also delays the control loop's ability to recognize a load upset.

Measurement time constants (e.g., filter times) give an attenuated version of the real process variability and an illusion of better control.

A controller that has too high a gain or rate action inflicts disturbances on itself by increasing the noise in the controller output. You can determine whether noise in the controller output is insufficiently smoothed out by the process time constant by determining whether the noise with the controller in automatic is greater than the noise when the controller is in manual. The solution to reduce noise introduced by automatic action is to first eliminate excessive gain or rate action and then, if necessary, increase signal filtering [5].

Noise introduced by automatic action should be first reduced by better tuning and then, if necessary, by adding a minimal filter on the process variable.

In reality, there is more than one time constant, which gives rise to a bend in the initial response, as shown in Figure 2-3. In the bend, the process variable accelerates. Subsequently, the response reaches an inflection point at which it decelerates. If there was a single time constant, the response immediately after the dead time would be the fastest and would form a right angle. For a first-order approximation, the small time constants effectively become dead time. This is determined by the intersection of the process variable's initial value by a line that is tangent to the inflection point, as shown in Figure 2-3.

In real applications, there are many small time constants that can be approximated as additional dead time.

Adding rate action can compensate for these small time constants if the measurement noise is small and a large process time constant exists to smooth out the reaction to noise in the controller output. Flow and speed loops have too much noise and no appreciable process time constant.

Composition loops may have too much sensor noise. The best candidates for rate action are generally temperature loops that have A/D noise. When one-way cooling is used (no heating via a tempered water system), then it is essential that derivative action be used to prevent overshoot of the set point.

The best candidates for rate action are generally temperature loops with minimal noise, in which the rate time is set to compensate for small time constants.

The open-loop gain is more linear for flow and speed loops that have a VSD because these loops do not have the flow characteristic of a control valve. The open-loop gain for gear pumps is relatively constant and does not vary much with operating point or process conditions. The steady-state process gain for a temperature loop that throttles coolant flow is inversely proportional to coolant flow. The equal percentage flow characteristic of a control valve can largely cancel out this process gain nonlinearity. If a VSD, linear valve, or secondary flow loop is used, then there is no inherent compensation. At low loads, the process gain increases dramatically. Also, at low flow, the transportation delay and heat transfer lag increase. This combination of a higher process gain and larger loop dead time can cause a limit cycle in a temperature control system. Whether it is noticeable or not depends on data compression, mixing, and tuning.

If the flow loop manipulates a VSD without velocity limits instead of using a control valve as the final element, then the time constant is dramatically faster. The loop's speed of response is mostly set by the module execution time, transmitter damping, and signal filters.

2.4 Integrating Processes

The process response is slow and integrating for the bioreactor primary loops for DO, pH, and substrate control. The concentration control of batch operations that have slow reaction rates resembles an integrating response because no discharge flow occurs for continuous control. Also, in batch operations, there is no steady state as evidenced by the batch profiles. After the dead time, the process slowly ramps in response to a manual change in the primary controller output, as shown in Figure 2-4. There is no steady state and consequently no steady-state open-loop gain for these processes. The process gain is an integrating gain that is the change in the ramp rate in percent per second per percent change in the controller output. The response is modeled as a loop dead time and an integrating process open-loop gain. Note also that the initial response is not flat for manual operation as it was for a self-regulating process. This is because a small imbalance (difference between load and controller output) causes the process variable to ramp. Thus, the open-loop gain must be calculated from the change in the ramp rate.

Primary loops typically have a slow integrating type of process response.

An integrating process does not line out when the controller is in manual.

If the process output's ramp rate (slope) is used for the controlled variable where the open-loop gain is the change in the process output's ramp rate for a change in the process

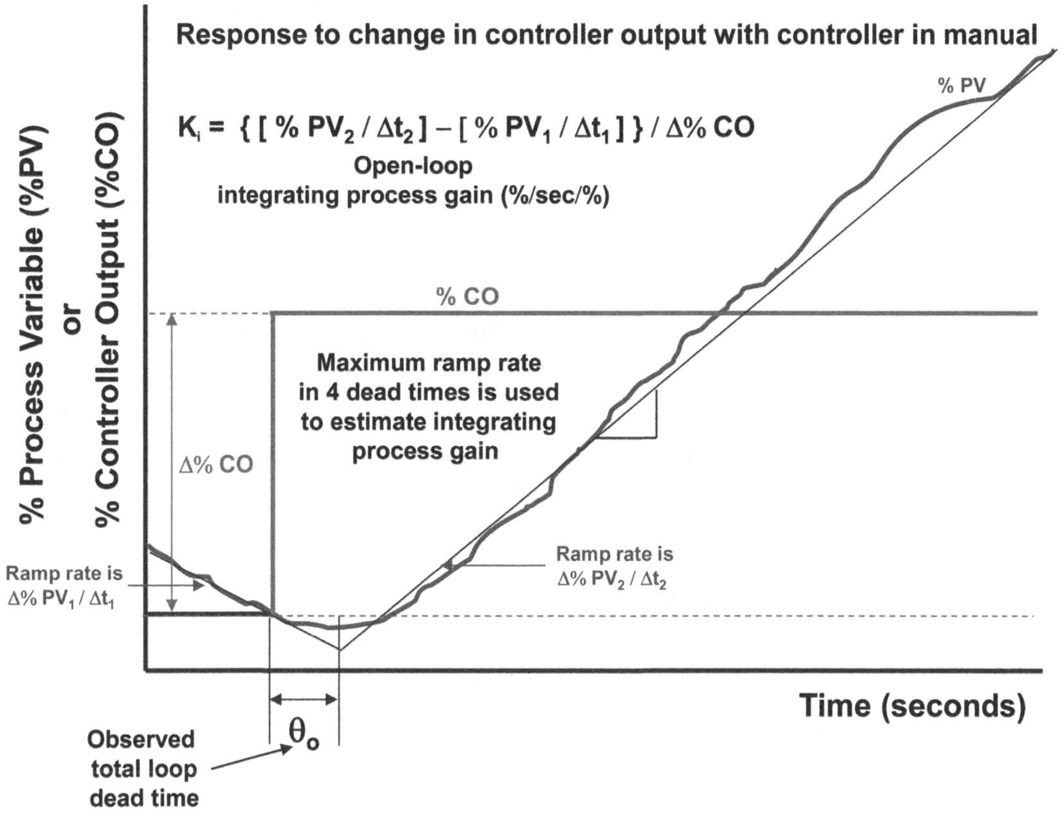

Figure 2-4. Integrating process response.

input, then the result can be modeled as a self-regulating response. When the ramp rate is constant for a given change in the controller output, there is effectively a steady state.

However, this ramp rate changes as the batch progresses.

An integrating process can be modeled as a self-regulating process with an interim steady state by computing the change of the process output's ramp rate in response to a change in a process input.

Integrating processes can be identified much faster because there is no steady state. The person or software must wait until the ramp rate is recognizable, which is usually possible after four dead times.

Integrating processes can be identified faster than self-regulating processes can.

Disturbances also change the ramp rate. The dead time and open-loop gain change in relation to operating point and process conditions. The dead time may also change as a function of the direction of the change. For these and other reasons, tests should be repeated at least twice in both directions at different conditions.

Tests to identify process dynamics should be repeated at least twice in both directions at different operating points, process conditions, and batch times.

Though the bioreactor temperature loop may have a self-regulating response, the time required to reach steady state is so long and the self-regulation so weak because of a lack of continuous discharge flow that it is best to model and tune the response as an integrating process. This can be visualized by noting that the first portion of the response in Figure 2-3 after the dead time is similar to the response in Figure 2-4. Processes that do not have a true integrating response but ramp in the time frame of interest are called *near* integrating processes [3, 5].

Equation 2-4a shows the conversion of the steady-state open-loop gain (often just referred to as *process gain*) and the open-loop time constant of a self-regulating process into the equivalent integrating process open-loop gain of a pseudo integrating process [3, 5]. Modeling a slow self-regulating loop as a pseudo integrating process can save considerable test time. Chapter 3, on basic feedback control, discusses how to apply the tuning rules for integrators to these processes.

Batch operations tend to have an integrating type of concentration and temperature response because there is no continuous discharge flow or steady state.

$$K_i = K_o / \tau_o \tag{2-4a}$$

$$K_o = K_v \bullet K_p \bullet K_m \tag{2-4b}$$

where

K_i = open-loop integrating process gain (PV percent per second per CO percent) (1/seconds)
K_m = measurement gain (PV percent/PV engineering units)
K_o = open-loop self-regulating process gain (PV percent/CO percent) (dimensionless)
K_p = steady-state process gain (PV engineering units/flow engineering units)
K_v = valve or VSD gain (slope of installed flow characteristic) (flow engineering units/CO percent)
τ_o = open-loop time constant (primary time constant) (seconds)

The bioreactor pressure loop has an integrating type of response if the loop manipulates a fan VSD or if the pressure drop across the control valve is so large that a change in vessel pressure does not translate into an appreciable change in vent flow. In other words, a higher vessel head space pressure does not force out more off-gas flow through the vent system.

As was the case for the self-regulating response, the initial bend in the response is associated with small time constants. If the process could be modeled by a loop dead time and an integrating process gain, then the process variable would abruptly reverse immediately after the dead time. In real applications, the transition is smoothed by small time constants. A rate time can be set to compensate for these time constants if the measurement noise is not excessive and the integrating process open-loop gain is low, which, according to Equation 2-4a, is equivalent to a large process time constant.

2.5 Wireless Devices and Analyzers

Often misunderstood is the effect of discrete updates on process control. Existing studies have typically shown a misleading, greatly diminished effect from sluggish tuning settings and mistakenly timing the set point change or load change right before an update. Frequency response and common sense indicate the dead time should be considered to be the result of an average timing where the change arrives in the middle of the update or cycle time interval. Wireless devices tend to have slow update rates and large trigger levels to help preserve battery life, not realizing that the additional dead time that is half the update rate hurts loop performance. [11]

One must estimate the dead time from the update rate or analyzer cycle time and then determine if it is appreciable compared to the implied loop dead time potentially causing oscillation. The tuning of a controller is almost always less aggressive than possible per the actual dead time and in many cases causes a sluggish response. An implied dead time larger than the actual dead time can be back calculated from the tuning settings. This innovative term is based on the observation that a slowly tuned controller behaves like a loop with a larger than actual dead time [2].

We can estimate the amount of measurement delay that causes no appreciable need to retune the controller based on the original loop dead time (without the additional measurement delay) and implied dead time (θ_i). Equation 2-5a gives the implied dead time as a function of the original dead time (θ_0) and the tuning parameter lambda (λ) detailed in Chapter 3:

$$\theta_i = 0.5 \bullet (\lambda + \theta_o) \tag{2-5a}$$

The amount of additional dead time available is the implied dead time minus the original dead time:

$$\theta_m = \theta_i - \theta_o \tag{2-5b}$$

Lambda is typically set to be three times the open-loop dead time to achieve robustness to deal with unknowns and nonlinearities. If the dynamics are well known and aggressive albeit oscillatory control is desired, lambda may be set as small as the loop dead time. In this case, the ratio of the new to the original peak error is proportional to the ratio of the increased total loop dead time to the implied dead time. The ratio of the new to the original integrated error is proportional to the ratio of these same dead times squared.

$$\theta_m < 0.5 \bullet (\lambda + \theta_o) - \theta_o \tag{2-5c}$$

Now that we know the maximum allowable dead time from the controller, we must know what this corresponds to in terms of measurement from a multitude of sources.

Measurement dead time has a negligible effect unless it is larger than the implied dead time that is larger than the actual dead time due to tuning for robustness and to prevent oscillations.

Equation 2-5d gives the measurement dead time (θ_m) from measurement lags 1 (τ_{m1}) and 2 (τ_{m2}), delays 1 (θ_{m1}) and 2 (θ_{m2}), wireless devices (θ_w), analyzers (θ_a), and threshold sensitivity limits. Note that the term *update rate* is a time interval between updates rather than the number of updates per second.

$$\theta_m = \theta_w + \theta_a + Y \bullet \tau_{m1} + Y \bullet \tau_{m2} + \theta_{m1} + \theta_{m2} + \theta_{mx} \tag{2-5d}$$

The many sources of measurement dead time should be tracked down.

The dead time from noncontinuous updates is 1/2 the wireless update time (Δt_{wx}) plus the wireless latency (Δt_{wz}), which is the time from the start of an update time interval to when a result is communicated as an output. Normally, this latency is negligible.

$$\theta_w = 0.5 \bullet \Delta t_{wx} + \Delta t_{wz} \tag{2-5e}$$

If a change in measurement is greater than the wireless trigger level, an update will occur before the next scheduled transmission set by the default update rate. Fortunately, the enhanced PID controller developed for wireless (described in Chapter 3 and Appendix C) can handle extremely large and variable update times by executing a single feedback correction when a change in the process variable is detected.

If the total loop dead time from a wireless device or analyzer exceeds the implied dead time by more than 10%, the peak and integrated errors will be noticeably degraded.

For self-regulating processes, the use of an enhanced PID controller may eliminate the need for the controller to be retuned as the wireless update time or analyzer cycle time and the multiplex time increases [10, 11]. For loops where the implied dead time is greater than the primary process time constant, the PID tuning is no longer a function of the discontinuous update time. If the total loop dead time is much greater than the 63% process response time (process dead time plus process time constant), the enhanced PID gain can be set as large as the inverse of the open-loop self-regulating process gain. While the enhanced PID controller prevents the loop from becoming oscillatory, the errors for a load disturbance are still increased. Equation 2-2a can be used for a rough estimate of the integrated error by adding the update time interval between wireless results to the numerator.

For near-integrating, true-integrating, and runaway processes, the enhanced PID controller cannot prevent the degradation in loop performance, and the need for retuning if the total loop dead time becomes larger than the implied dead time. The enhanced PID controller can still help in terms of dealing with a large variable update time and a communication failure or sensor failure (e.g., coated or broken pH electrode) [11].

The latency for an analyzer is the cycle time because the analysis result is available at the end of the cycle time. Therefore, the dead time is 11/2 times the analyzer cycle time ($0.5 \bullet \Delta t_{ax}$).

Because analyzers are a large investment in capital and maintenance, multiple sample points are serviced by a single analyzer, particularly for parallel trains of unit operations. The time required for the analyzer to take another sample from a given sample point is the multiplex time. The analyzer dead time is increased by 1/2 of the multiplex time ($0.5 \bullet \Delta t_{az}$).

$$\theta_a = 1.5 \bullet \Delta t_{ax} + 0.5 \bullet \Delta t_{az} \tag{2-5f}$$

At-line analyzers often serve several process units resulting in a dead time from multiplexing.

At-line analyzers can have quite a range of analysis cycle times, from a few seconds for a mass spectrometer to 30 minutes for a chromatograph. These analyzers also have a sample transportation delay, that is, the sample tubing volume divided by the sample flow rate. Because the flow rate needed by the analyzer is extremely small, transportation delays, often simply known as *sample delays*, can be quite large, particularly if the analyzer is located in a central analyzer house to provide a better climate for analyzer operation and maintenance. Often a much higher flow recirculation line is used so that the sample line is as short as possible to reduce the sample delay.

The dead time from discrete updates is 1/2 the time interval plus a latency that is generally zero for wireless and equal to cycle time for analyzers.

Off-line analyzers have extremely large and variable analysis cycle times because the operator may not immediately take a sample to the lab, and the technician in the lab may not do the analysis right away and quickly enter the result into the data historian. In some cases, the sample must be sent to a special lab. The off-line analysis time can vary from hours to days.

The effect of analyzer cycle time and multiplex time in terms of causing oscillations can be mitigated by using the enhanced PID controller developed for wireless. Results can be obtained that are similar to those discussed for wireless transmitters at the end of the previous section. However, the dead time per Equation 2-2k still limits the ultimate possible performance of the control loop.

An enhanced PID controller for self-regulating processes can make tuning independent of large cycle times from at-line analyzers and even off-line analyzers.

Off-line analyzers can only be used if the open-loop integrating process gain is so slow that the excursion is acceptable between worst-case analysis intervals. Off-line analyzers generally are not used for closed-loop control of runaway processes due to the risk of excessive delay, human error, and missing results.

where

Δt_{ax} = analyzer cycle time (seconds)
Δt_{az} = analyzer multiplex time (seconds)
Δt_{wx} = time interval between wireless updates (wireless default update rate) (seconds)
Δt_{wz} = latency (e.g., communication delay after an update) (seconds)
λ = lambda (closed-loop time constant for self-regulating processes or arrest time for near-integrating and true-integrating processes) (seconds)
θ_a = analyzer dead time (seconds)

θ_i = implied loop dead time from PID tuning (seconds)
θ_m = total measurement dead time (seconds)
θ_{m1} = measurement delay 1 (seconds)
θ_{m2} = measurement delay 2 (seconds)
θ_{mx} = measurement delay from resolution or threshold sensitivity limit (seconds)
θ_o = original loop dead time (seconds)
θ_w = wireless dead time (seconds)
τ_{m1} = measurement time constant 1 (seconds)
τ_{m2} = measurement time constant 2 (seconds)
Y = fraction of a small time constant converted to dead time seen in Equation 3-5d where the fraction approaches 1.0 as the ratio of small to large time constant becomes very small (Y = 0.88 for ratio = 0.01)

2.6 Best Practices

1. Make sure sensor lags are not excessive due to low velocities and fouling.
2. Select control valves designed for throttling with the least deadband and best resolution and invertors with high-resolution input cards and minimum deadband and rate limiting settings.
3. Use a smart positioner without integral action tuned with higher gain for valves and speed to torque cascade control in pulse-width modulated inverters for variable frequency drives.
4. Select and size control valves and variable-speed drives (VSDs) to provide a good rangeability and loop linearity.
5. Minimize sources of dead time, particularly from piping and dip tube transportation delays.
6. Tune the controllers with the largest gain and smallest reset time that does not cause any oscillation for worst-case conditions (e.g., largest dead time and highest process gain).
7. To prevent set point overshoot, use a proportional-integral-derivative (PID) structure of I on Error and proportional plus derivative (PD) on process variable (PV) as specified in Chapter 3.
8. To convert a batch integrating response to a self-regulating process, use the slope of the batch profile (e.g., composition) with a good signal-to-noise ratio as the PV detailed in Chapter 3.
9. Realize the many sources of measurement dead time.

10. For wireless devices and analyzers, use the enhanced PID controller detailed in Chapter 3 and Appendix C.

References

1. Joseph S. Alford, email correspondence, 2006.

2. Gregory K. McMillan, *Essentials of Modern Measurements and Final Elements in the Process Industry* (Research Triangle Park, NC: ISA [International Society of Automation], 2010).

3. Gregory K. McMillan, *Tuning and Control Loop Performance,* 4th ed. (New York: Momentum Press, 2015).

4. Terrence L. Blevins, Gregory K. McMillan, Willy K. Wojsznis, and Michael W. Brown, *Advanced Control Unleashed: Plant Performance Management for Optimum Benefits* (Research Triangle Park, NC: ISA [International Society of Automation], 2003).

5. Gregory K. McMillan, *Good Tuning: A Pocket Guide,* 4th ed. (Research Triangle Park, NC: ISA [International Society of Automation], 2015).

6. Gregory K. McMillan, and Robert Cameron, *Advanced pH Measurement and Control,* 3rd ed. (Research Triangle Park, NC: ISA [International Society of Automation], 2005).

7. Gregory K. McMillan, Mark Sowell, and Peter W. Wojsznis, "The Next Generation Adaptive Control Takes a Leap Forward," *Chemical Processing* (September 2004).

8. Tore Hägglund, "A Control Loop Performance Monitor," *Control Engineering Practice* 3, no. 11 (1995): 1543–51.

9. Gregory K. McMillan, "What Is Your Control Valve Telling You?" *Control Design* (May 2004): 43–48.

10. Gregory K. McMillan and P. Hunter Vegas, *Process/Industrial Instruments and Controls Handbook,* 6th ed. (New York: McGraw-Hill Education, 2019).

11. Gregory K. McMillan, "Wireless: Overcoming Challenges in PID Control & Analyzer Applications," *InTech* 57 (July–August 2010).

3

Basic Control

3.1 Introduction

To ensure cells are healthy and productive, the medium in which they live must be at the right temperature, pH, and pressure and have sufficient access to oxygen, food, and nutrients. The basic control system must be designed, configured, and adjusted to provide these fundamental requirements. As cells and products increase in sophistication, the need for tight control at the right operating point greatly increases. For mammalian cells, deviations of a tenth of a pH and a tenth of a degree centigrade from the optimum can cause a significant deterioration in cell growth rate or product formation rate. While the sensitivity to dissolved oxygen (DO) concentration is lower, the rangeability requirement is incredible as cells go from the pre-exponential growth phase at seed culture concentration through the exponential growth and product formation phases to the stationary phase at maximum cell concentration. This extreme rangeability requirement and possible interactions with pH control can lead to DO being a limiting factor to cell performance.

Loops for pH control, and possibly for DO and temperature control, tend to have oscillations that persist across the split-range point, most notably due to nonlinearities, higher backlash, and stiction in control valves, and deadband inadvertently introduced into variable frequency drives (VFDs) and proportional-integral-derivative (PID) configuration. The control valve, VFD designs, and PID designs that minimize these sources of oscillations are detailed.

This chapter provides the knowledge and guidance to maximize the effective use of accurate measurements, precise final control

elements, and tight robust basic control, including what is needed for the best PID configuration, settings, and control system design.

The sources of most of the information presented in this chapter are *Essentials of Modern Measurements and Final Elements in the Process Industry* [1], *Advances in Reactor Measurement and Control* [2], and the *Process/Industrial Instruments and Controls Handbook* [3]. See these references for a more detailed understanding.

Learning Objectives

- Be able to specify the best flow, pH, and temperature measurements
- Be aware of the common mistakes made with control valves and VFDs
- Develop the ability to specify the best control valves and variable frequency drives
- Understand the PID algorithm used in industry and how to better apply its many features
- Learn how PID forms and structures can increase flexibility
- Appreciate how to estimate and improve performance for load and set-point changes
- Learn how to tune for the desired objective
- Become familiar with power of external-reset feedback
- Be able to design the best control system for state-of-the-art bioreactors

3.2 Key Process Measurements

The best basic process control starts with the best measurements in terms of accuracy and rangeability. Accuracy statements include drift, repeatability, sensitivity, and linearity in addition to error over the entire stated range. For loops in the chemical industry, repeatability might be sufficient because the minimization of variability is not as stringent, and often analyzers and advanced control systems are available to correct the set points of the basic loops.

Measurement accuracy and rangeability requirements are extraordinary for biological processes.

The repeatability and accuracy (including drift) are two orders of magnitude better and the signal strength is four orders of magnitude larger for a resistance temperature detector (RTD) compared to a thermocouple (TC). The operating temperature ranges of biological processes are well within the RTD range for exceptional accuracy. The 1°C to

20°C drift per year of a TC is of particular concern because of the profound effect on product quality from control at the wrong operating point. The already exceptional accuracy for a *Class A* RTD of 0.1°C can be improved to 0.02°C by "sensor matching," that is, entering the four constants of a Callendar–Van Dusen (CVD) equation (provided by the supplier for the sensor) into the transmitter. The main limit to accuracy of an RTD is the transmitter and wiring. The use of three extension lead wires between the sensor and transmitter or input card can enable the measurement to compensate for changes in resistance in the lead wires due to temperature, assuming the change is exactly the same for both lead wires. The use of four extension lead wires enables total compensation that accounts for the inevitable uncertainty in resistance of lead wires. Standard lead wires have a tolerance of 10% in resistance. For 500 ft of 20-gauge lead wire, the error could be as large as 26°C for a two-wire RTD and 2.6°C for a three-wire RTD. The best practice is to use a four-wire RTD unless the transmitter is located close to the sensor, preferably on the sensor.

RTDs offer orders of magnitude greater accuracy.

The air in the annular clearance between the thermowell interior wall and sensor sheath is the dominant source of slowness in the temperature response. For the same reason, a spring-loaded, sheathed TC or RTD is used to ensure the tip of the sheath is touching the bottom of the thermowell. A stepped thermowell minimizes the additional lag associated with metal mass near the tip and reduces heat conduction error.

Spring-loaded sheathed RTDs in stepped thermowells increase the speed of response and accuracy.

Spherical glass electrodes and flowing junction reference electrodes offer the greatest pH measurement accuracy. If a flowing junction is not used, the reference electrode potential should be made as small and constant as possible by liquid junction design. Solid references and flat glass electrodes should not be used in bioreactors because of accuracy, rangeability, and calibration concerns.

Spherical glass electrodes with a reference designed to provide a constant small junction potential are essential to meet accuracy and speed of response requirements for biological processes.

Three electrodes and middle signal selection should be used to inherently ignore a single failure of any type, decrease drift and offset, decrease vulnerability to a slow electrode, ignore noise from electromagnetic fields (EMFs) or nonideal mixing, and to provide intelligent diagnostics that reduce maintenance. The life cycle cost is greatly reduced and savings can be enormous if you include the effect of process efficiency and capacity. When you consider the cost of an electrode versus the cost of a bad batch, the decision is easy.

Three pH electrodes with middle signal selection are essential to increase accuracy, reliability, speed of response, and intelligence and to reduce noise and maintenance.

Coriolis flowmeters have an order of magnitude better accuracy and rangeability (e.g., 250:1 and 0.05% of reading) than other flowmeters. It is the only flow measurement that is not affected by changes in concentration that affect density. In fact, the accompanying highly accurate and sensitive density measurement can provide an inference of concentration. The installation and maintenance requirements are by far the least of any flowmeter.

Coriolis flowmeters offer extraordinary accuracy and rangeability with by far the fewest installation and maintenance requirements.

3.3 Control Valves and VFDs

The most common mistake is choosing a control valve to minimize leakage. This leads to the wrong choice of valve and actuator type. These valves have too much capacity and excessive friction and backlash. The piston, positioner, and shaft-to-valve-stem connection design in the associated actuator tends to aggravate the problem of rangeability, poor resolution, and deadband. Unfortunately, these on-off rotary valves posing as throttling valves have an attractive price, high flow capacity, the simplest internals, and the clearest flow path in addition to the most familiarized maintainability. These rotary valves are often used throughout the plant and are already in the piping spec. The positioner readback of the actuator shaft often lies about response, not realizing that the internal disk or ball is not moving because of backlash in linkages and stiction from seal friction. Because so many suppliers of true throttling valves have bought companies that make these seemingly attractive on-off rotary valves, the user may not be able to distinguish between a true throttling valve and a rotary valve posing as one [1, 7].

Control valve specification sheets emphasize capacity and leakage but not response requirements, and bids emphasize price, leading to on-off valves posing as throttling valves.

The next most common mistakes are undersizing actuators, selecting stem packing with an unnecessarily high temperature rating and excessive friction (e.g., graphoil), using boosters instead of positioners, and putting integral action in positioners. Widely publicized rules originating from theoretical Nyquist plot studies recommending the use of boosters instead of positioners on fast loops introduce great safety concerns. As a result of this mistake, the rotary valve can dangerously slam shut from fluid forces due to positive feedback between the booster and diaphragm actuator, which is caused by flexure of the diaphragm and high booster outlet port sensitivity. Additionally, this mistake causes poor precision due to a booster's poor inlet port sensitivity and a great delay followed by large overshoot when opening due to the lack of position feedback to overcome breakaway torque and seal friction. Not realized is that positioners have a faster execution and tuning than most PID controllers, eliminating violation of the cascade rule where the inner controller must be five times faster than the outer controller (the positioner is the inner controller, and the process PID is the outer controller). Finally, digital positioners have integral action turned on as a default to make the response step changes larger than 1% look better. Unfortunately, this reduces the proportional action allowed, causing the response time for small step changes to be much worse and creating limit cycles from backlash. The best solution is to use a sensitive digital positioner tuned aggressively without integral action combined with a sensitive actuator, tight shaft and stem connections (no linkages), and low-friction packing. If this is not fast enough, a booster with a slightly open bypass valve can be put on the positioner output. For exceptionally fast loops, a variable speed pump or fan with a VFD should be considered [1, 8].

There are many misleading articles that cause the incorrect use of booster and positioners.

The most common mistakes for VFDs include:

- putting the speed control in the distributed control system (DCS) rather than keeping it in the VFD control system, slowing down speed control;
- adding a deadband in the invertor setup to reduce reaction to noise, creating delay;
- adding rate limiting in the inverter setup to prevent motor overload, slowing down response;
- using 8-bit I/O input cards, deteriorating resolution to 0.4%;
- not using "inverter rated" motors with a service factor rating of at least 1.15, leading to motor overload;
- not increasing frame size or not using totally enclosed fan cooled (TEFC) with line power;
- not inserting a valve to create more system drop when static pressure approaches discharge pressure, causing a quick opening type of installed flow characteristic;
- using older inverter technologies such as six-step voltage instead of pulse-width modulation (PWM), causing cogging, bearing damage, and electromagnetic interference (EMI);

- not using the most applicable PWM control strategy, decreasing response time or rangeability; and
- omitting inverter input chokes and necessary cable shielding and spacing causing EMI. Cascade control in the inverter of speed to torque provides the greatest rangeability [1].

VFDs can have more nonlinearity and less rangeability and precision than a well-designed control valve unless the right motor, frame, inverter, settings, input card, cable, and control strategy are used.

Often not realized is that deadband introduced in a PID configuration or a VFD setup, or from backlash in a control valve, will result in a limit cycle if there are two or more integrators in the control loop. Integral action can occur in many important processes (e.g., gas pressure control, batch composition, pH and temperature control), digital positioners, and each PID controller in a cascade control system. Stiction requires only one integrator to develop a limit cycle. Oscillations also occur when actuator or positioner sensitivity is poor (e.g., spool positioners and piston actuators). The good news is the oscillation period and amplitude from deadband and poor sensitivity can be reduced by more aggressive tuning, and the oscillation from a splitter configuration deadband stops when the output is far enough away from the split-range point.

A control valve's installed flow characteristic with a linear trim (inherent flow characteristic) becomes increasingly like a quick opening characteristic when the ratio of valve pressure drop to total system pressure drop becomes less than 0.25 in a misguided attempt to save energy. The minimum flow for an equal percentage inherent flow characteristic significantly increases as well. In both cases there is a significant nonlinearity and loss of rangeability that particularly affects split-range control. There is a less recognized corollary situation with a VFD's installed flow characteristic normally linear distorting to be more like a quick opening characteristic when the ratio of static head to total system pressure drop becomes greater than 0.25 in a misguided attempt to save energy by not inserting a valve to create some pressure drop.

The Baumann 84000 roller mechanism diaphragm valves shown in Figures 3-3a and 3-3b offer exceptional throttling capabilities and are designed to satisfy the stringent demands of the pharmaceutical and biotechnology industries. These valves are in compliance with the 3-A Sanitary Standards, Inc., requirements for clean-in-place (CIP) and sanitize-in-place (SIP) service. The angle body is self-draining. The diaphragm eliminates the need for stem packing and provides a small, smooth path for fine adjustment of extremely small flows that meets sanitary valve requirements. The low friction of the packless and roller bearing linkage design enables the valve to

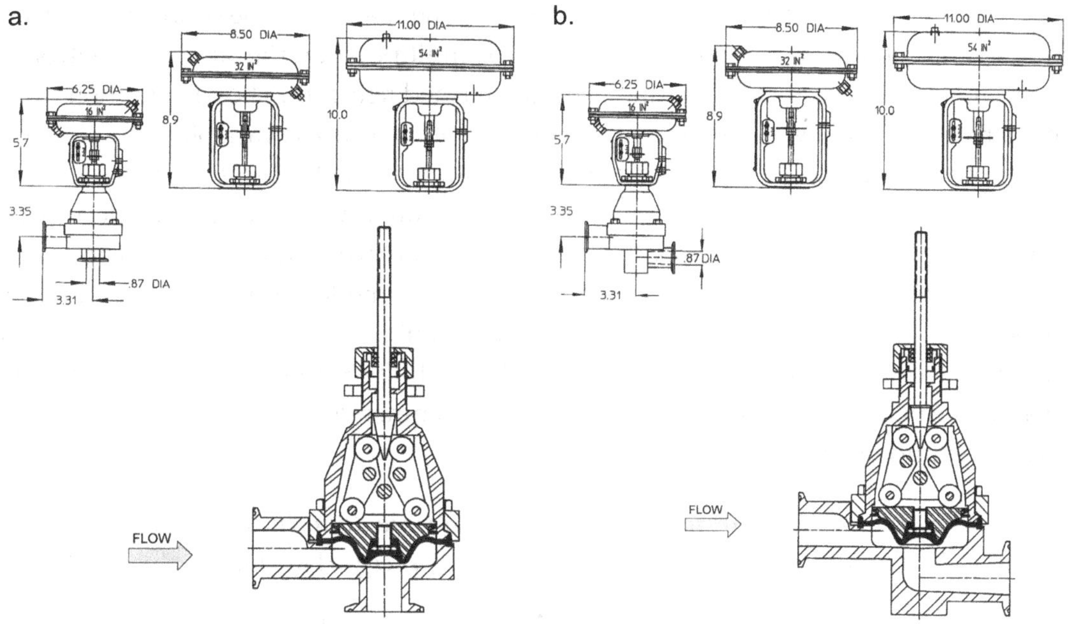

Figure 3-3. Angle (a) and inline-roller diaphragm (b) valves offer incredible rangeability and precision.

respond to incredibly small changes in signal. Digital positioners and sensitive diaphragm actuators are used. These valves can be employed for fine control of additives and nutrients in food and drug production. The flow is forced to be laminar. The pressure drop is small compared to the pressure available. As a result, the installed characteristic is close to the inherent characteristic, enabling an accurate linearization of the controller output in the DCS provided there is at least a 16-bit output card [1].

3.4 PID Fundamentals

For more than 95 years, it has been demonstrated, and even been theoretically proven, that a PID controller provides near optimal control for rejecting unmeasured load disturbances at the process input. The main challenges are how to select and effectively use the most applicable tuning rules and the many features and options in an industrial PID controller.

The three modes in a PID controller are proportional, integral, and derivative. There are many ways of computing and combining the contributions of each mode. The PID block used in industry within a modern DCS or programmable logic controller (PLC) differs from what publications show.

In analog and early DCS and PLC controllers, the contribution from the proportional mode was simply the error between the set point (*SP*) and process variable (*PV*) multiplied by the control action sign (S) and controller gain (K_c). The contribution from the integral mode was the integral of the error multiplied by the sign, the controller gain, and the inverse of integral time (T_i), often termed *reset time*. The contribution from the derivative mode was a filtered rate of change of the process variable multiplied by sign, controller gain, and derivative time (T_d), often termed *rate time*. The filter time was 1/10 to 1/8 the rate time. A *Series Form* was commonly used where the derivative mode was computed first and became the input to the proportional mode. The Series Form offered an advantage by ensuring the derivative mode contribution could never exceed the integral mode contribution by an inherent interaction between the modes in the time domain via interaction factors. Although the Series Form is still available in a modern DCS or PLC, the *Standard Form* is recommended to enable improved performance, understanding, and communication.

The Series Form provides interaction in a time domain that prevents derivative action from exceeding integral action.

In the Standard Form, the derivative time setting should be limited to less than 1/4 the integral time setting to prevent oscillations.

The PID control algorithms in older supervisory computers and in the literature showed a velocity PID algorithm where for a constant error, the contribution from the proportional and derivative modes would be moving with a velocity based on the tuning settings. In industry, a positional PID algorithm is used where, for a constant error (constant SP and PV), the contribution from the proportional mode is constant and from the derivative mode is zero. The integral mode is the only mode driving the controller output when the error is constant. The positional algorithm is important not only for understanding but also for better tuning and performance, including the ability to effectively add adaptive control, override control, valve position control, wireless control, and seamless automation of operator actions for start-up and abnormal conditions. A further simplification occurs by use of the Standard Form, where the contribution from the three modes are computed in parallel and simply added together along with a bias. The bias is normally zero but is adjustable when there is no integral mode.

In the literature and some PLCs, a Parallel Form is offered where the controller gain is not shown as affecting all three modes. Instead of a reset time or rate time, an integral gain and derivative gain are specified, causing considerable confusion in the use of tuning rules.

A Parallel Form with an integral gain and derivative gain must be converted to forms used in industry where the proportional mode gain setting affects all modes.

Older PID controllers often had the integral mode setting as *repeats per minute*. This term originates from the concept that the setting defines how many times the contribution of the proportional mode is repeated per minute. Many modern PID controllers use reset time in seconds, which is the inverse of repeats per minute after converting the time units from minutes to seconds. While technically the equivalent term would be *seconds per repeat*, the "per repeat" is dropped in most systems. The tuning setting for the proportional mode in older PID controllers and in some PID controllers today is *proportional band*. This term originates from the concept that the setting defines what percent change in error causes a 100% change from the proportional mode contribution. The equivalent controller gain is 100% divided by the proportional band. Fortunately for the derivative mode, rate time is the same except often the time units are seconds in modern PID controllers instead of minutes. The bottom line is knowledge of the tuning setting units besides the PID Form and structure is essential when using tuning software or performing tuning calculations.

It is critical that a PID controller tuning settings be understood as to whether gain or proportional band is used for proportional action, whether repeats per unit time or simply unit time is used for integral action, and what the time units are for integral and derivative action.

Today, eight different PID structures (Table 3-4) are possible by introducing simple set-point weight factors for the proportional and derivative modes and turning off different modes in the Standard Form. The set point in the proportional and derivative mode contributions is multiplied by the set-point weight factors beta (β) and gamma (γ) as seen in Equations 3-4a and 3-4c, respectively. These factors are set to 0 or 1 to achieve different structures or are adjustable in a two degrees of freedom (2DOF) structure. The measurement gain for all loops (100% divided by the PV scale span) and a manipulated variable gain for a primary loop (manipulated variable scale span divided by 100%) must be factored into the open-loop gain calculation. The scale spans must be in engineering units. Industrial PID tuning software does this automatically when you enter the *PV* and controller output (*CO*) scales in engineering units.

$$\text{Proportional mode } P = K_c \bullet S \bullet (PV - \beta \bullet SP) \tag{3-4a}$$

$$\text{Integral mode } I = K_c \bullet \frac{1}{T_i} \bullet S \bullet \int (PV - SP) \bullet dt \tag{3-4b}$$

$$\text{Derivative mode } D = K_c \bullet T_d \bullet S \bullet filter\left[d(PV - \gamma \bullet SP)/dt\right] \tag{3-4c}$$

$$\text{Controller output } CO = max\left[min(P + I + D + B,\ CO_{max}), CO_{min}\right] \tag{3-4d}$$

Table 3-4. Major PID structure choices.

1	PID action on error ($\beta = 1$ and $\gamma = 1$)
2	PI action on error, D action on PV ($\beta = 1$ and $\gamma = 0$)
3	I action on error, PD action on PV ($\beta = 0$ and $\gamma = 0$)
4	PD action on error, no I action ($\beta = 1$ and $\gamma = 1$)
5	P action on error, D action on PV, no I action ($\beta = 1$ and $\gamma = 0$)
6	ID action on error, no P action ($\gamma = 1$)
7	I action on error, D action on PV, no P action ($\gamma = 0$)
8	2DOF controller (β and γ adjustable 0 to 1)

where

P = proportional mode contribution to PID output (percent)
I = integral mode contribution to PID output (percent)
D = derivative mode contribution to PID output (percent)
B = bias contribution to PID output (normally zero but is adjustable for no I mode (percent))
PV = process variable (percent)
S = sign of control action (+1 for direct action and –1 for reverse action)—the control action is the opposite of process action if valve action is correctly set in configuration and positioner.
SP = set point (percent)
CO = controller output (percent)
CO_{max} = controller high output limit (percent)
CO_{min} = controller low output limit (percent)
β = set-point weight factor for proportional mode (0 to 1)
γ = set-point weight factor for derivative mode (0 to 1)
K_c = controller gain (dimensionless)
T_i = integral time (reset time) (seconds)
T_d = derivative time (rate time) (seconds) with filter time that is 1/10 to 1/8 rate time

A less recognized aspect of PID control is that the contribution from the integral mode originally developed for pneumatic controllers had to use a bellows and adjustable restrictor acting as an adjustable filter to provide integral action. The reset time was the filter time. Integral action was achieved by a positive feedback implementation where the input to the filter was the controller output and the output of the filter was added to the contribution of the proportional and derivative mode in series to give the PID output. The term *positive feedback* is the result of realizing that changes in the filter output show up as changes in the filter input and then again as changes in the filter output. It helps to recognize that when the error is zero, the filter input and output are equal. A Laplace transform shows the positive feedback implementation is equivalent to integral action given by an integrator. When electronic analog controllers and digital control systems came on the scene, integrators were possible. The positive feedback implementation was replaced by an integrator providing what was thought to be a modern solution, not realizing what was lost. Most notably, there was no longer the ability to inherently prevent the integral mode from changing faster than the manipulated valve, VFD, or secondary PID controller could respond by turning on external-reset feedback (ERF). Figure 3-4 is the first correct block diagram to date in the time domain that shows the positive feedback implementation for a Standard Form with eight different structures possible. Presently, the literature shows a simpler frequency domain block diagram of a Series Form algorithm without any structure options. In Figure 3-4, a switch enabling ERF illustrates how the input to the reset time filter can be a signal representing the variable being manipulated instead of the controller output.

Integral action is favored over proportional action because it works to decrease the error in the direction more readily understood by humans. For example, if a temperature is below the set point, operators want a valve for heating to be open. A valve for cooling must be open if the temperature is rising quickly and is close enough to the set point that if cooling has not started, the temperature will overshoot the set point.

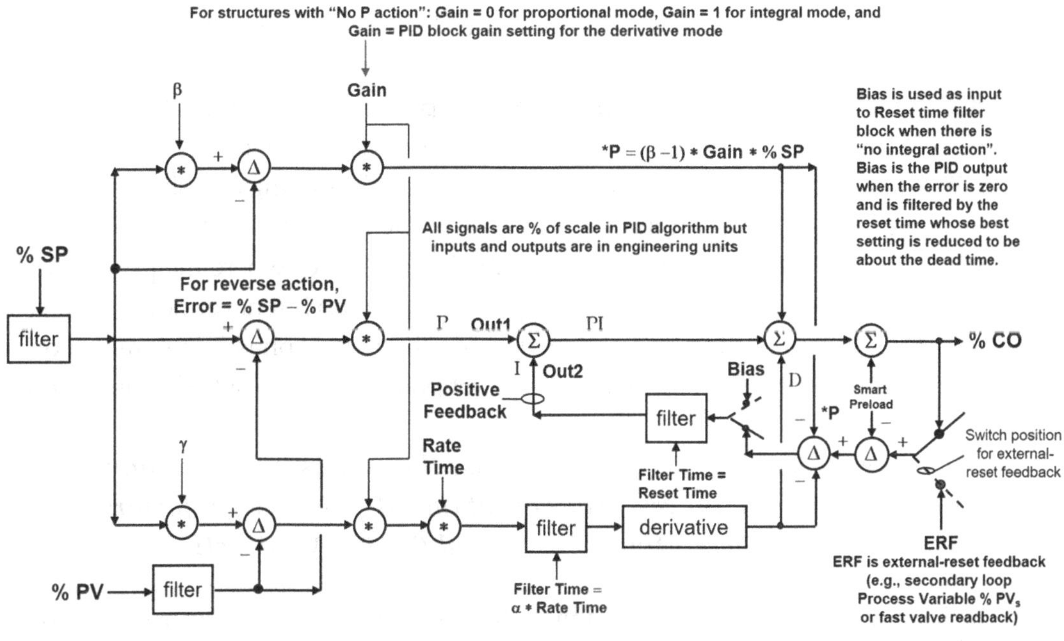

Figure 3-4. Standard Form with structure options and external-reset feedback.

Integral action will continue to try to increase the amount of heat until the temperature is greater than the set point, which is too late. The use of digital displays aggravates the problem in that a temperature a few hundredths of a degree below the set point will be perceived by the operator as requiring heating. Integral action is never satisfied, has no sense of direction, and tries to work off slight offsets from resolution limits by ramping the controller output. Proportional action decreases the request for heat and seeks cooling before the temperature reaches the set point. Proportional action also stops changing the output for an uncorrectable offset. PID controllers for composition, level, pH, pressure, and temperature loops on vessels and columns not tuned by the rules in Section 3.7 or by appropriate tuning software typically have integral action that is too large (reset time too small) and proportional action that is too small (gain too small) often by an order of magnitude or more.

3.5 PID Form and Structure

Use the Standard Form to reduce confusion on tuning rules and enable a greater focus on the application. Select the best PID structure for the Standard Form based on whether or not a mode is needed and how the set-point weigh factors β and γ are set (Figure 3-4) for the application. Table 3-4 gives eight different structure choices. If the proportional mode set-point weight factor β is set to 1 or 0, there is proportional action on error or PV, respectively. If the derivative mode set-point weight factor γ is set to 1 or 0, there is derivative action on error or PV, respectively. The default settings for most PID applications to provide some proportional action for set-point changes—but eliminate a kick in PID output for a set-point change from derivative action—are $\beta = 1$ and $\gamma = 0$ giving a structure of proportional and integral action on error (E) and derivative action on PV (PI on E and D on PV structure). To make the load and set-point response equivalent to a filter set point equal to reset time, set $\beta = 0$ and $\gamma = 0$ giving I on E and proportional plus derivative (PD) on PV. To provide a faster set-point response while minimizing overshoot similar to what is achieved by a lead-lag on the set point, use a 2DOF structure where you adjust β and γ between 0 and 1 (e.g., $\beta = 0.5$ and $\gamma = 0.25$). For processes where the process variable only responds in one direction, (e.g., temperature goes up but never down for a vessel with heating, but no cooling and no reactions or changes in phase), integral action is omitted because recovery from overshoot of the set point is not possible (PD on E, no I).

To make the load and set-point response the same, set the set-point filter time equal to the reset time or set the set-point weight factors $\beta = 0$ and $\gamma = 0$ creating a "I on E and PD on PV" structure.

If the integral mode is turned off, a bias (*B*) is adjusted to be the desired controller output when the error is nearly zero. The bias is an effective way of determining the controller output when an objective has been achieved, such as reaching the temperature and pH set point in a batch where there is not both heating or cooling and both acidic or basic reagents, respectively. The integral mode is not used in these types of applications where there is no ability to reverse direction as a result of overshoot caused by integral action. The integral mode is also omitted in a PD controller that is used to manipulate a trim valve to provide a faster and more precise response than a larger valve manipulated by a PID controller with the same set point. In this case, the bias is the desirable midrange position of the trim valve. The industrial PID algorithm uses percent signals.

If the integral mode is turned off, a bias is set that is the expected or desired PID output when the PV settles at the set point.

Finally, the PID controller must be tuned to meet loop objectives, such as minimizing peak and integrated errors for load disturbances, minimizing overshoot or time to reach the set point for set-point changes, minimizing overshoot of final resting value for heat integration, maximizing absorption of variability for level control, and minimizing resonance and interaction of interrelated loops.

3.6 PID Tuning

Because the omission of the sources, units, and types of open-loop gains in the literature has led to misconceptions and misunderstandings, this section goes into considerable detail in giving the basic equations that are used to define factors contributing to the open-loop gains. This section assumes the PID algorithm works with percent signals, which is the case for nearly all PID controllers in the process industry. The PID controller is used in many university courses, in MATLAB for simulation, and in a MeasureX system for sheet-line-thickness control work in engineering units. The conversion of tuning settings is extremely difficult.

The open-loop gains define what the PID controller sees as a percent change in PV ($\Delta\%PV$) for a percent change in controller output ($\Delta\%CO$). Consider a loop with a control valve or VFD. The open-loop steady-state gain is the product of the final control element, flow ratio, process, and measurement gains. For small changes, the final control element gain (change in flow divided by the change in percent PID output) is the slope of the valve or VFD installed flow characteristic curve at the current operating point (valve position or speed). The flow ratio gain is the inverse of the feed flow for composition, pH, and temperature control. For flow, level, and pressure control, the flow ratio gain is simply one. For small changes, the process gain is the slope of a plot of the process variable versus the ratio of the manipulated flow to the feed flow near the set point. The measurement gain (change in percent PID input divided by the change in the process variable) is 100% divided by the measurement span. Because the PID algorithm uses percent signals, the open-loop, steady-state gain must end up being the change in percent in the PID input divided by the percent change in the PID output. Consequently, the open loop gain is dimensionless despite the fact that the PID block and graphics show the PID process variable, and in some DCSs, the PID output in engineering units.

Because the PID algorithm works with percent input and output signals, the open-loop steady-state gain must be dimensionless and the open-loop integrating process gain has units of 1/s.

The open-loop steady-state gain for a self-regulating process is the final change in the percent process variable input ($\Delta\%PV$) after the process has reached a new steady state for a step change in the percent controller output ($\Delta\%CO$):

$$K_o = \frac{\Delta\%PV}{\Delta\%CO} \tag{3-6a}$$

The open-loop steady-state gain (K_o) is the product of the gains for the valve or VFD (K_v), flow ratio (K_r), process (K_p), and measurement (K_m). The product of the flow ratio gain and process gain defined here is most often simply expressed as a process gain [8]:

$$K_o = K_v \bullet K_r \bullet K_p \bullet K_m \tag{3-6b}$$

For small changes in flow, the valve or VFD gain is the slope of the installed flow characteristic near the current operating point that is the change in flow in engineering units (ΔF_v) divided by the change in controller output ($\Delta\%CO$):

$$K_v = \frac{\Delta F_v}{\Delta\%CO} \tag{3-6c}$$

For composition, temperature, and pH loops, the flow ratio gain (K_r) is inversely proportional to the main process feed flow (F_p):

$$K_r = \frac{\Delta R}{\Delta F_v} = \frac{\Delta F_v / F_p}{\Delta F_v} = \frac{1}{F_p} \tag{3-6d}$$

The process gain is the change in the process variable in engineering units (ΔPV) divided by the change in ratio (ΔR):

$$K_p = \frac{\Delta PV}{\Delta R} \tag{3-6e}$$

For flow and pressure loops, the product of the flow ratio and process gain simply becomes the change in the process variable per the change in flow (both in engineering units):

$$K_r \bullet K_p = \frac{\Delta PV}{\Delta F_v} \tag{3-6f}$$

It is not well recognized that loops for continuous composition, pH, and temperature control have an open-loop steady-state gain that is inversely proportional to feed rate.

The measurement gain (K_m) is 100% divided by the calibration span of the transmitter:

$$K_m = \frac{100\%}{\left(PV_{\max} - PV_{\min}\right)} \tag{3-6g}$$

Decreasing the measurement span will increase the open-loop gain via an increase in the measurement gain; this requires the PID gain setting to be decreased to prevent oscillations if the setting was already near the maximum allowable gain.

For the upper (primary) loop in a cascade control system, the PID output becomes the set point of a lower (secondary) loop rather than the set point of a valve or VFD. In Equation (3-6b), the set-point scale gain as shown in Equation (3-6h) for the manipulation of a set point is substituted for the gain for the manipulation of a valve or VFD. The nonlinearity of the installed flow characteristic that depends largely on pressure drop for a valve and on static pressure for a VFD has been eliminated from the view of the upper loop. The gain for manipulation of a set point is computed as detailed in Equation 3-6i as the span of the set-point scale of the lower loop divided by 100%. The set-point scale normally matches the measurement scale of the lower loop. If the DCS output scale is in percent, this match happens automatically. If the PID output scale is in engineering units, the user must set the upper PID output scale to match the lower PID scale. The most common lower (secondary) loop is flow, in which case there is still the flow ratio gain (K_r) for composition, pH, and temperature loops [8].

For cascade control where an upper composition, pH, or temperature loop is manipulating a lower flow loop, the valve or VFD gain is replaced with a set-point gain in the open-loop steady-state gain giving:

$$K_o = K_{sp} \bullet K_r \bullet K_p \bullet K_m \tag{3-6h}$$

The set-point gain is the span of the lower loop set-point scale divided by 100%:

$$K_{sp} = \frac{(SP_{\max} - SP_{\min})}{100\%} \tag{3-6i}$$

An important higher cascade control loop is an upper vessel temperature loop manipulating the set point of a heating or cooling coil or jacket. In this case, the flow ratio gain has been eliminated. Additionally, the process gain simplifies to about 1.0, giving Equation 3-6j for the heating or cooling of a batch vessel when there is no reaction or feed. Equation 3-6j is Equation F-3d in Appendix F with a zero feed flow and zero heat of reaction. The simplification and linearization of the open-loop gain is a significant advantage in addition to the compensation of cooling water or steam pressure or temperature changes by the coil or jacket loop before they appreciably affect vessel temperature.

For cascade control where an upper vessel temperature loop is manipulating the lower coil or jacket temperature loop, the open-loop gain simplifies to the product of the set point and measurement scale gains for zero feed flow and zero heat of reaction yielding:

$$K_o = K_{sp} \bullet K_m \tag{3-6j}$$

Note that an important check as to the correctness of a calculation in terms of units and even expressions is that the open-loop steady-state gain is dimensionless.

In cascade control where the vessel temperature is controlling the coil or jacket temperature, the open-loop steady-state gain can simplify to being linear and a function of temperature scale spans.

For integrating processes, the open-loop self-regulating (steady-state) gain (K_o) is replaced with an open-loop integrating process gain (K_i) with inverse time units (1/s), which results from the process gain term units not cancelling out the flow time units as seen in Equation 3-6k through 3-6m for level. For level, we can eliminate the ratio gain term defined by Equation 3-6d in the open-loop gain (K_o) for a self-regulating process giving an equation for the open-loop integrating process gain (K_i) with the distinction of no longer being dimensionless but having units of 1/s [8].

$$K_i = K_v \bullet K_p \bullet K_m \tag{3-6k}$$

$$K_v = \frac{\Delta F_v}{\Delta \%CO} \tag{3-6l}$$

$$K_p = \frac{1}{A \bullet \rho} \tag{3-6m}$$

$$K_m = \frac{\Delta\%PV}{\Delta L} \tag{3-6n}$$

One can convert an open-loop self-regulating process (steady-state) gain (K_o) for a process with a primary time constant (τ_o) much larger than the dead time (θ_o) to an open-loop integrating process gain (K_i) by the use of Equation 3-6o, which was developed for the near-integrator approximation to enable the use of tuning rules for integrating processes to improve disturbance rejection. The equation can also save a huge amount of time in the identification of the process dynamics because the largest persistent ramp rate, $Max(\Delta\%PV / \Delta t)$ within two to four dead times after a step change in the PID controller output can be used in Equation 3-6p to compute the integrating process gain instead of waiting for the process to essentially reach a steady state (i.e., 98% response time) to identify the primary time constant and the open-loop steady-state gain. For a time constant that is 20 times the dead time (not uncommon for well-mixed liquid reactors), the identification time is 20 times faster.

The conversion of the open-loop steady-state gain and primary time constant of a self-regulating process to an open-loop integrating process gain has many benefits in PID tuning and in the identification and analysis of process dynamics.

$$K_i = \frac{K_o}{\tau_o} \tag{3-6o}$$

$$K_i = \frac{Max(\Delta\%PV / \Delta t)}{\Delta\%CO} \tag{3-6p}$$

For integrating processes, the product of the controller gain (K_c) and the integral time (reset time) setting (T_i) must be greater than twice the inverse of the open-loop integrating process gain (K_i) to prevent the start of slow rolling oscillations [8].

$$K_c \bullet T_i > \frac{2}{K_i} \tag{3-6q}$$

Because most PID controllers on integrating processes have a controller gain much less than the maximum allowed, the equation is reformulated to show the minimum integral time.

$$T_i > \frac{2}{K_c \bullet K_i} \tag{3-6r}$$

The oscillations will decay slowly unless the following inequality is enforced:

$$T_i > \frac{1}{2 \bullet K_c \bullet K_i} \tag{3-6s}$$

Recent test results show that the previous rules apply to near-integrating and runaway processes as well. As one goes from near-integrating to true integrating to runaway processes, the consequences of violating these rules get more severe in that the oscillations are larger and slower to decay.

If the PID gain is much less than the maximum allowed, the reset time must be increased to prevent the start of slow, nearly undamped oscillations for near-integrating, true integrating, and runaway processes.

Most of the tuning problems associated with level control, gas pressure control of large volumes and composition, pH, and temperature control of large well-mixed liquid volumes stems from a reset time that is orders of magnitude too small (e.g., reset time is 100 s instead of 10,000 s). The reset time is too small because reset provides the type of response a human would provide (no sense of direction or anticipation, as noted in the Appendix B example). Gain settings are an order of magnitude smaller than the maximum allowed because gain settings above 10 are not described in the literature and the operator is concerned about full-scale PID output changes for set-point changes. Simulations can show the value of higher gain settings. Set-point rate limits can be added to the signal manipulating a valve, VFD, or secondary flow loop (i.e., effectively move suppression) to make these changes less disruptive to other loops if the PID controller has external-reset feedback.

The reset time setting for level, gas pressure, and well-mixed liquid volume composition, pH, and temperature control is frequently orders of magnitude too small.

Nomenclature for PID Gain and Reset Time Calculations

A = cross-sectional area of vessel (square meter)

K_c = controller gain (dimensionless)

K_m = measurement gain (PV percent/PV engineering units)

K_i = open-loop integrating process gain (PV percent per second per CO percent) (1/seconds)

K_o = open-loop self-regulating process gain (PV percent/CO percent) (dimensionless)

K_p = process gain (PV engineering units)

K_r = flow ratio gain often imbedded in process gain (1/flow engineering units)

K_v = valve or VSD gain (slope of installed flow characteristic) (flow engineering units/CO percent)

$\Delta\%CO$ = change in controller output converted to percent of controller output scale (CO percent)

ΔF_v = change in valve or VFD flow (flow engineering units)

F_p = process flow at current production rate (flow engineering units)

ΔL = change in level (meters)

$\Delta\%PV$ = change in the process variable converted to percent of controller input scale (percent)

PV_{max} = maximum process variable scale value (PV engineering units)

PV_{min} = minimum process variable scale value (PV engineering units)

ΔR = change in ratio of manipulated flow to process flow (dimensionless)

SP_{max} = maximum set-point scale value (SP engineering units)

SP_{min} = minimum set-point scale value (SP engineering units)

T_i = integral time (reset time) (seconds)

λ = lambda for closed-loop time constant or arrest time (seconds)

θ_o = total loop dead time (seconds)

τ_o = open-loop time constant (primary time constant) (seconds)

τ_s = secondary time constant (second largest time constant in loop) (seconds)

ρ = liquid density in vessel (kilograms/cubic meter)

The lambda tuning method for self-regulating processes is advocated by the author when the open-loop time constant is less than four times the total loop dead time. For self-regulating processes that are not near-integrating ($\tau_o < 4 \bullet \theta_o$), we have the following series of equations.

The self-regulating process tuning rules with limits are used for self-regulating processes where the open-loop time constant to dead time ratio is less than 4.

The reset time is set equal to the open-loop time constant (Equation 3-6t). The reset time steadily decreases as the ratio of the time constant to dead time decreases from its maximum of four dead times if this tuning method is used when the open-loop time constant is less than four times the total loop dead time. For systems with an open-loop time constant much less than the dead time (dead time dominant), the PID action becomes essentially integral only due to a steady decrease in the controller gain and the reset time. This type of control helps deal with the abrupt, almost step changes and noise in a dead time dominant system. Thus, the PID controller takes over the role of providing a gradual and smoothing response that is missing in the process. The best scenario is to have a large process time constant to fulfill this role, but for plug flow and sheet line processes, this is not possible.

$$T_i = \tau_o \qquad (3\text{-}6t)$$

A low limit is added to traditional equations to prevent a reset time smaller than 1/4 the dead time when the time constant is much smaller than the dead time (Equation 3-6u). This low limit provides more proportional mode action (Equation 3-6v). Keeping the PID gain from approaching zero gives a more immediate response to disturbances and set-point changes and helps the PID output get through valve or VFD deadband and resolution limits faster.

$$T_i = Max\left[0.25 \bullet \theta_o , \tau_o\right] \qquad (3\text{-}6u)$$

For dead time dominant processes, a low limit of 1/4 the dead time is placed on the reset time to provide some proportional action unless nearly integral-only control is desired.

$$K_c = \frac{T_i}{K_o \bullet (\lambda_f \bullet \tau_o + \theta_o)} \qquad (3\text{-}6v)$$

In industrial applications, the lambda factor multiplication of the open-loop time constant is replaced with lambda, the closed-loop time constant ($\lambda = \lambda_f \times \tau_o$), which is the time after the dead time to reach 63% of a set-point change. The use of lambda rather than a lambda factor provides the recognition that lambda is set relative to the dead time to minimize variability in the PID input or output and to provide the gain margin needed to deal with extenuating circumstances (e.g., interaction, inverse response, and resonance).

$$K_c = \frac{T_i}{K_o \bullet (\lambda + \theta_o)} \qquad (3\text{-}6w)$$

For maximum unmeasured disturbance rejection, a lambda equal to the dead time is used, giving Equation 3-6x for the PID gain setting when the reset time is not limited. For maximum disturbance rejection of a self-regulating process, a lambda equal to 0.6 times the dead time is used, giving nearly the same PID gain as the original

Ziegler-Nichols reaction curve method defined in the landmark paper "Process Lags in Automatic Control Circuits" [9].

$$K_c = 0.5 \bullet \frac{\tau_o}{K_o \bullet \theta_o} \tag{3-6x}$$

A rate setting is not normally included as part of the lambda tuning for self-regulating processes. If the primary time constant is greater than 1/2 the dead time, rate action may be beneficial. If a secondary time constant (the next largest time constant) can be identified, the rate time is set equal to the secondary time constant. The rate time should be larger than 1/2 the dead time but not be greater than 1/2 the reset time for Standard Form. The derivative mode must have a built-in filter whose time constant is a fraction of the rate time. The fraction is the inverse of the alpha setting (e.g., 8).

When derivative action is beneficial, the rate time should be the secondary time constant or 1/2 the dead time, whichever is largest, but not be greater than 1/2 the reset time for a Standard Form with a built-in filter whose time constant is a fraction of the rate time. The fraction is the inverse of the alpha setting (e.g., 8).

If the primary time constant is greater than 1/2 the dead time ($\tau_o > 0.5 \bullet \theta_o$):

$$T_d = Min\left[0.5 \bullet T_i, Max\left(0.5 \bullet \theta_o, \tau_s\right)\right] \tag{3-6y}$$

For maximum disturbance rejection of a self-regulating process, a lambda equal to 0.6 times the dead time is used.

The lambda tuning method for integrating processes is advocated by the author for use on self-regulating processes when the open-loop time constant is greater than four times the total loop dead time (near-integrating processes). This method is used to ensure sufficient gain action to provide the self-regulation that is appreciably missing in near-integrating processes and totally missing in true-integrating and runaway processes, and to prevent the violation of the window of allowable gains. The minimum PID gain is defined in Equation 3-6q for all of these processes where the ramp rate in the first four dead times gives an effective integrating process gain.

Integrating process tuning rules with limits are used for near-integrating processes where the open-loop time constant to dead time ratio is greater than 4 and for true integrating processes.

Near-integrating, true integrating, and runaway processes should use integrating process tuning rules with a low limit of four dead times for the reset time.

The reset time is twice the arrest time plus the dead time.

$$T_i = 2 \bullet \lambda + \theta_o \tag{3-6z}$$

A low limit is added so that the reset time is not less than four dead times.

$$T_i = Max\left[4 \bullet \theta_o, 2 \bullet \lambda + \theta_o\right] \tag{3-6aa}$$

The original equation for controller gain:

$$K_c = \frac{T_i}{K_i \bullet \left(\frac{\lambda_f}{K_i} + \theta_o\right)^2} \tag{3-6ab}$$

In industrial applications, the lambda factor multiplication of the inverse of the integrating process gain is replaced with lambda, the closed-loop arrest time ($\lambda = \lambda f/K_i$), which is the time to stop an excursion for an unmeasured disturbance (time to peak error). Lambda is minimized as a multiple of the dead time to minimize variability in the PID input or is maximized to maximize absorption of variability to reduce upsets to other loops (e.g., minimize disruption to downstream users when a level loop manipulates a discharge flow feeding these users).

The numerator uses the unlimited reset time.

$$K_c = \frac{2 \bullet \lambda + \theta_o}{K_i \bullet (\lambda + \theta_o)^2} \tag{3-6ac}$$

Near-integrating, true integrating, and runaway processes should use integrating process tuning rules with a low limit of four dead times for the reset time. The unlimited reset time is used in the numerator of the lambda calculation for the PID gain.

For maximum unmeasured disturbance rejection, a lambda equal to the dead time is used:

$$K_c = 0.75 \bullet \frac{1}{K_i \bullet \theta_o} \tag{3-6ad}$$

A lambda equal to three dead times minimizes the consequences of nonlinearities, inverse response, and resonance.

Equation 1.3f for the rate time is used for integrating processes including derivative action when the secondary time constant is not known:

$$T_d = Min\left[0.5 \bullet T_i, Max\left(0.5 \bullet \theta_{o,}\tau_s\right)\right] \quad \text{(3-6ae)}$$

3.7 PID Options

In cascade control loops, the secondary loop must be at least five times faster than the primary loop to prevent oscillations. For slow secondary loops that violate this rule, simply flipping the ERF switch and configuring the process variable of the secondary loop as the primary's ERF inlet signal can prevent such oscillations. Similarly, flipping the switch and using a fast readback of actual valve position or VFD speed can prevent the PID controller output from changing faster than the valve or VFD can respond. The movement of the PID output will also stop for deadband and stiction, reducing oscillations.

Simply putting a set point up or down rate limit on an analog output block or via a rate limit block on the splitter bock input and turning on ERF (e.g., dynamic reset limit) can slow down the integral action providing directional move suppression (DMS) that can enable better optimization by the valve position controller or prevent unnecessary crossings of the split-range point. Valve action to maintain process performance and safety would not be slowed down (e.g., coolant valves and vent valves would not be slowed down). For bioreactors, where pH is controlled by manipulating carbon dioxide and sodium bicarbonate, DMS would be set to slow down any increases in the sodium bicarbonate. This is critical for cell health because high sodium concentrations that can easily build up over the weeklong batch causes an increase in cell osmotic pressure, damaging cell membranes and leading to cell rupture. The many valuable uses of DMS will be noted in the recommendations and best practices for the best PID control.

ERF is also the key to an enhanced PID controller where the integral mode update is only done when there is an update of the PV measurement. The update of the integral mode is the computed first-order response of the reset filter for the time since the last measurement update. This capability is extremely valuable for wireless transmitters and analyzers. For self-regulating processes where the dead time from the wireless update time or analyzer cycle time is greater than the sum of the process dead time and the primary time constant, the PID controller does not have to be retuned as the wireless update time or analyzer cycle time increases. Amazingly, an off-line analyzer can be used with an incredibly large and variable time between lab results without retuning the PID controller. The PID gain calculation simplifies to the inverse of the largest open-loop gain. While the control is stable, the peak and integrated errors increase for unmeasured disturbances as the time interval between lab results increases.

ERF enables an often-neglected increase in PID gain. The proportional mode gives an immediate response for minimizing peak error and provides the missing negative feedback in the process. A common mistake of not using a high enough PID gain leads to more oscillations from many processes lacking self-regulation and from valves or VFDs with excessive deadband or poor sensitivity. Operators do not like sudden changes in controller output, and engineers are taught that the cause of oscillations and possible instability is the result of too high a PID gain. Subsequently, when oscillations occur, the tendency is to incorrectly decrease the PID gain, making the oscillation amplitude and period larger. If the PID gain and reset time were both increased, the oscillations would be smaller and faster and possibly stop.

3.8 PID Performance

The primary metrics for load disturbances (disturbances at the process input) are peak error and integrated error (accumulated error). Figure 3-8a shows that the minimum

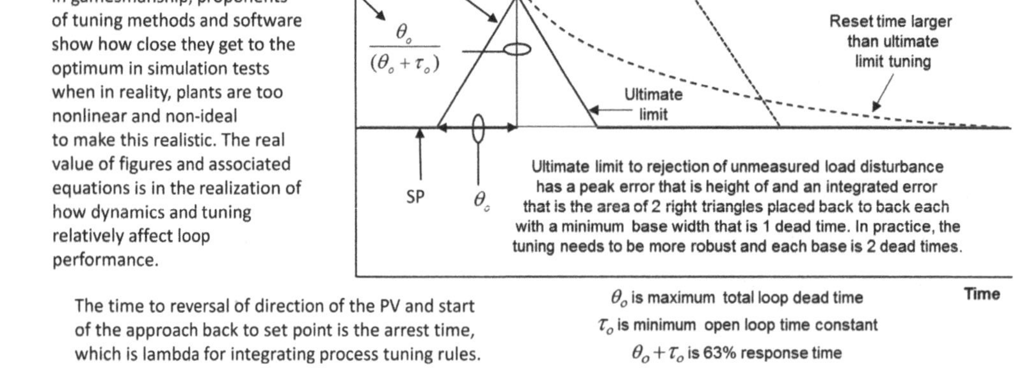

Figure 3-8a. Peak and integrated error metrics for load response.

peak error obtainable from feedback control for the most aggressive tuning is reached in one loop dead time. The minimum integrated error is the area of two right triangles placed back-to-back. A smaller PID gain increases the peak error. A slower integral action (larger reset time) most notably slows the return to the set point.

Internal model control and lambda tuning methods have been criticized for too low of a PID gain and too large of a reset time for loops with a large primary process time constant that resulted from an original emphasis on set-point response. The use of a lambda arrest time equal to the dead time and a switch from self-regulating to integrating process tuning for near-integrating processes with a low limit for reset time of four dead times and a rate time of 1/2 the dead time results in tuning settings that are close to the optimum, eliminating this criticism [3].

For the rejection of load disturbances, the metrics are peak error and integrated error.

Accumulated error is the totalized deviation of the controlled variable from the set point. For a composition control loop (e.g., a distillation column temperature loop or a pH control loop), the accumulated error multiplied by the average product flow provides a measure of the total amount of product that deviates from the optimum product purity specification. For a flow control loop (e.g., a reactor feed loop or a furnace fuel loop), the accumulated error provides a measure of the total amount of feed that deviates from the stoichiometric ratio specification that may be important not only for product purity, but also for safety (if the excess feed is flammable, explosive, exothermically reactive, or toxic) or emission control (if the excess feed or associated by-products are environmentally restricted). If the controlled variable is a utility flow (e.g., steam flow or cooling water flow), the accumulated error is representative of energy usage in excess of the set point.

The accumulated error is the integrated error where the positive and negative errors are canceled by the volume of the system to provide a net accumulated positive or negative error. Integrated absolute error (IAE) is equal to the accumulated error for an overdamped or critically damped (non-oscillatory) response. Integrated squared error (ISE) can be approximated by a combination of accumulated error and peak error. The value of calculating the IAE and ISE is usually not worth the additional effort by the design or maintenance engineer for loop performance analysis. The exception occurs for split-ranged flows where oscillations across the split-range point can cause wasted energy resulting from alternating between cooling and heating and wasted reagent resulting from alternating between acid and base addition. However, a small measured accumulated error does not necessarily indicate a well-tuned stable loop.

A small measured accumulated error can result from a loop that is marginally stable because the positive and negative errors will cancel for sustained oscillations. For a limit cycle, the accumulated error will generally steadily increase with time because the oscillation is rarely centered on the set point. The accumulated error can be accurately calculated by the use of a relatively simple algebraic equation if the oscillations are decaying in amplitude (Equation 3-8g).

The accumulated error of the process variable with respect to the set point provides a measure of the total quantity of reactants deviating from stoichiometric ratio, off-spec material produced, and the total quantity of excess energy, raw material, and reagent used.

The following equations for a self-regulating process can also be used for integrating and runaway processes. The open-loop error is the ramp rate if the controller is in manual multiplied by the closed-loop arrest time (lambda for integrating process tuning rules), which is a multiple of the dead time. A good learning tool is an Excel spreadsheet with fields to enter loop gains, scales, dead times, and time constants computing tuning settings and associated gain and phase margins for lambda as a multiple of dead time and resulting performance metrics.

Equation 3-8a shows how the accumulated (integrated) error for a closed loop can be calculated for a given size load disturbance from the PID gain, reset setting, execution time, and filter time [1, 2, 3, 4]. The measurement gain (100% divided by the measurement span) is included to get the error in percent of PID scale. The disturbance size is expressed as the percent change in PID output required ($\Delta\%CO$). For dead time compensators and the enhanced PID controller for wireless, the reset time can be decreased to the point where these equations will predict too low of an error.

$$E_i = \frac{T_i + \Delta t_x + \tau_f}{K_m \bullet K_c} \bullet \Delta\%CO \tag{3-8a}$$

A more useful formulation of the equation uses the size of the open-loop error (the error if the PID controller was in manual) from the disturbance. The change in valve or VFD flow, which is the change in controller output ($\Delta\%CO$) multiplied by the valve or VFD gain (K_v), must balance out the change in the disturbance variable (ΔDV) multiplied by the disturbance gain (K_d). The disturbance gain converts the load upset to the required change in manipulated flow:

$$\Delta\%CO \bullet K_v = K_d \bullet \Delta DV \tag{3-8b}$$

The load multiplied by the gains gives the open-loop error in engineering units:

$$E_o = K_d \bullet K_r \bullet K_p \bullet \Delta DV \tag{3-8c}$$

Solving Equation 3-8b for ΔDV and substituting for ΔDV in Equation 3-8c cancels out K_d:

$$E_o = K_v \bullet K_r \bullet K_p \bullet \Delta\%CO \tag{3-8d}$$

Solving Equation 3-8d for $\Delta\%CO$ and substituting the result into Equation 3-8a gives the integrated error in terms of the open-loop error.

$$E_i = \frac{T_i + \Delta t_x + \tau_f}{K_c \bullet K_v \bullet K_r \bullet K_p \bullet K_m} \bullet E_o \tag{3-8e}$$

$$K_o = K_v \bullet K_r \bullet K_p \bullet K_m \tag{3-8f}$$

$$E_i = \frac{T_i + \Delta t_x + \tau_f}{K_c \bullet K_o} \bullet E_o \tag{3-8g}$$

The accumulated error is proportional to the reset time and inversely proportional to the product of the PID gain and the open-loop self-regulating process gain.

The effect of the PID execution time or signal filter time on the accumulated error is negligible if the sum of these times is less than 10% of the reset time.

Nomenclature for Accumulated Error Metric

E_i = integrated error from load disturbance (PV percent seconds)

E_o = open-loop error (load disturbance error with controller in manual) (PV engineering units)

$\Delta\%CO$ = change in controller output converted to percent of controller output scale (CO percent)

ΔDV = change in disturbance variable (load) (disturbance engineering units)

K_c = controller gain (dimensionless)

K_d = disturbance gain (flow engineering units/disturbance engineering units)

K_m = measurement gain (PV percent/ PV engineering units)

K_o = open-loop self-regulating process gain (PV percent/CO percent) (dimensionless)

K_p = process gain (PV engineering units/flow engineering units or flow ratio)
K_r = flow ratio gain often imbedded in process gain (1/feed flow engineering units)
K_v = valve or VSD gain (slope of installed flow characteristic) (manipulated flow engineering units/CO percent)
T_i = integral time (reset time) (seconds)
τ_f = signal filter time constant (seconds)
Δt_x = controller execution time (seconds)

Equation 3-8g for the accumulated error leads to four conclusions that are important enough to emphasize. First, if the disturbances are nearly zero in magnitude ($E_o \approx 0.0$), even the most difficult loop will perform excellently. Therefore, before one can decide whether a difficult loop justifies the additional expense of special equipment, special instruments, or advanced control algorithms, knowledge of the size of the disturbance is necessary.

Second, if the controller was tuned with too small a gain (too large a proportional band) or too large a reset time (too slow a reset action), an easy loop will perform as poorly as a more difficult loop that required these mode settings. Any special efforts or expense during design to reduce dead time or increase a primary time constant will be wasted if overly conservative controller tuning is used in the field.

Third, if the PID execution time or filter time is large compared to the reset time, the accumulated error could be too large for even the best tuning.

Fourth, if the process gain is increased, the open-loop error, and therefore the accumulated error, increases per Equation 3-8g for a given disturbance. To recognize this relationship, realize that the K_p in the denominator is canceled by the K_p in the expression for PID gain K_c because K_c is inversely proportional to K_p. It is therefore important that the process engineer and control engineer review the effect of equipment design and operating conditions on the process gain at an early stage of the project (preferably before the project appropriation request is made final).

In applications where the measurement is an inference of a process condition (e.g., temperature is an inferential measurement of composition), a high process gain (high process sensitivity) provides better recognition of the disturbance, ensuring a small change in the process condition is above the threshold sensitivity limit of the sensor and the resolution limit of the PID. The resolution limit of wide-range thermocouple input cards in 1980s vintage DCSs was about 0.25°C, causing steps in the PID measurement that prevented the use of high PID gain and rate time settings. An increase in PID execution time and an increase in PID filter time settings helped to smooth the steps and increase the signal-to-noise ratio, but the accumulated error per Equation 3-8g was adversely affected. In a modern DCS, the resolution of the input cards is better than 0.02°C, which exceeds the capability of the sensor.

A PID controller tuned sluggishly will perform as badly as a loop with much more dead time. Any special efforts or expense during design to reduce dead time or increase a primary time constant will be wasted if overly conservative controller tuning is used in the field.

Peak error is the maximum deviation of the controlled variable from the set point. For a reactor temperature or reactor pH loop, the maximum deviation might have to be limited to prevent the start of an undesirable secondary reaction. For a pressure control loop, the maximum deviation might have to be limited to prevent actuation of relief valves or vacuum breakers. For a Claus furnace combustion control loop, the maximum emission level of sulfur dioxide must meet environmental regulations. (A Claus furnace produces sulfur from sulfur dioxide waste gas.) For a waste treatment neutralizer, the pH must always be between low and high environmental limits to prevent a recordable incident. The peak error is equal to approximately 1.5 times the steady-state offset for a proportional-only controller per Equation 3-8h developed by Peter Harriott in the book *Process Control* [5].

Limiting the peak error is important for preventing the triggering of undesirable reactions, high temperature and high-pressure shutdowns, environmental violations, and the blowing of relief valves and rupture discs.

For a proportional-only (P-only) controller:

$$E_x = \frac{1.5}{K_c \bullet K_o + 1.0} \bullet E_o \tag{3-8h}$$

The peak error is reached in about one dead time. The integral mode contribution is the number of repeats of the proportional mode contribution until the peak error is reached, giving Equation 3-8i. For processes with a small dead-time-to-time-constant ratio, the integral time is much larger than the dead time and the peak error is mostly affected by the proportional mode. For processes with a large dead-time-to-time-constant ratio (dead time dominant), the integral time for most tuning methods is about 1/2 the dead time, and the integral mode contribution approaches the proportional mode contribution for minimizing peak error. For lambda tuning of dead time dominant self-regulating processes, the integral time is set equal to the time constant, which makes the integral mode contribution much larger than the proportional mode contribution. The addition of the low limit of 1/4 the dead time for the reset time helps give more proportional action. Note that the controller gain affects the contribution from both modes. For dead time dominant processes, the controller gain is less than the inverse of the open-loop gain resulting in the peak error approaching the open-loop error. Figure 3-8a shows how the ultimate limit to the peak error depends on the ratio of the dead time to the 63% response time (dead time plus primary time constant).

For a proportional-integral (PI) controller:

$$E_x = \frac{1.5}{K_c \bullet K_o \bullet \left(1.0 + \frac{0.5 \bullet \theta_o}{(T_i + \Delta t_x + \tau_f)}\right) + 1.0} \bullet E_o \tag{3-8i}$$

For processes with a small time constant to dead time ratio (dead time dominant), the product of the PID gain and open-loop steady-state gain is about 0.2 and the reset time is less than 1/2 the dead time. The denominator ends up being about equal to the numerator. The result is that the peak error is about equal to the open-loop error, which is consistent with the observation that there is little if any reduction in open-loop error by feedback control for an unmeasured step load disturbance to a dead time dominant loop. There is more integral action than proportional action, and the process has reached close to its final value in one dead time, the minimum time needed for seeing and correcting for an unmeasured disturbance.

For dead time dominant processes, the peak error for a step disturbance approaches the open-loop error (error if there was no feedback control).

For processes with a large primary time constant to dead time ratio, a feedback controller is potentially capable of making the peak error a small fraction of the open-loop error if the controller is aggressively tuned to make the most of the proportional and derivative mode to provide an immediate response. The numerator in Equation 3-8i reduces from 1.5 to 1.25 from the derivative mode contribution, giving Equation 3-8j [8].

The peak error is best minimized by the more immediate action provided by the proportional and derivative modes possible with processes that have a large time constant to dead time ratio.

For a PID controller:

$$E_x = \frac{1.25}{K_c \bullet K_o \bullet \left(1.0 + \frac{0.5 \bullet \theta_o}{(T_i + \Delta t_x + \tau_f)}\right) + 1.0} \bullet E_o \tag{3-8j}$$

For PID control of processes dominated by a large primary time constant tuned with an aggressive gain as a near-integrator to minimize peak error, the equation simplifies to:

$$E_x = \frac{1.1}{K_c \bullet K_o} \bullet E_o \qquad (3\text{-}8k)$$

The size of the peak error for near-integrating, true integrating, and runaway processes is inversely proportional to the PID gain.

Nomenclature for Peak Error Metrics

E_x = peak error from load disturbance (PV engineering units)

E_o = open-loop error (load disturbance error with controller in manual) (PV engineering units)

K_c = controller gain (dimensionless)

K_o = open-loop self-regulating process gain (PV percent/CO percent) (dimensionless)

T_i = integral time (reset time) (seconds)

τ_f = signal filter time constant (seconds)

Δt_x = controller execution time (seconds)

The time to reach a new set point (rise time) in batch and continuous processes is minimized to reduce the time to reach a new operating condition. For near-integrating, true integrating, and runaway processes, the effect on start-up, transition, and batch times is significant. However, the rise time is much less important for bioreactors because the batch cycle times are long (e.g., 10 days) and fixed, and the consequences of a temperature or pH overshoot are so severe.

The time to reach a new set point (rise time) is important for minimizing the start-up and transition time of continuous unit operations and the cycle time of chemical batch operations.

The maximum deviation after the first crossing of the set point (overshoot) must be minimized for many of the same reasons as for minimizing peak error. Overshoot is independent of direction and is the maximum deviation above and below the set point for an increase and decrease in the set point, respectively. Bioreactors are particularly sensitive to overshoot in batch temperature or pH. An overshoot can result in cell death. In highly exothermic chemical reactors, an overshoot can trigger a runaway reaction.

The maximum deviation after a second crossing of the set point (undershoot) increases the settling time to reach the more optimum set point possibly lowering reaction yield from reduced growth and product formation rates for bioreactors and low reaction rates for chemical reactors. Figure 3-8b shows the rise time, overshoot, and undershoot for a step increase in the set point.

Overshoot and undershoot tend to have more of an effect on process efficiency than rise time.

The minimum total rise time (T_r) to go from 0% to 100% of set-point change can be approximated as the change in percent set point ($\Delta\%SP$) divided by the maximum rate of change of the process variable plus the total loop dead time. For an integrating or "near integrating" process, the maximum percent process variable ramp rate is the integrating process (K_i) gain multiplied by the change in controller output. The change in controller output is the minimum of the maximum allowable change per output limit ($\Delta\%CO_{\max}$) and the change as the result of the controller gain (K_c) and the feed-forward gain (K_{ff}) as detailed in the denominator of Equation 3-8l. If the step change in controller output from the proportional mode for a structure of proportional action on error is less than the maximum available output change ($\Delta\%CO_{\max}$) (i.e., the difference between the current output and the output limit), Equation 3-8l simplifies to Equation 3-8m for feedback control. In general, the tuning and PID structure that minimizes rise time increases the propensity for overshoot and undershoot [8].

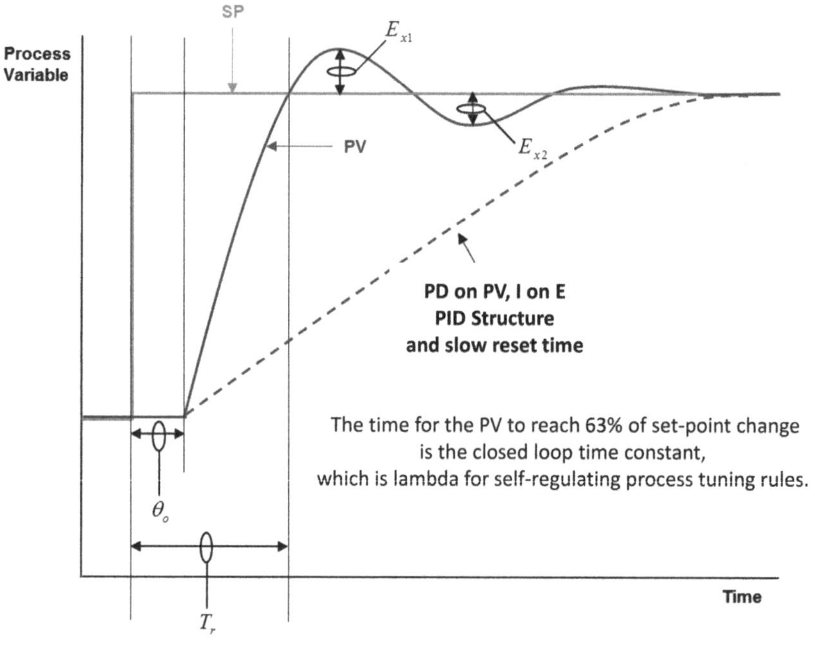

Figure 3-8b. 100% of set-point change rise time (T_r), overshoot (E_{x1}), and undershoot (E_{x2}) metrics.

The output change can be proactively set to make the set-point response faster. For set-point feedforward, the step change in output is a combination of the feedforward and feedback action. The bang-bang logic detailed by the author in the article "Full Throttle Batch and Startup Response" [6] minimizes the rise time by making a step output change that is the maximum available output change and then switching to a final resting value when a future value of the process variable is projected to reach the set point. These methods offer the opportunity to reduce rise time without increasing overshoot. Set-point rate limits on the manipulated secondary flow controller with external-reset feedback to the temperature controller provide a directional move suppression that can be used to smooth out step changes in flow from utility and raw material systems.

$$T_r = \frac{\Delta\%SP}{K_i \bullet \min(|\Delta\%CO_{\max}|, (K_c + K_{ff}) \bullet \Delta\%SP)} + \theta_o \quad (3\text{-}8l)$$

For a maximum available output change larger than the step from the proportional mode ($|\Delta\%CO_{\max}| > K_c \bullet \Delta\%SP$), the changes in the set point in the numerator and denominator cancel out, yielding a simpler equation for feedback control:

$$T_r = \frac{1}{(K_i \bullet K_c)} + \theta_o \quad (3\text{-}8m)$$

For the near-integrating process response seen in vessel and column temperature loops where the process time constant is significantly larger than the total loop dead time, the integrating gain is the open-loop gain (K_o) divided by the open-loop time constant (τ_o), yielding Equation 3-8n.

$$T_r = \frac{\tau_o}{(K_o \bullet K_c)} + \theta_o \quad (3\text{-}8n)$$

The equations assume that there is a step change in PID output that is the step change in the set point multiplied by the PID gain. However, these step changes in the PID output only occur if there is no set-point filter, rate limit, or lead-lag and a PID structure with full proportional action on error is used (e.g., "PI on error and D on PV" structure) or the beta factor is set equal to 1.0 in a 2DOF. A set-point lead-lag with the lead time set to zero is the same as a set-point filter. The promotion of a step change in PID output increases overshoot and undershoot unless the reset time is increased enough for the proportional mode to dominate correcting the step change in PID output as the process variable approaches the set point.

A PID structure with derivative action on error can help the PID output get through a large valve or VFD deadband and resolution limit faster for small set-point changes and small gains by a kick in the PID output (step change followed by a rapid exponential decay similar to what is seen when the lead is larger than the lag in a lead-lag block). The result can be an increase in the initial rate of approach to the set point. Otherwise, the impact on rise time from derivative action on error is minimal. Normally, a PID structure with derivative action on PV (e.g., "PI on error and D on PV") or a small gamma in a 2DOF PID structure is chosen to minimize kicking in the PID output from the derivative mode for set-point changes.

PID structures, tuning, and features applied to reduce overshoot tend to increase rise time.

Set-point rise time and overshoot can be minimized for near-integrating, integrating, and runaway processes by smart bang-bang logic. The disruption to the loops is minimized by the use of a set-point rate limit for the manipulated flow and external-reset feedback.

Nomenclature for Rise Time Metrics

$\Delta\%CO_{\max}$ = maximum change in the controller output converted to percent of controller output scale (CO percent)

$\Delta\%SP$ = step change in the set point converted to percent of PID input scale (PV percent)

K_c = controller gain (dimensionless)

K_{ff} = feedforward gain (percent/percent) (dimensionless)

K_i = open-loop integrating process gain (PV percent per second per CO percent) (1/seconds)

K_o = open-loop self-regulating process gain (PV percent/CO percent) (dimensionless)

T_r = rise time (time to reach a new set point) (seconds)

τ_o = open-loop time constant (primary time constant) (seconds)

θ_o = total loop dead time (seconds)

Step changes and oscillations in the PID output can be disruptive to other loops that are relying on a constant pressure in the same volume (e.g., a utility header or inline system) that is the source or destination of the flow manipulated by the PID. Some tuning techniques are designed to make the changes in the PID output as smooth as possible. Some go so far as to choose to minimize overshoot of the PID output beyond its final resting value (FRV). Although this objective may work for volumes with no back mixing and short residence times (e.g., catalyst bed gas reactors and plug flow liquid reactors), preventing step changes in the PID output and overshoot of the FRV is not feasible for integrating processes (e.g., level or batch temperature) and downright dangerous for runaway processes because an excursion in the process variable will not reverse direction until the PID output correction goes beyond the FRV.

While prevention of PID output overshoot is possible for self-regulating processes, it is not advisable for well-mixed volumes with large residence times that consequently have a large primary process time constant. The rise time can become orders of magnitude larger for these near-integrating processes. A more prevalent goal is to focus on the minimization of process variable overshoot (peak error after first crossing of the set point) and undershoot (peak error after second crossing of the set point). Undershoot is in the opposite direction of overshoot.

PID output overshoot of the final resting value of the manipulated variable (e.g., jacket temperature or coolant flow) is necessary for an integrating process or runaway process to reach a new set point.

PID output overshoot is of practical necessity for near-integrating processes to reach a new set point in a matter of minutes rather than hours.

The objective of minimizing PID output overshoot is only reasonable for moderate self-regulating or dead time dominant processes (e.g., plug flow unit operations).

In general, a decrease in rise time causes an increase in overshoot for feedback control. Overshoot and undershoot of the process variable can be reduced by making the contribution from the proportional and derivative modes optimally larger than the contribution from the integral mode. The integral mode is the principal cause of overshoot because this mode does not reverse direction of the change in PID output until the process variable crosses the set point. The integral mode has no sense of direction.

Overshoot and undershoot can be eliminated by a PID structure with no proportional action on error or a set-point filter time equal to the reset time at the expense of a considerable increase in rise time. The trade-off between rise time and overshoot can be optimized by the use of a 2DOF structure and adjustment of the beta and gamma factors or the use of a set-point lead-lag and an adjustment of the lead time to lag time ratio given that the lag time is set equal to the reset time.

The trade-off between rise time and overshoot from feedback control can be optimized by the use of a 2DOF PID structure or a set-point lead-lag.

For batch temperature control where there is only heating or only cooling (no split range), and there is no significant heat generation from reaction or heat loss from evaporation in the process, the response is in one direction only. For heating, the temperature can only increase, and for cooling, the temperature can only decrease. A one-sided response also occurs in batch pH control where there is only an acid or only a base reagent and there is no significant reagent consumption from reaction. For these one-sided responses, integral action will cause overshoot. The structure must have proportional action on error, derivative action on the process variable, and no integral action. When integral action is not used, an output bias must be specified. For these one-sided batch applications, the output bias is generally the PID output when the process variable is at the set point.

For batch control loops where the process variable can only respond in one direction, a PID structure with P on error and D on PV with no I must be used to prevent overshoot. The PID output bias is the PID output when the PV is at the set point.

3.9 Control Strategies

Mammalian bioreactors have the most stringent control requirements, particularly for pH and temperature. The liquid termed *media* consists of water, nutrients, and a substrate of sugar (glucose) and amino acid (glutamine). Agitation is kept at a low rate because mammalian cells have no cell walls (they have membranes) and are easily damaged by impeller shear. Mammalian bioreactors have all the process control for bacterial bioreactors but with more complex DO and pH loops (Figure 3-9a). Sparger design is critical to provide dispersion of gases and mixing and to reduce interaction between the DO and pH controllers.

Mammalian cell bioreactors have more stringent control requirements and a greater opportunity for process analytical technology (PAT) initiatives and process optimization due to the sensitivity of the cells to operating conditions and the sophistication of the cell functions.

The DO controller output is split-ranged to manipulate air and oxygen sparge and air overlay (vapor space addition). The pH controller output is split-ranged between carbon dioxide addition (carbonic acid) and sodium bicarbonate (base). These are weak

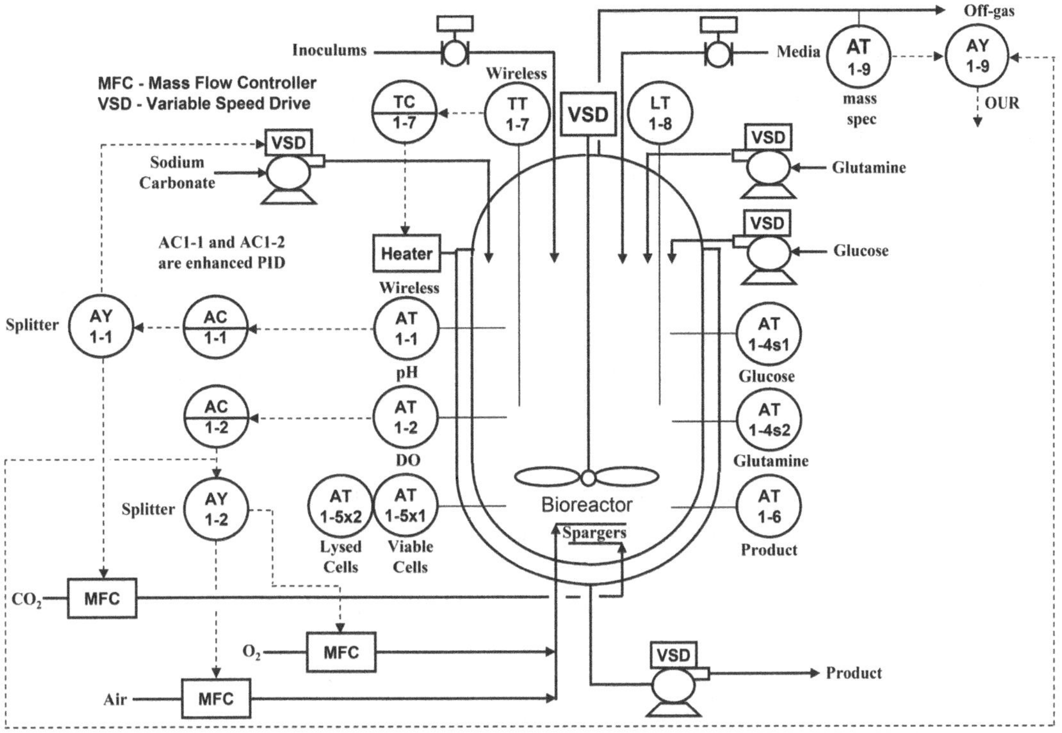

Figure 3-9a. Mammalian bioreactors have all the process control of bacterial reactors but with more complex DO and pH loops with decoupling and smarter split-ranged control.

acids and bases, providing a great moderating (buffering) effect on the titration curve slope and enabling much tighter pH control.

Weak acids and bases, and the use of a tiny aperture for the reference electrode junction, reduce the liquid junction potential, providing a more accurate pH measurement. The disturbances from changes in cell concentration and function take days to develop (slower than concentration disturbances in any other type of reactor). With proper tuning, pH control within 0.02 pH is possible. Instead of dealing with cell disturbances, the focus is on eliminating overshoot for changes in the pH set point, reducing interaction with the DO control, and preventing high carbon dioxide concentrations that can suffocate cells and the excessive accumulation of sodium bicarbonate concentrations that can increase the osmotic pressure (cell internal pressure).

Incredibly tight pH control is possible because of buffering and extremely slow disturbances.

Interaction between DO and pH control can be greatly reduced by sparger design. The remaining interaction can be dealt with by an enhanced PID controller for DO and pH control with a threshold sensitivity setting for ignoring noise and a decoupling feedforward signal that is the opposing controller's gas addition rate. Because tight pH control is more important than DO control, a feedforward of DO PID output (e.g., airflow set point) can be added to the pH PID output to help the pH controller immediately correct the carbon dioxide (CO_2) flow set point.

The feedforward does not need to correct the pH PID output when manipulating the sodium bicarbonate set point. This simple half decoupler of air to CO_2 flow set point is usually sufficient unless there is a sparger design problem. Problems occur when air and CO_2 flow share the same sparger, the spargers are too close to each other, or there is no tight pressure control of air or CO_2.

Good sparger design can eliminate most of the interaction between pH and DO control. What remains can be reduced by a half decoupler where the airflow set point from the DO PID output is added to the pH PID output that is the CO_2 flow set point.

High concentrations of CO_2 can be avoided by preventing overreaction of the pH controller to pH shifts (pH set-point changes to promote product formation) by using an enhanced PID controller and a secondary CO_2 flow PID controller with set-point rate limits. High CO_2 concentrations will dissipate with time. Inert gases are sometimes added to help sweep CO_2 out of the batch.

Mammalian bioreactor batches take 10 or more days. The exceptionally slow kinetics translate to extremely slow changes in reagent, oxygen, and cooling demand. Exceptionally slow load disturbances and slow integrating process gains result in negligible errors from disturbances. The objectives of tuning and control-loop performance for pH, DO, and temperature control are in terms of set-point response and optimization rather than disturbance rejection. The recognized opportunity is getting to a set point with no overshoot for a temperature shift and pH shift (set-point change) to move from enhancing cell growth to promoting product formation. A PID structure that has proportional action on the process variable rather than on error and no derivative is useful for preventing overshoot. Getting to the set point faster is not a consideration because the minutes saved are inconsequential in terms of the batch cycle time.

Mammalian cell kinetics are extremely slow, with batch cycle times of 10 or more days. The errors from load disturbances are so slow that tuning for disturbance rejection or getting to the set point faster is not the goal. Most of the focus is on the prevention of overshoot.

As previously mentioned, excessive accumulation of sodium bicarbonate causes a high sodium concentration leading to a high osmotic pressure that can contribute to cell membrane rupture and cell death. Unlike CO_2 that can escape in the vent gas, sodium bicarbonate is trapped in the media. Excessive crossing back and forth of the split-range point for pH control over the course of the batch can lead to high osmotic pressure. A split-range gap or deadband complicates tuning and does not stop the eventual excursion from the integrating process response. ERF (dynamic reset limit) in the primary pH PID controller and set-point rate limits in the secondary flow PID controller can provide the directional move suppression needed to slow down a movement into the direction of adding sodium bicarbonate, reducing unnecessary split-range excursions without requiring special tuning. New at-line analyzers can measure osmolality (osmotic pressure) by freeze point depression to help in optimization of the set-point rate limits.

External-reset feedback in the pH PID controller and set-point rate limits in the CO_2 and the sodium bicarbonate flow controllers can provide directional move suppression to eliminate unnecessary crossings of the split-range point and an increase in osmotic pressure.

New online and at-line analyzers for measuring glucose and glutamine open the opportunity for improved concentration control of this sugar and this amino acid. At present, glucose and glutamine addition are scheduled in anticipation of utilization rates. A move from sequenced batch charges to fed-batch closed-loop control can compensate for disturbances and consistently maintain a concentration (Figure 3-9b). The glutamine feed set point can be ratioed to the glucose feed set point. A feedforward signal for the glucose addition rate based on the oxygen uptake rate (OUR) can provide preemptive correction for the acceleration of substrate utilization in the exponential growth phase. The sampling rate for at-line analyzers does not need to be more frequent than once per 4 hours because the changes in mammalian cell metabolism are slow. An enhanced PID controller can provide tight control for relatively large sample times and can suppress reaction to noise and feedforward timing errors by a deadband (threshold sensitivity) setting.

New online and at-line analyzers of glucose and glutamine concentration offer the opportunity for optimizing the ratio of these concentrations for better cell performance (higher cell growth rate and product formation rate and better product quality).

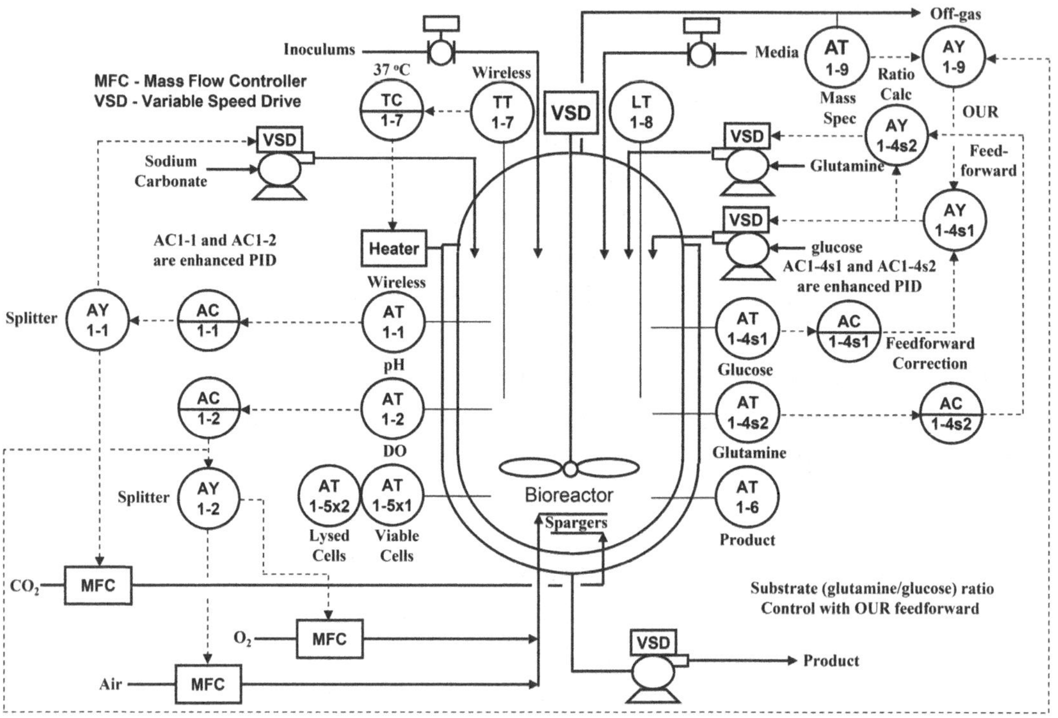

Figure 3-9b. Online and at-line analyzers enable substrate concentration control with glutamine feed ratioed to glucose. An OUR feedforward anticipates changes in glucose utilization rate.

3.10 Best Practices

RTD

1. Use an immersion length that is long enough to minimize heat conduction errors (e.g., L/D > 5).

2. Use an immersion length that is short enough to prevent vibration failure (e.g., L/D < 20).

3. Ensure velocity is fast enough to provide a fast response (e.g., > 0.5 feet per second [fps]) and to prevent fouling for sticky fluids and solids (e.g., > 5 fps).

4. For pipes, locate the tip near the centerline.

5. For vessels, extend the tip sufficiently past the baffles (e.g., L/D > 5).

6. Use a spring-loaded sheathed RTD sensor that fits tightly in the thermowell to minimize lag from air acting as an insulator (e.g., annular clearance < 0.02 in.).

7. To increase RTD reliability, use dual RTD elements except when vibration failure is more likely due to smaller gauge sensor, in which case use redundant separate thermowells.

8. Use a Class A element and four lead wires to minimize RTD measurement uncertainty.

9. Use integral mounted temperature transmitters in accessible locations to eliminate extension wire and lead wire errors and to reduce noise.

10. Use "sensor matching" and proper linearization tables in the transmitter for primary control loops and in safety instrumented systems achieving accuracy better than 0.1°C.

pH

1. Select the best glass, best reference design, and a fill that will prevent an appreciable increase in drift, error, failure rate, and the 86% response time for the worst case.

2. Use a spherical glass bulb for maximum accuracy and minimum response time.

3. Use a removable reference junction to enable an electrode to be rejuvenated (reference junction and fill quickly replaced) before plugging/poisoning becomes a problem.

4. If the deterioration in accuracy and equilibration speed requirement from the reference electrode causes poor continuous or batch performance or excessive maintenance, use a pressurized flowing junction to provide the most constant reference potential and the fastest junction equilibration rate to minimize the size and variability of the liquid junction potential. When inserting these electrodes in samples, ensure the junction flow rate is small and the sample size is large and stirred so that the sample pH does not change due to reference electrolyte entering the sample.

5. Use a Varipol (VP) or equivalent connector to easily disconnect the electrode cable, eliminating the need to disconnect the transmitter and retract cable through conduit.

6. Use a smart electrode with a stored calibration record.

7. Use a solution ground for impedance diagnostics and ground potential elimination.

8. Use a smart transmitter to detect glass and reference problems.

9. Ensure the electrode is installed with the tip pointing down at about a 45° angle to prevent bubbles from residing in the tip or being caught on the internal electrode.

10. Ensure electrodes are always wetted (a common problem for batching and shutdown).

11. Use middle signal selection to eliminate response to single failure and reduce the effect of noise, coatings, and contamination and to provide more intelligent diagnostics of a deviant electrode. This reduces life-cycle costs.

12. Ensure the stream velocity and tip protective shroud design and orientation is the best for the process conditions. Use a low velocity (0.1 to 1 fps) and protect the tip from flow impingement to reduce abrasion or contamination and a high velocity (5 to 10 fps) and sweeping of action at the tip to reduce coatings.

13. Ensure the electrode tip extends into the pipe centerline and past vessel baffles.

14. Choose an electrode location free from bubbles (e.g., not near the sparger ring).

15. Locate the electrode sufficiently downstream from a pump or static mixer to reduce concentration fluctuations but not so far as to increase dead time by more than 6 s.

16. Put insertion electrodes in series in the same line (at least 10 pipe diameters apart) rather than in parallel lines so each electrode sees the same velocity and same composition.

17. Choose an electrode location free from flashing (e.g., not in the pump suction or valve discharge).

18. Ensure the insertion type of electrode assembly has a ball valve and a restraining strap to safely withdraw (retract) the electrode from a pipe or vessel and prevent blowout.

19. Ensure the electrode and transmitter location is safely accessible for maintenance.

20. Ensure the electrode and transmitter signal connections are always dry by location (i.e., that connections are not exposed to process fluid or moisture in air) and type of connection seals and, if necessary, by the use of a dry air purge. Avoid submersion assemblies and transmitter enclosure or electrode holder cable entries that are into the top or are not sealed.

Coriolis

1. Select a Coriolis sensor material that is chemically compatible with all expected process fluids and cleaning/flushing agents.

2. Orient the meter properly to keep the meter full and avoid two-phase conditions—for liquids: tubes down or flag position flowing up; for gases: tubes up or flag position flowing down.
3. Size the meter properly to maintain the best measurement performance across the full process flow range within an acceptable pressure loss range.
4. Insulate and/or heat-trace the meter for the best density, concentration, and volume measurement performance, or if temperature change will cause undesirable effects such as flashing, condensation, or freezing.
5. Verify the calibration constants from the sensor are reflected properly in the electronics.
6. Locate all control valves downstream of the meter to provide back pressure (and to keep the tubes full).
7. Perform a zero calibration if the normal process temperatures or pressure is far from the manufacturer's calibration conditions. Circulate fluid at the process pressure until all meter components are brought to thermal equilibrium, then block the meter in with no flow, and zero the meter.
8. Determine a field baseline for the tube integrity diagnostic test values (e.g., smart meter verification) if required by the manufacturer.

Valves and VFDs

1. Use sizing software with physical properties for worst-case operating conditions.
2. Include the effect of the piping reducer factor on the effective flow coefficient.
3. Select the valve location and type to eliminate or reduce damage from flashing.
4. Preferably, use a sliding stem valve to minimize backlash and stiction.
5. Use a roller diaphragm valve achieving best precision and compliance with the 3-A Sanitary Standards, Inc., for clean-in-place (CIP) and sanitize-in-place (SIP) service.
6. Use Teflon, and for higher temperature ranges, use ultra low friction (ULF) packing.
7. Compute the installed valve flow characteristic for worst-case operating conditions.

8. Size the actuator to deliver more than 150% of the maximum torque or thrust required.
9. Select an actuator and positioner with threshold sensitivities of 0.1% or better.
10. Ensure the total valve assembly deadband is less than 0.4% over the entire throttle range.
11. Ensure the total valve assembly resolution is better than 0.2% over the entire throttle range.
12. Choose an inherent flow characteristic and a valve-to-system pressure drop ratio that does not cause the product of the valve and process gain divided by the process time constant to change more than 4:1 over the entire process operating point range and flow range.
13. Tune the positioner aggressively for application without integral action with readback that indicates actual plug, disk, or ball travel instead of only actuator shaft movement.
14. Use volume boosters on the positioner output with the booster bypass valve opened enough to assure stability o reduce valve 86% response time for large signal changes.
15. Use small (0.2%) as well as large (20%) step changes to test valve 86% response time.
16. Use International Society of Automation standard and technical report relaxing expectations on travel gain and 86% response time for small and large signal changes, respectively.
17. Counterintuitively increase PID gain to reduce the oscillation period and/or amplitude from backlash, stiction, and poor actuator or positioner sensitivity.
18. Use external-reset feedback of accurate and fast valve position readback to stop oscillations from poor valve precision and slow response time.
19. Use input and output chokes and isolation transformers to prevent electromagnetic interference (EMI) from the inverter.
20. Use pulse-width modulation (PWM) to reduce torque pulsation (cogging) at low speeds.
21. Use an inverter duty motor with Class F insulation, a 1.15 service factor, and a totally enclosed fan-cooled (TEFC) motor with a constant speed fan or booster

fan or a totally enclosed water-cooled (TEWC) motor for high temperatures to prevent overheating.

22. Use a National Electrical Manufacturers Association (NEMA) Design B instead of Design A motor to prevent a steep torque curve.
23. Size the pump to prevent operation on the flat part of the pump curve.
24. Use a recycle valve to keep the pump discharge pressure well above static head at low flow and a low speed limit to prevent reverse flow for the highest destination pressure.
25. Use at least 12-bit signal input cards to improve the resolution limit to 0.05% or better.
26. Use a drive and motor with a generous amount of torque for the application so that speed rate-of-change limits in the variable frequency drive (VFD) setup do not prevent changes in speed from being fast enough to compensate for the fastest possible disturbance.
27. Minimize deadband introduced into the drive setup, as it would cause delay and limit cycling.
28. For tachometer control, use magnetic or optical pickup with enough pulses per shaft revolution to meet the speed resolution requirement.
29. For tachometer control, keep speed control in the VFD to prevent cascade rule violation where the secondary speed loop is not five times faster than the primary process loop.
30. To increase rangeability to 80:1, use fast cascade control of speed to torque in the VFD to provide closed-loop slip control.
31. Use external-reset feedback of accurate and fast speed readback to stop oscillations from poor VFD resolution and excessive deadband and rate limiting.
32. Use foil-braided shield and armored cable for VFD output spaced at least 1 ft from the signal wires with never any crossing of signal wires, ideally via separate cable trays.

PID Controller

1. Realize that the PID algorithm works with percent signals introducing a measurement gain that is 100% divided by the PV scale span (Equation 3-6g) and an output gain that is the manipulated flow span or secondary loop PV scale span divided by 100% (Equation 3-6i).

2. Realize that the open-loop gain is the product of the manipulated valve or VFD gain or secondary loop scale gain, the flow ratio gain for composition, pH, and temperature control, process gain, and measurement scale gain.
3. Use the PID Standard Form and make sure units of tuning settings are addressed.
4. Use the PID structure that best meets application requirements.
5. Match the PID output limits and engineering units to the set-point limits and engineering units of the secondary loop. Set any anti-reset windup limits to match the output limits except for special cases where you set the anti-reset windup (ARW) limits inside the output limits when the PID controller must move quickly through stiction at a low limit or recover from a high limit.
6. Use lambda self-regulating process tuning rules when the primary time constant is less than four times the total loop dead time for self-regulating processes.
7. Use lambda integrating process tuning rules for runaway, true-integrating, and near-integrating processes when the primary time constant is greater than four times the total loop dead time for self-regulating processes.
8. To eliminate abrupt changes in PID output from high PID gain, put set-point rate limits on the analog output block or secondary PID set point and use external-reset feedback to prevent the PID output from changing faster than the desired rate of change.
9. Use more aggressive gain to deal with deadband, stiction, and poor sensitivity.
10. Use fast external-reset feedback (ERF) from readback of the actual valve position to stop limit cycles from deadband in cascade loops or in single loops on integrating processes or with integral action in the positioner and to stop limit cycles from stiction in all loops.
11. Use fast ERF from readback of the actual valve position and speed to stop oscillations from a slow valve or slow VFD, respectively.
12. If the process is near-integrating, true integrating, or runaway, ensure the product of the PID gain and reset time is greater than twice the inverse of the integrating process gain to prevent the start of slow oscillations (lambda greater than the dead time).
13. Use fast ERF of the secondary loop PV to stop oscillations from violating the cascade rule that the secondary loop must be five times faster than the primary loop.

14. Use slower rate limiting of the analog output or the secondary loop set point in the safe direction of approaching the split-range point with ERF of the manipulated valve position or flow to prevent unnecessary crossings of the split-range point.

Bioreactor Control

1. For cascade control, make sure the primary loop is tuned with a lambda that is at least five times greater than the lambda for the secondary loop.
2. If there is interaction between DO and pH loops, use a decoupler.
3. Use ERF and directional set-point limits (section 8.1 in reference 3) to prevent rapid changes in manipulated flows and to prevent unnecessary crossings of the split-range point (especially in the direction of adding sodium bicarbonate that increases sodium ion concentrations and, consequently, osmotic pressure).
4. For DO control, use a split range between air for the pre-exponential growth phase and oxygen for the exponential growth phase supplemented by air overlay as needed. Use adaptive control for changes in the growth rate (slope of viable cell concentration) and changes in the gain of manipulated flow (e.g., gain for oxygen is five times that for airflow).

References

1. Gregory K. McMillan, *Essentials of Modern Measurements and Final Elements in the Process Industry* (Research Triangle Park, NC: ISA [International Society of Automation], 2010).
2. Gregory K. McMillan, *Advances in Reactor Measurement and Control* (Research Triangle Park, NC: ISA [International Society of Automation], 2015).
3. Gregory K. McMillan and P. Hunter Vegas, *Industrial/Process Instruments and Controls Handbook*, 6th ed. (New York: McGraw-Hill Education, 2019).
4. F. Greg Shinskey, *Process Control Systems: Application, Design, and Tuning*, 4th ed. (New York: McGraw-Hill Companies, Inc., 1996).
5. Peter Harriott, *Process Control* (New York: McGraw-Hill Companies, Inc., 1964).
6. Gregory K. McMillan, "Full Throttle Batch and Startup Response," *Control* (May 2006).

7. Gregory K. McMillan and P. Wu, "Valve Response—Truth or Consequences," *Control* (March 2016).

8. Gregory K. McMillan, *Tuning and Control Loop Performance,* 4th ed. (New York: Momentum Press, 2014).

9. John G. Ziegler and Nathaniel B. Nichols, "Process Lags in Automatic Control Circuits," *ASME Transactions* (1943): 433–444.

4 Model Predictive Control

4.1 Introduction

Because model predictive control (MPC) uses an experimental model, it can create a future trajectory of the process response based on multiple measured process inputs. Unknown disturbances, noise, limit cycles, random error, and incorrect model parameters result in a bias correction and a shift of the trajectory. Because MPC seeks to minimize the squared error of the trajectory over the time horizon, the short-term effects of unknowns and erratic signals are minimized. In contrast, a proportional-integral-derivative (PID) control block only knows what it sees as a change or rate of change for one process feedback measurement in the current scan. Additionally, MPC can simultaneously manipulate multiple variables, whereas a PID control block is restricted to one controller output and one feedback measurement. Consequently, PID controller must use various downstream blocks, such as the *splitter* block for split-range control of multiple manipulated variables (MVs) and the *selector* block for override control of multiple controlled variables (CVs). With all these PID techniques, a single CV is matched up with a single MV at any given time with no inherent knowledge of the dynamics of the pairing. A PID controller in override control uses sequential pairing, whereas MPC offers simultaneous manipulation of multiple process inputs for control of multiple process outputs. Finally, MPC is inherently better at optimization because it can predict future violations of constraints; has built-in features for maximization, minimization, and setting priorities; has a tuning adjustment to smooth out the optimization; and has hooks to a linear program.

Section 4.2 of this chapter addresses how process gains and an inaccurate model dead time prevent MPC from doing its job. MPC becomes more sensitive to an inaccurate dead time if the process is dominated by dead time or the process gains indicate that the process inputs are not actually independent variables, that the process outputs do not strictly depend on the process inputs, or that some process inputs have too great or too little of an effect. This section discusses the performance advantages of MPC in terms of its functional capability and concludes with a summary of its advantages.

Section 4.3 describes the setup, tuning, and performance of MPC for the simultaneous manipulation of multiple process inputs. In particular, the section details MPC's ability to optimize process inputs while rejecting load disturbances and eliminating split-ranged control. A standard MPC 9×9 block has been offered in distributed control systems for more than a decade.

Section 4.4 begins with an overview of the optimum relationships between biomass growth rate and process conditions, such as temperature, pH, dissolved oxygen (DO), and substrate concentration. The section alerts the user to a potential change in the sign of the process gain because of a peak in the optimum process variable that necessitates a change in controller action. The section also shows how translating CVs is essential for optimization of the batch profiles of biomass and product concentrations. It ends with an example of the MPC models and setup.

Section 4.5 will give an example of using MPC to control and optimize mammalian cell growth and production rates by manipulation of glucose and glutamine feed rates with an oxygen uptake rate (OUR) feedforward.

Section 4.6 will list best practices for MPC design and implementation.

Learning Objectives

- Understand how the MPC algorithm gains knowledge of the future from the past
- Recognize the implications of MPC's long-term view versus the PID controller's short-term view
- Appreciate MPC's ability to deal with erratic responses and interaction
- Know how to check whether an MPC solution is viable
- Be able to summarize the advantages of MPC
- Know how to set up an MPC block for optimizing a process input

- Be able to set up an MPC block to replace split-range control
- Recognize the nature of the optimization problem
- Determine the requirements for a change in control action for optimization
- Understand the translation of CVs needed for profile control
- Be aware of the types of models and MPC setup used for profile control

4.2 Capabilities and Limitations

MPC uses incremental models of the process in which the change in controlled or constraint variables is predicted from a history of changes in the MVs, called *moves*, and changes in the disturbance variables. The initial values of the controlled, constraint, manipulated, and disturbance variables are set to match those in the plant when MPC is in the initialization mode. MPC can be run in manual and the trajectories monitored. If the trajectories drastically change or flip-flop from one execution to another, then the models must be improved or the model's time horizon, which is how far trajectories into the future are seen by the model and operator (projected time to steady state), increased. Often, the significant delay and slowness in the observed response helps operations and process technology personnel appreciate the implications of the process dynamics. While in manual, a back-calculated signal from the downstream function blocks is used. This makes it possible for the MVs to track the appropriate set points so MPC is ready for a bumpless transition to automatic. When a new MPC application is commissioned, the MPC block maximum and minimum output limits are narrowed until experience and confidence are gained with the model accuracy and optimization strategy [1, 2].

The trend in the future response to current disturbances and past moves helps operations and process technology personnel appreciate the implications of process dynamics.

A simplified review of MPC's functionality will help us gain the understanding essential to our later discussion of MPC's capabilities and limitations. Figure 4-2a shows the response of a change in the CV to a step change in each of two MVs at time zero. If the step change in the MVs were twice as large, the individual responses of the CV would be predicted to be twice as large. The bottom plot shows the linear combination of the two responses. Nonlinearities and interdependencies can cause this principle of linear superposition of responses to be inaccurate. Any error in the modeled-versus-actual process response shows up as an error between the predicted value and actual value of the CV, as shown in the upper plot of Figure 4-2b. A portion of this error is then used to bias the process vector, as shown in the middle plot. The

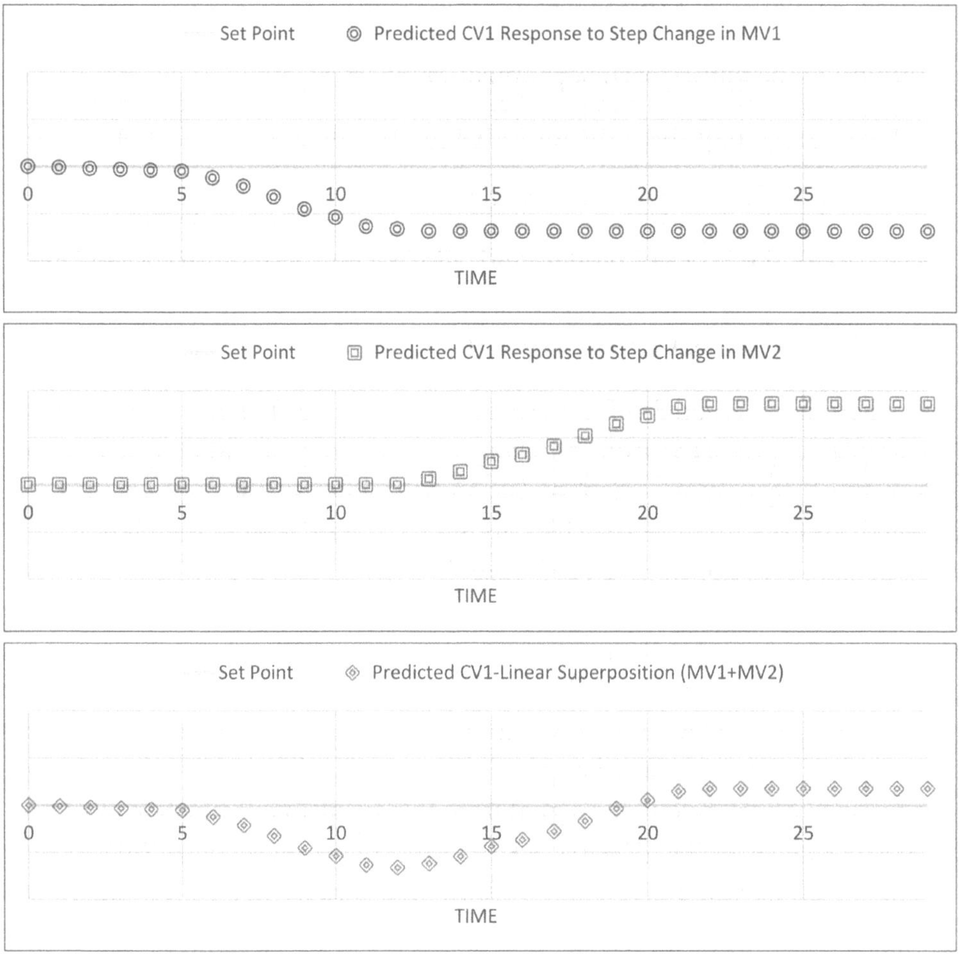

Figure 4-2a. Linear superposition of MPC response.

MPC algorithm then effectively calculates a series of moves in the MVs that will provide a control vector that is the mirror image of the process vector about the set point. This is shown in the bottom plot of Figure 4-2b. MPC has a built-in knowledge of the combined inverse response where the initial reaction of the CV is in the opposite direction of the final change in the CV. If there are no nonlinearities, load upsets, or model mismatch, the predicted response and its mirror image should cancel out, with the CV ending up at its set point. How quickly the CV reaches the set point depends on the process dynamics, penalty on move (PM, penalty for changes to a MV), move size limit, and number of moves set for the MVs [1, 2].

MPC can handle interactions better than PID control because it can simultaneously manipulate multiple process inputs to control the predicted response of multiple process outputs to the combined effects of disturbances and previous process inputs.

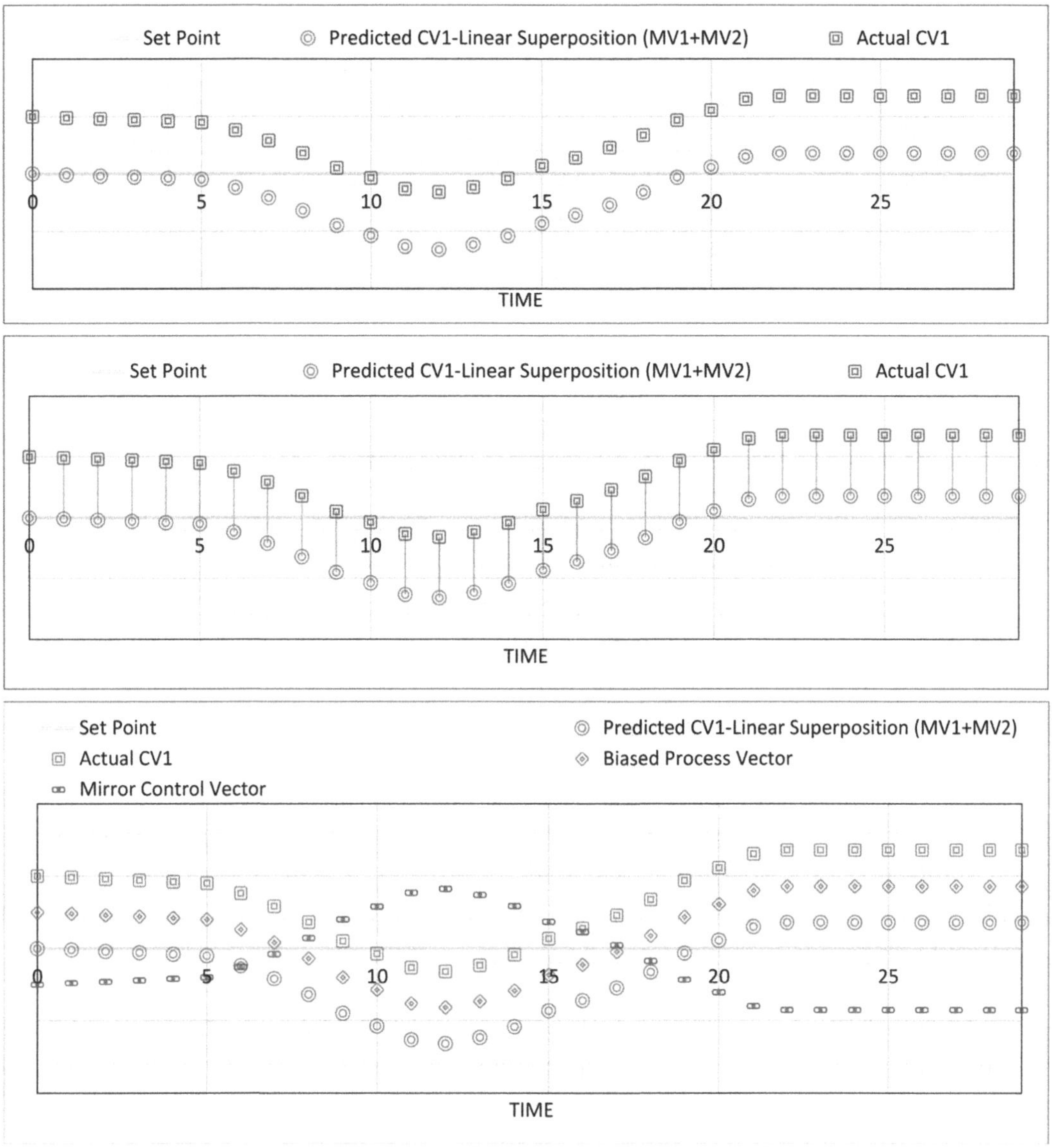

Figure 4-2b. Shift of process vector and mirror image control vector.

Based on the process model, the MPC predicts the process behavior (CV response) based on projected future moves over a time horizon. The controller, however, executes only the first projected move at the current time. Based on this first move, the CV experiences a change that might differ from the expected or projected response. This difference is the result of disturbances, noise, limit cycles, random error, and incorrect model parameters. A portion of the difference is used to provide a bias correction to shift the trajectory. The MPC executes its algorithms again to compute a new MV move based on the biased response.

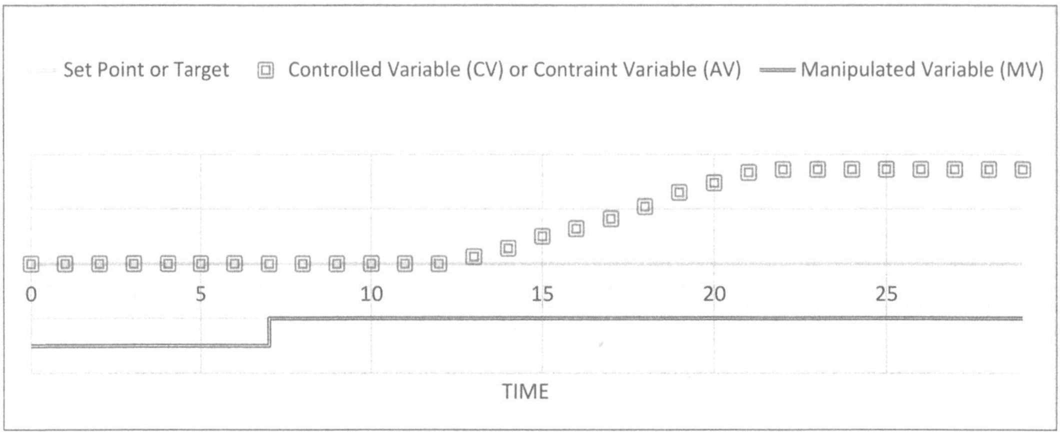

Figure 4-2c. MPC and PID controller views of the process response.

Each move in the MV is actually the linear summation of all the first of a projected series of moves based on the controlled, disturbance, and constraint variables. Figure 4-2c shows how MPC has a view of the trajectory of each controlled and constraint variable based on changes in the manipulated and disturbance variables. The prediction horizon is the time to steady state, which is approximately the time to 98% of the final value (T_{98}), as defined by Equation 2-3 in Chapter 2.

In contrast, a PID controller only sees the current value and the rate of change of its CV. Adding a dead-time compensator only extends the controller's view of the future to the end of the dead time, as shown in Figure 4-2c. The derivative (rate) mode provides some anticipation based on the CV's slope of response, and the proportional (gain) mode gives an immediate and abrupt action.

Feedforward and decoupling can be added, but the addition of these signals to the controller output is based on current values, has no projected future effect, and is designed to achieve the goal of returning a single CV back to its set point [1, 2].

MPC takes a long-term view of the process response, whereas the PID controller reacts to the immediate situation. Consequently, abrupt, erratic, or nonrepresentative changes, such as steps or a scatter of readings from a chromatograph, resolution limits, process and measurement noise, and inverse responses, are more problematic for a PID controller.

MPC can handle noise, resolution limits, steps, random errors, and erratic responses better than a PID controller because MPC has a longer-term view of the process response.

Tuning an MPC application can often be reduced to PM, also known as *move suppression,* which sets the degree of transfer of variability from the CV to the MV. The

PM combined with a move size limit offered by MPC prevents large and fast changes in the MVs.

The performance advantages of MPC that directly relate to economic benefits are its inherent capability to meet process objectives, manipulate multiple outputs, and deal with interactions. Because it also computes the trajectories of constrained variables and has built-in capabilities for suppressing moves and maximizing or minimizing an MV, it is well suited to multivariable control and optimization. Move suppression provides the smoothness desired for optimization.

MPC's built-in ability to handle interactions, predict future violations of constraints, and smoothly maximize or minimize an MV is important for optimization.

As with the PID controller, a trade-off exists between an MPC's performance (minimum peak and integrated error in the CV) and its robustness (maximum allowable unknown change in the process gain, dead time, or time constant). Higher performance corresponds to lower robustness. However, the direction of the dead-time change that most destabilizes MPC is the opposite of the change that destabilizes a PID controller [1].

MPC's robustness is more sensitive to a decrease than an increase in dead time. This is especially true in dead-time-dominant processes, where the total dead time is much greater than the process time constant. A decrease in plant dead time can rapidly lead to growing oscillations that are much faster than the ultimate period. A large increase in dead time shows up as much slower oscillations, with a superimposed high-frequency limit cycle. In contrast, a PID controller can become unstable for an increase in total loop dead time. A decrease in dead time for a PID controller translates to lost opportunity in performance because the PID controller becomes less aggressive than is possible in terms of its reset and gain settings [1, 2].

Unlike PID standard control without dead time compensation, MPC is oscillatory for an actual plant dead time that is either smaller or larger than the identified dead time. MPC is also more sensitive to and destabilized by a decrease in plant dead time in dead-time-dominant processes.

PID controllers thrive on the steady and gradual response of a process time constant or process integrating gain that is much slower than the total dead time. If the PID controller has a single MV and no interactions, PID control can achieve a level of fast, unmeasured load disturbance rejection that is difficult to beat. However, for bioreactors, the disturbances are so slow and the PID controller is so detuned that moving to MPC with a good model may improve control even in cases where the PID controller normally excels.

For measured disturbances, MPC generally has a better dynamic disturbance model than a PID controller with feedforward control. This is primarily because it is difficult for the user to properly identify the feedforward lead-lag times for PID control. Often, the feedforward dynamic compensation for PID controllers is omitted or adjusted by trial and error.

For constraints, MPC anticipates a future violation by looking at the final value of a trajectory versus the limit. MPC can simultaneously handle multiple constraints. PID override controllers handle constraints one at a time through the low or high signal selection of PID controller outputs.

For interactions, MPC is much better than PID control. The decoupling signals sometimes added to a PID controller are generally exclusively computed based on steady-state gains. This PID decoupling ignores the effects of the relative speed of loops on interaction. However, the benefits of MPC over detuned or decoupled PID controllers deteriorate as the condition number of the matrix increases.

The steady-state gains of the 2×2 matrix in Equation 4-2a show that each MV has about the same effect on the CVs. The inputs to the process are linearly related. The determinant is nearly zero and provides a warning that MPC is not a viable solution [1, 2].

$$\begin{bmatrix} \Delta CV_1 \\ \Delta CV_2 \end{bmatrix} = \begin{bmatrix} 4.1 & 6.0 \\ 4.4 & 6.2 \end{bmatrix} \bullet \begin{bmatrix} \Delta MV_1 \\ \Delta MV_2 \end{bmatrix} \tag{4-2a}$$

The steady-state gains of a CV for each MV in Equation 4-2b are not equal but exhibit a ratio. The outputs of the process are linearly related. Such systems are called *stiff* because the CVs move together. The system lacks the flexibility to move them independently to achieve their respective set points. Again, the determinant is nearly zero and provides a warning that MPC is not a viable solution [1, 2].

$$\begin{bmatrix} \Delta CV_1 \\ \Delta CV_2 \end{bmatrix} = \begin{bmatrix} 4.1 & 6.0 \\ 2.2 & 3.1 \end{bmatrix} \bullet \begin{bmatrix} \Delta MV_1 \\ \Delta MV_2 \end{bmatrix} \tag{4-2b}$$

In Equation 4-2c, the steady-state gains for the first MV (MV_1) are several orders of magnitude larger than for the second MV (MV_2). Essentially, there is just one MV, MV_1, because the effect of MV_2 is negligible in comparison. Unfortunately, the determinant is 0.9, which is far enough above zero to provide a false sense of security. The condition number of the matrix computed from the singular values of the matrix provides a more universal indication of a potential problem than either the determinant or relative gain matrix. A higher condition number indicates a larger problem. At some point, the condition number becomes too high. For Equation 4-2c, the condition number exceeds 10,000 [1, 2].

$$\begin{bmatrix} \Delta CV_1 \\ \Delta CV_2 \end{bmatrix} = \begin{bmatrix} 1 & 0.001 \\ 100 & 1 \end{bmatrix} \bullet \begin{bmatrix} \Delta MV_1 \\ \Delta MV_2 \end{bmatrix} \tag{4-2c}$$

The matrix condition number provides a numerical assessment of potential MPC performance problems.

The matrix can be visually inspected for indications of possible MPC performance problems by looking for gains in a column that has the same sign and size, gains that differ by an order of magnitude or more, and gains in a row that are a ratio of gains in another row. Very high process gains may cause the change in the MV to be too close to a control valve's deadband and resolution limits, and very low process gains may cause an MV to hit its output limit [1, 2].

The desired relationship for MPC is for process outputs to depend only on measured process inputs that are independent variables. If the process inputs affect each other, they are not independent. If the process outputs affect each other, they are not solely dependent on the process inputs. If the process gains associated with an MV are more than an order of magnitude greater or smaller than the process gains of other MVs, the effect of process inputs cannot be sufficiently balanced for control.

MPC performance relies on process inputs that are independent of each other, process outputs that depend on process inputs rather than other process outputs, and the set of process gains for each MV having a comparable magnitude.

MPC software packages compute the condition number after the models have been identified for the variables included in the control matrix [3]. The setup screen gives a relative index of the importance of each MV and computes the condition number for the CVs and MVs selected. The user can move variables in and out of the control matrix and see the effects on the condition number [4].

A condition number above 100 indicates a serious degradation in both performance and robustness. MPC may work well enough up to a point if the PM is increased and the changes in set points are coordinated. However, the best solution is to come up with different CVs or MVs. The problems associated with a condition number that is greater than 1000 cannot be sufficiently mitigated by tuning. One innovative possibility is to use principal component analysis, as described in Chapter 8, to provide new independent latent variables to replace existing collinear CVs (i.e., CVs with linear interrelationships).

Some MPC software helps guide users in selecting MVs and CVs in the control matrix in order to reduce the condition number.

To summarize, if the condition number is not too high, then MPC is useful for the following situations:

- Process and measurement noise
- Erratic or stepped measurement response
- Inverse response
- Complex (irregular) responses
- Move size limits
- Measured disturbances
- Multiple MVs
- Interactions
- Constraints
- Optimization

The remaining sections of this chapter will focus on applying MPC to situations involving multiple MVs and optimization.

The following business and practical issues must be addressed for MPC to be applied to industrial biopharmaceutical processes [5]:

- Today, most basic control loops in major pharmaceutical companies are satisfactorily handled as single input, single output loops.
- Perturbing the system to generate the model, such as changing temperature and pH in a running fermentation for the sake of identifying a model, is not something most plants want to do. This is because even small excursions from the set point, especially involving temperature and pH, can negatively affect the culture.

- The perception is that MPC is so difficult to set up that an outside firm or consultant is typically needed.
- The literature lacks documentation of research showing how MPC in bioprocesses has provided a significant financial benefit.
- Many automation and process control support engineers are not familiar with MPC.
- MPC software comes at an extra cost. With many control systems, PID control is standard off-the-shelf functionality that comes with the system, but MPC is an often somewhat expensive add-on software package.
- The perception is that MPC was developed for and is best suited for steady-state continuous processes, such as those found in the petrochemical industry. Most pharmaceutical processes are non-steady-state batch processes with several step-to-step transitions.
- There is discomfort among development and manufacturing engineers, as well as quality control folks, over the formal validation of MPC applications, which is a Current Good Manufacturing Practices (CGMPs) requirement for systems making medicines for human use. Specifically, how does the user make regulatory professionals comfortable that a controller is robust enough to handle contingencies and disturbances that it had not been exposed to when the model was created and when the nature and structure of the model equations are not disclosed by the MPC supplier?

The benefits from optimization can provide the motivation to advance beyond single input and output control noted in item (1) to multivariable control. Solving the remaining issues is facilitated by using relatively low-cost and easy-to-use software for advanced control algorithms that has been developed and prototyped in a digital twin and demonstrated in benchtop and pilot-plant systems with industrial control systems. Item 8 is not as serious an issue as it used to be in that proprietary algorithms have been validated for use (e.g., freeze thaw skids algorithms).

4.3 Multiple Manipulated Variables

The relative precision, cost, and speed of each MV depend on the process and the equipment, piping, and automation system design. For this first case involving situation 7, we consider the control of DO by manipulating airflow and the agitator speed for a viscous bacterial cell culture where the relative cost of agitation power is greater than airflow. Several examples have demonstrated the successful use of MPC to control

DO by manipulating multiple variables. In one case, an MPC application controlled the DO in a bioreactor by manipulating airflow, agitation speed, and vessel pressure [6]. In another case, an MPC application for a bioreaction controlled DO by manipulating air, oxygen, and nitrogen to the headspace and oxygen sparge to the media, and it controlled pH by manipulating carbon dioxide to the headspace and a sodium carbonate solution to the media [7].

High- and Low-Cost MV

For maximum process efficiency, it is important to reject unmeasured load disturbances and achieve a smooth optimization. In some cases, the MV with the fastest effect on the process is the most expensive. This section considers the use of MPC with one high-cost fast MV and one low-cost slow MV to achieve the dual objective of load rejection and optimization for a single CV. Many ways to set up the MPC application are available, but one method that has been demonstrated to work well is to configure the MPC block for maximization of the low-cost MV and make a simple tuning adjustment to reduce the importance of the optimization compared to load rejection [8].

In general, it is best to optimize the slow MV and keep the fast MV free for rapid response to disturbances and set points.

MPC can be configured and tuned to maintain a critical CV (e.g., dissolved oxygen) at its target and to maximize the low-cost slow MV (airflow) set point as an optimization variable. The best load and set-point response for the critical CV is obtained through a short-term trade-off in efficiency: reducing the penalty on error (PE) for the optimization variable rather than increasing the PM of either MV. When riding the low-cost MV maximum set point, such as the airflow that can cause foaming, or when riding the high-cost MV low set-point limit, such as agitator speed , this adjustment of the PE rather than the PM lets both the slow and fast MV move to improve the load and set-point response of the critical CV. Agitator speed provides a fast response by immediately changing the mass transfer coefficient, but there is a cost penalty in terms of potential cell damage from the impellers. Only the response of the optimization variable is slowed down. This is consistent with the general theme that disturbance rejection (regulatory control) should be faster than optimization. Figure 4-3a shows the general setup of a 2 × 2 MPC with a critical CV that has a normal PE and an optimization variable that has a lowered PE [8]. The generic exponential responses indicate models with a primary and secondary time constant, gain, and dead time. The horizontal line indicates a model with a unity gain, and "Null" indicates no model is needed.

Reducing the PE of the optimization variable makes regulatory control more important than optimization.

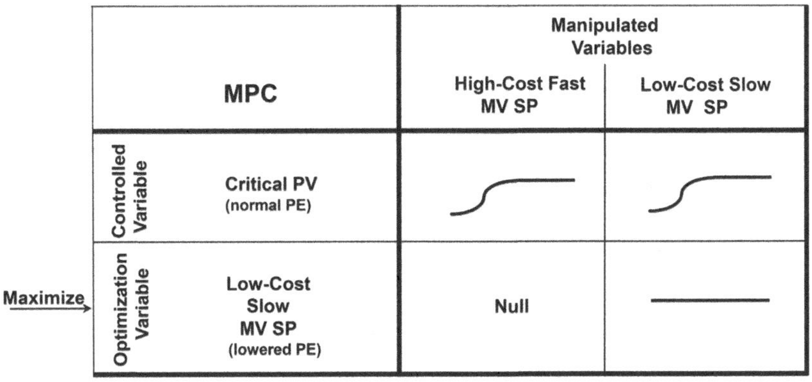

Figure 4-3a. Setup of MPC for tight control of critical PV with maximization of low-cost slow MV.

The optimization could also have been slowed down by increasing the PM on the low-cost slow MV. However, a large PM forces the high-cost fast MV to take most of the correction, particularly if you are riding the low-cost MV maximum set point. Additionally, if you are riding the high-cost MV low set-point limit and the upset or set-point change is in the direction that requires a decrease in this high-cost MV, then the large PM makes the load and set-point response noticeably slower because the response depends totally on the slow MV with its high PM [8].

Leaving the PM of the MVs at the values that are best for performance and robustness enables all of the MVs to help with regulatory control.

The test results shown in Figure 4-3b are for a process where the dead time is about as large as the time constant. This causes the peak error to be almost the full-load upset because of the high degree of dead time. The improvement is mostly seen in the integrated absolute error (IAE). For processes where the time constant is much larger than the dead time, the reduction in peak error and IAE would be greater [8].

The trend chart in Figure 4-3b shows the response for load disturbances and set point changes in both directions. In the first operating mode, the low-cost MV is riding its maximum set point, which leaves the fast-cost MV above its low set-point limit, and thus free to respond in the negative as well as a positive direction. In the second operating mode, the maximum for the low-cost MV has been increased to the point where it is no longer achievable, which drives the high-cost MV to its low set-point limit. The succession of load upsets can be so frequent that the high-cost MV does not always ride its hard low limit, which is the case shown in Figure 4-3b. The test sequence also displays the slow response of the optimization variable to large changes in its maximum set point. It is important for operators to realize that the

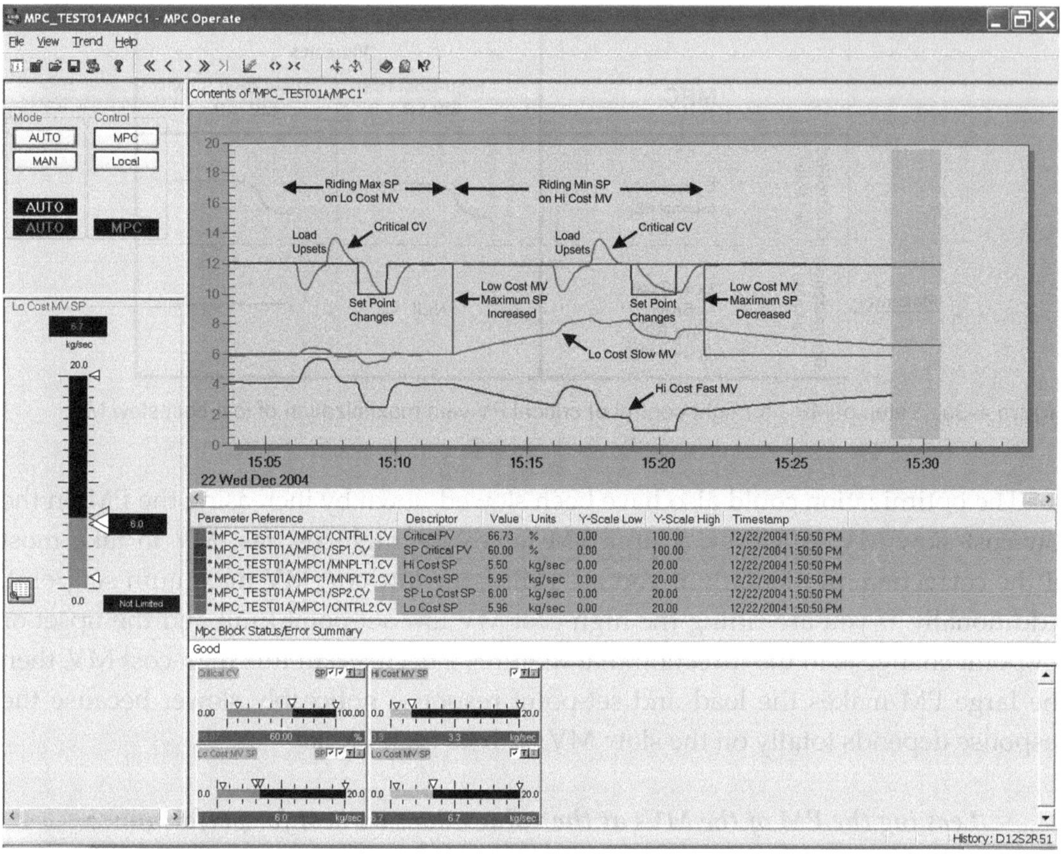

Figure 4-3b. Response of high- and low-cost MV MPC to disturbances and set points.

optimization is purposely designed to be slow and smooth, and in the long term, it does a great job [8].

The optimization algorithm has a tight grip on the optimization variable. A large decrease in the PE helped free up the associated MV to move when it was riding its maximum set point. Setting the set point for maximizing the low-cost MV at a level higher than is obtainable freed up the slow MV for upsets and set-point changes that demanded an increase in the MV. However, a higher-than-obtainable slow MV maximum forces the fast MV to its low set-point limit where it is unavailable for corrections that require a decrease in the fast MV [8].

The alternate strategy of minimizing the high-cost fast MV is not advisable for best load rejection because it makes it difficult to free up this fast MV sufficiently once it becomes an optimization variable. However, this might be a viable alternative if it is desirable that the minimum for the fast MV be a soft limit via an economic objective rather than a hard limit via a low set-point limit and if it is sufficient to turn over most of the load rejection to the slow MV. Again, the PE must be reduced for the optimization variable [8].

Coarse and Fine MV

Two or more MVs are often used to extend the range of loads that loops can handle. Typically, the variables are split-ranged, where the variable with smallest effect (low-gain MV) is throttled for low loads (low controller outputs) and the variable with the largest effect (high-gain MV) is throttled for high loads (high controller outputs). A standard Fieldbus "splitter" block is used on the controller output to split a single controller output into set points for analog outputs. Ideally, to help linearize the loop, the split-range point or point of transition from the variables that have small to large effects is based on the process gain. For example, if the high-gain MV has nine times the effect of the low-gain MV, then the split-range point should be 10% so the low-gain MV and high-gain MV move from minimum to maximum values for a controller output of 0% to 10% and 10% to 100%, respectively. However, this practice of compensating for gain by selecting a split-range point is uncommon. For split-ranging two MVs, the split-range point is traditionally set at 50% even though "splitter" blocks, position read-back, speed measurements, and secondary loops have eliminated both special calibrations and operator misunderstandings over actual valve position or drive speed [9].

Besides the process gain, the process dead time and time constant are generally different for each split-ranged output. If there is no compensation for the changes in process dynamics with each MV, then the controller tuning must be compromised to ensure stability for the worst-case MV.

The controller tuning is compromised to deal with the changes in process dynamics that are associated with split-ranged output.

Unfortunately, the high-gain MV resolution limit, which is expressed in terms of percentage of full scale, will have a bigger effect on the process because of its higher process gain. Also, if the high-gain MV is a rotary valve, the percent resolution limit is bigger to begin with than it is for sliding stem valves or variable speed drives (VSDs) [10]. From here on, the high- and low-gain MV will be referred to as the *coarse* and *fine* (trim) MV, respectively.

Most loops exhibit a "limit cycle" across the split-range point from the gain discontinuity caused by the different MVs, the rangeability limits of valves and drives, and the additional friction from both the seating of the plug in sliding stem valves and the sealing of the ball or disc in rotary valves [9].

Most control loops oscillate around the split-range point because of discontinuities, rangeability limits, and stick slip at the split-range point.

To address these issues, "integral only" (integral-only) controllers are sometimes used to eliminate the split-ranging. Because these controllers are normally implemented for big and small valves, they are called *valve position* controllers. However, tuning rules are scarce and the term *valve position* causes confusion between these new controllers and the traditional valve positioners mounted on the valves; it also unnecessarily excludes other final control elements, such as VSDs.

The original process PID controller output now goes to just the trim MV. An "integral only" *valve position* controller is added with the fine MV signal as its CV; a mid-range value, such as 50%, as its set point; and an output that only goes to the coarse MV. The "integral only" controller reset time is set larger than five times the product of the PID controller gain and reset time settings in order to make the interaction between the controllers negligible. Unfortunately, this detuning of the "integral only" controller is too slow to handle disturbances and causes a slow limit cycle from the resolution limit of the coarse MV [9]. Also, a study of the application of dual input (MV) temperature and pressure processes shows that the tuning of these controllers is much more critical than was suspected [11].

Tuning "integral only" controllers to minimize interactions but still ensure responsive manipulation of the coarse MV is critical but difficult.

MPC can be configured to simultaneously manipulate a coarse and fine MV, eliminating the problems inherent in split-ranged and "integral only" controllers. This provides the precision of control offered by the fine MV with the range of control possible from combining fine and coarse MV. Because MPC has the process gain and dynamic response of each MV built in to its model of the process, MPC knows how much to move each MV and inherently eliminate the interactions from the multiple process inputs.

MPC can provide simultaneous fast fine and coarse MV control from its knowledge of the process dynamics for each MV.

The MPC Block in Figure 4-3c has two MVs (fine MV and coarse MV). These are process inputs, one critical CV (e.g., DO or pH), which is a process output, and one optimization variable (fine MV). The MPC optimization mode selected is "Observe Limit," which will push the fine MV from either direction toward its set point (e.g., 50%). The PE for the optimization variable is decreased from 1.0 to 0.1 to increase the relative importance of the control of the critical variable. The tuning setting of the

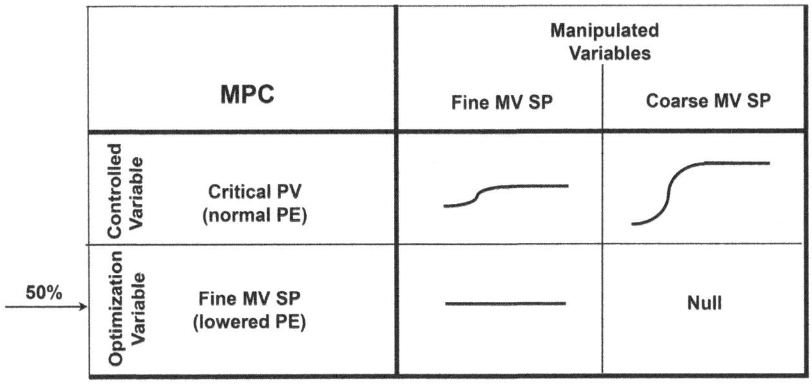

Figure 4-3c. Setup of MPC for rapid simultaneous manipulation of dual process inputs.

PM computed by MPC was used because the PM was correctly initialized to be much larger (PM = 20) for the coarse MV than for the fine MV (PM = 5). Using this setting ensures that the coarse MV does not move too fast. If MPC does not automatically tune the PM correctly, the user must manually increase the PM of the coarse MV [9, 10].

The PM for optimizing the fine MV at midrange is reduced to make this optimization less important than regulatory control.

The MPC application could have been set up to optimize the coarse MV to a minimum. However, the set point (target) for this minimization would need to be calculated based on the installed characteristics and capacities in order to keep the fine MV from running near its output limit. This complication is unnecessary with the MPC setup shown in Figure 4-3c, which works to ensure the fine MV is fully available by pushing it to the middle of its range. The generic exponential responses indicate models with a primary and secondary time constant, gain, and dead time. The horizontal line indicates a model with a unity gain, and "Null" indicates no model is needed.

The trend chart in Figure 4-3d shows the response for large steps in load and set points for both the critical PV and the desired optimum position for the fine (trim) MV. Notice that for the successive load upsets, the coarse MV moves rather quickly to a new position, which enables the fine MV to return to its optimum. The load rejection is smooth and fast. The details, such as signal values and tag name, are not relevant. Only the general trend appearances in Figure 4-3d are important.

Similarly, for a change in the set point of the critical PV from 50% to 70%, the coarse MV moves to take care of the long-term need at the new set point. Finally, a change of the optimum set point of the fine MV from 50% to 60% shows only a small bump to the process.

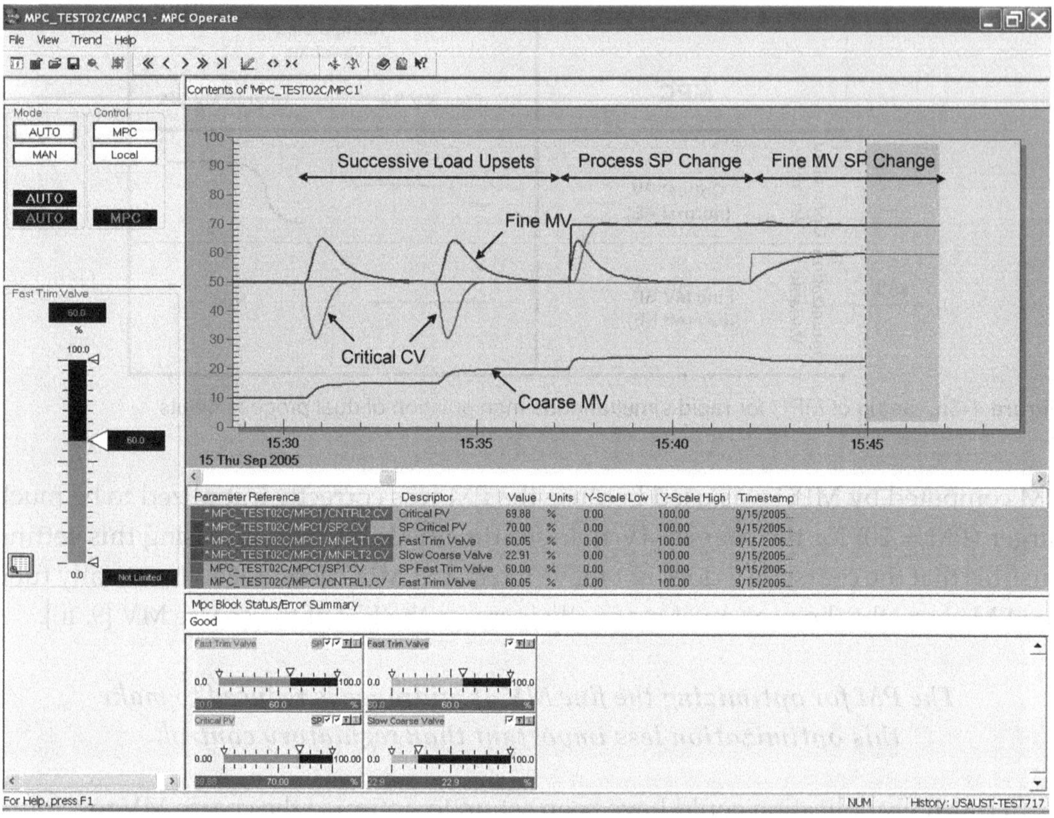

Figure 4-3d. Response of fine and coarse MV MPC to disturbances and set points.

Normally, the operator would not be changing the optimum, but a set-point filter could be added for this optimization variable so as to prevent even a small bump [9, 10].

The simultaneous throttling of both valves by MPC will reduce the limit cycle from the coarse MV resolution limitations. If the "Maximum MV Rate" parameter in the MPC setup is written to zero when a separate filtered value of the "critical CV" in Figure 4-3d stays within a band around its set point, then the limit cycle from the coarse MV is eliminated. The width of the band can be calculated as the resolution limit multiplied by the process gain in the MPC model for the coarse MV [9, 10].

The limit cycle from the resolution limit of the coarse MV can be eliminated by setting its "Max MV Rate" to zero when the critical PV is close to its set point.

For those applications in which the MVs have an opposite effect on the process—such as steam and coolant, oxygen and nitrogen, and acid and base reagents—a different setup of the model predictive controller is needed to keep both MVs from being fed at the same time and wasting energy or raw materials [9, 10].

4.4 Optimization

The optimum values of the cell environment, such as temperature or pH, depend on process, product, and in some cases, the batch phase. Chapter 6 describes the equations for the kinetics employed in first-principle models to quantify the effects of operating conditions on biomass growth rate and product formation rate. Chapter 5 describes how model predictive control can be used to adapt kinetic parameters and expressions to changing batch conditions and seed cultures.

The plot of biomass growth rate versus temperature in Figure 4-4a has a rather sharp peak. As the temperature approaches the optimum, the growth rate may double for every 10°C per an Arrhenius relationship over a narrow range. Microbiologists responsible for defining the parameters do an experimental design involving several different temperatures and then a standard curve fit to the experimental results. When the temperature gets high enough to start denaturing proteins and killing cells, the rate of kill is such that a 10°C change in temperature results in an order-of-magnitude increase in kill rate. Therefore, there are lots of different activities occurring within cells at different temperature-dependent rates. The location of the peak may shift and the slope may change for the product formation rate [12].

There is a sharp drop in growth rate at temperatures greater than the optimum.

The plot of biomass growth rate versus pH in Figure 4-4b has a more symmetrical shape. It may be possible to widen the width of the optimum for some bacterial cells by adapting the culture such as by making small changes in the pH of the seed culture prior to its transfer to the bioreactor [13].

As with temperature, the optimum pH for the product formation rate may be different from the optimum pH for growth rate. Many microorganisms can maintain the

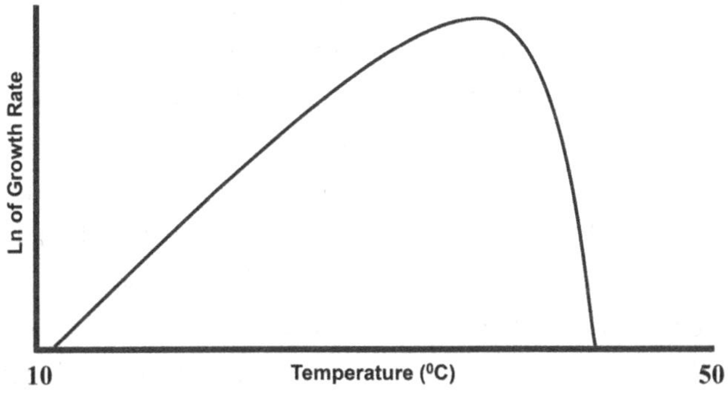

Figure 4-4a. Temperature effect on growth rate for bacterial cells.

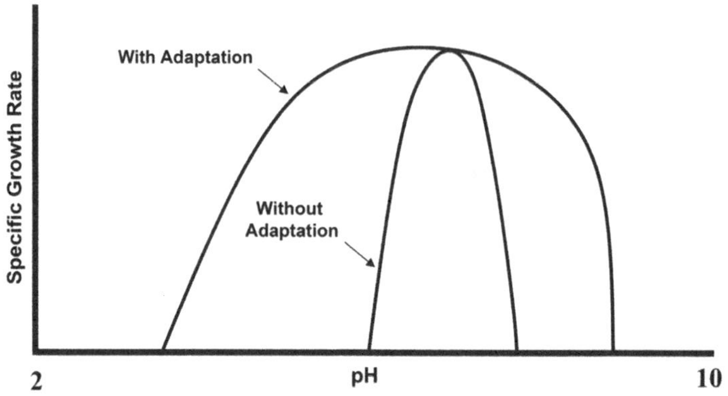

Figure 4-4b. pH effect on growth rate for bacterial cells with and without adaptation.

intracellular pH to some extent for changes in the pH of their environment. However, the shifts in the pH caused by the change in ammonium ions from the metabolism of nitrogen sources, the production of organic acids and bases, the utilization of amino acids, and the evolution or feed of carbon dioxide makes pH control important [14]. One major pharmaceutical company's experience in making 20 different products from bioprocesses (including ones made with bacteria, fungi, yeast, and mammalian cell cultures) is that tight pH control is very important and all the optimum pH set points are between 6 and 8. A drift of more than a few tenths of a pH unit away from the set point has negative consequences for the fermentation [15]. Ethanol processes generally run at a lower pH.

pH is important for proper cell metabolism.

The plot of biomass growth rate versus DO in Figure 4-4c shows a steep decline below a critical DO concentration (below 0.1 mg/L). This is because the cells for aerobic processes are deprived of oxygen. High viscosities may decrease mass transfer rate so that a higher aeration rate or agitation rate may be needed to achieve the critical concentration. Fermentations that are filamentous in nature (i.e., tangled strands of cell chains that look a little like spaghetti when viewed using a microscope) have tangled webs that grow in size. The broth viscosity will increase (perhaps up over 100 centipoise [cp]), and it may become more difficult for oxygen to get to the cells in the middle of these tangled masses of cells. In this case, it would be prudent to raise the DO set point. Between 0.2 and 0.8 mg/L, the growth rate is rather constant. Concentrations above 0.8 mg/L can be toxic, possibly because free radicals form, which accelerates cell aging and even cell death [15]. Foaming can sometimes be related to DO, especially when aeration is increased to help achieve the desired DO (foam creation is a function of the nutrient composition of the broth or media, aeration rate, and agitation rate) [15]. Foaming is particularly problematic when antifoam agents cannot be added. The cost of air, agitation, and oxygen means there is an economic optimum not shown in Figure 4-4c.

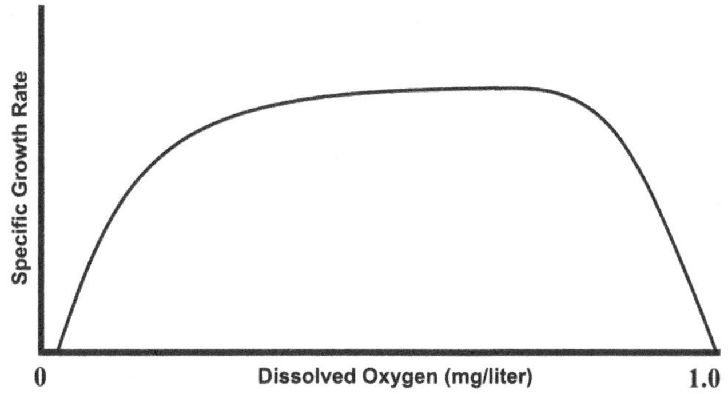

Figure 4-4c. DO effect on growth rate.

Most discussions and plots of the optimum DO do not include the effects of operating cost, foaming, and free radicals at high DO concentrations.

The plot of substrate saturation and inhibition effect on biomass growth in Figure 4-4d has a steep decline at low and high concentrations. Not shown is that at low biomass concentrations, such as might occur in the lag phase, the effect is negligible. Typically, there is no substrate inhibition until the substrate concentration gets to some specific value, with 20 to 25 g/L being typical for some glucose fermentations [15]. Therefore, a fermenter is operated at a high substrate concentration to maximize the specific growth rate in accordance with the kinetic equations presented in Chapter 6. However, the concentration is not so high as to cause substrate inhibition to become important. Then, near the end of the fermentation, the substrate concentration is typically lowered to avoid throwing away a lot of unused substrate when harvesting the fermentation. However, achieving substrate concentrations close to zero near the end is difficult, may increase the consumption rate of the product as food (hydrolysis rate),

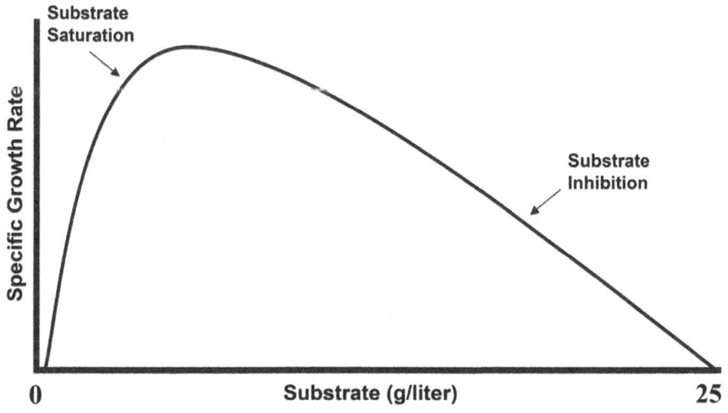

Figure 4-4d. Substrate saturation and inhibition effect on growth rate.

and may starve some cells as a result of localized depletion of food from the low agitation rates used for animal cell cultures.

The typical optimum profile for substrate concentration shows a decline near the end of the batch to improve yield.

Although the plots shown in Figures 4-4a through 4-4d are for biomass growth rate, similar plots may exist for product formation rates as a function of broth or media operating conditions (temperature, pH, DO, and substrate). In some processes, such as ethanol, there is a product inhibition, in which the product formation rate goes to zero when the product concentration reaches a critical value (e.g., 65 g/L of ethanol). If the kinetic parameters are known, the plots for biomass growth and product formation rates can be generated by plotting the results for the kinetic expressions for each broth or media operating condition of interest.

In Figures 4-4a through 4-4d, any change in the sign of the slope corresponds to a change in the sign of the process gain if the process variable on the horizontal axis was manipulated to optimize the biomass growth rate. The controller action must be the opposite of the process action or the controller output will walk to its output limit. For example, on the plot biomass versus pH, at low pH the slope ($\Delta G/\Delta pH$) and therefore the process gain is positive (direct action). The controller must have reverse action so the controller reacts to a decrease in growth rate by increasing the pH. However, at high pH the slope and process gain is negative (reverse action). The controller must now have direct action. The solution is to restrict the controller to working on one side or the other of the optimum.

The change of sign in the slope of key process variables with a relatively sharp peak, such as pH or temperature, as it goes from one side to the other of the optimum requires a change in the controller action.

The manipulation of temperature or pH as a secondary control loop for profile control requires a translation of the CV from rate of change (velocity) to acceleration and deceleration to deal with the peak in the plot of growth rate or formation rate versus temperature and pH. The change in the sign of the slope and process gain would confuse a controller whose process variable is the slope. The creation of a new CV for the secondary loop that is the change in the batch profile slope with respect to a change in temperature or pH eliminates the reversal of the sign of the process gain. The CV decreases as the operating point approaches the optimum in the plot of cell growth rate or product formation rate versus pH or temperature. The CV is zero at the peak and becomes more negative as the operating point moves further to the right of the peak [17].

The MV for this new secondary controller is the temperature or pH set point of a tertiary lower loop. Thus, there is a triple cascade of batch profile slope to batch profile acceleration to temperature or pH where MPC is used for the uppermost primary controller. Careful set point limits must be placed [17].

The triple cascade is an extension of the "hill climbing" strategy described by Shinskey. To create the innovative secondary loop, the new CV is scaled from a positive change in velocity (acceleration) to a negative change in velocity (deceleration) with respect to a change in temperature or pH. For a constant profile slope, the set point is zero. The acceleration and deceleration can be computed by passing the concentration rate of change through a dead time block to give a significant difference between a new and old value of the slope to compute a second derivative of the change in concentration. With this calculation, it is critical to choose a dead time interval that is long enough to provide a strong signal-to-noise ratio. The change in temperature or pH is first computed in the same manner by the use of a dead time block with the same dead time parameter. This change in temperature or pH must be dynamically compensated so that a change in profile is coincident with a change in temperature or pH. The dynamic compensation can be identified by the use of MPC software. The calculation must be protected against a divide by zero. When the triple cascade is turned on, a change in the manipulated temperature or pH in the most desirable direction should be initiated to create a change in slope. The tuning of the primary MPC must be slower than that of the secondary PID controller, which must be slower than the tertiary PID controller to manipulate reagent, heating, and/or cooling. Because this is an optimization and biological batch cycle times are extremely long, gradual corrections are in order; corrections should include exceptionally large reset times, low PID gains, and move suppression. The product of the PID gain and reset time must be greater than the inverse of the integrating process gain for the primary and secondary loops. Smooth and gradual action is also important from the standpoint of not reacting to noise. The numerator used in the calculation of the CV for the secondary PID controller is a second derivative of the batch profile with respect to time giving the acceleration or deceleration in the batch profile caused by a change in pH or temperature. The computation is consequently susceptible to noise [17].

For maximizing the cell growth rate or product formation rate, a triple cascade extension of the "hill climbing" technique can optimize the pH or temperature set point to be near the peak in the plot of cell growth rate or product formation rate versus pH or temperature. For temperature, the target is just to the left of the peak because of the risk of cell death from temperatures to the right of the peak from sensor offset or set-point overshoot.

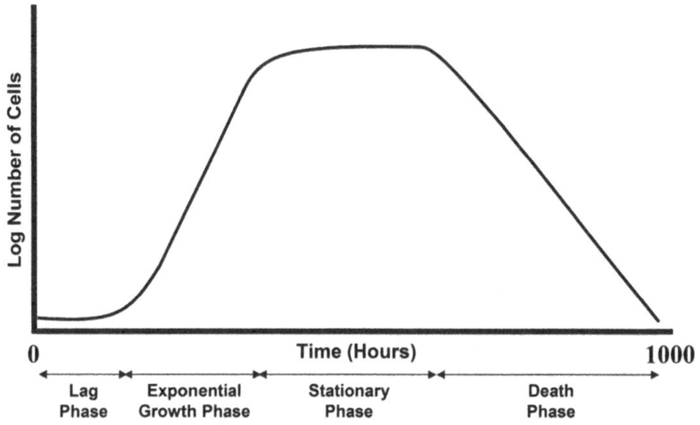

Figure 4-4e. Idealized phases of batch cell cultivation.

A bioreactor batch can be characterized by a lag phase, an exponential growth phase, a stationary phase, and a death phase, as shown in Figure 4-4e [12]. The plot is an idealized representation of the biomass taken as a whole if the batch cycle was allowed to continue. The phases of individual cells will vary in that some cells may already be dying near the end of the exponential growth phase. Optimization of the batch would have the goal of minimizing the lag phase and maximizing the exponential growth phase. This phase is also known as the *logarithmic growth phase* because the plot has a nearly constant slope when the vertical axis is the log of the biomass. The batch cycle would end near the start of the stationary phase, and the death phase would be avoided.

In the lag phase, an increase in the number of cells is not noticeable. The length of the lag phase depends on the age, number, and acclimation of the seed cells, and the nutrient content of the fresh medium. The seed cells go through a lag and exponential growth phase in a seed fermenter that can be optimized to harvest cells at the best age in a medium and at conditions similar to the main bioreactor. To deal with a different nutrient or higher concentration of the same nutrient, cells take time to produce new or more enzymes, respectively, which increases the duration of the lag phase [12]. The serial use of carbon sources by cells can cause a second lag phase as the preferential carbon source is depleted and the cell diverts its metabolic process from growth to the adjustment of enzymes in order to use the next carbon source [12].

In the exponential growth phase, the response of biomass or product concentration is a one-sided integrator because it always increases as it ramps up. Because it is undesirable and impossible to decrease the biomass and product in this phase, it is not possible to control these concentrations directly. However, if the CV is translated from a concentration to a growth or product formation rate, then the one-sided integrating response becomes a two-sided self-regulating response where a steady state can be approached from either direction. The growth rate or product formation rate would be computed as previously described in this section for the purpose of monitoring the process sign.

Batch profile control requires that the CVs be translated from concentration to a rate of change of concentration in order to transform the one-sided integrating response to a two-sided self-regulating response.

Figure 4-4f shows the self-regulating process responses identified for 2×2 MPC where the CVs are biomass growth rate and product formation rates and the MVs are

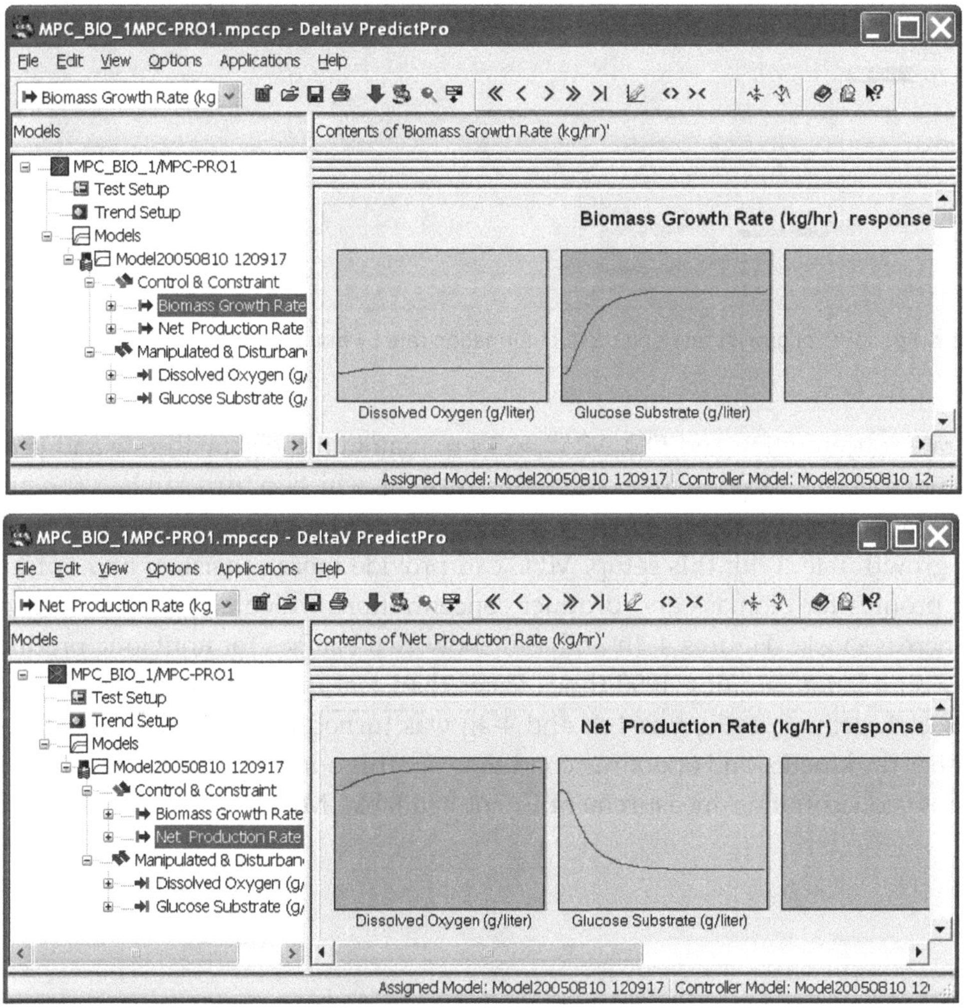

Figure 4-4f. Identified responses for batch profile control.

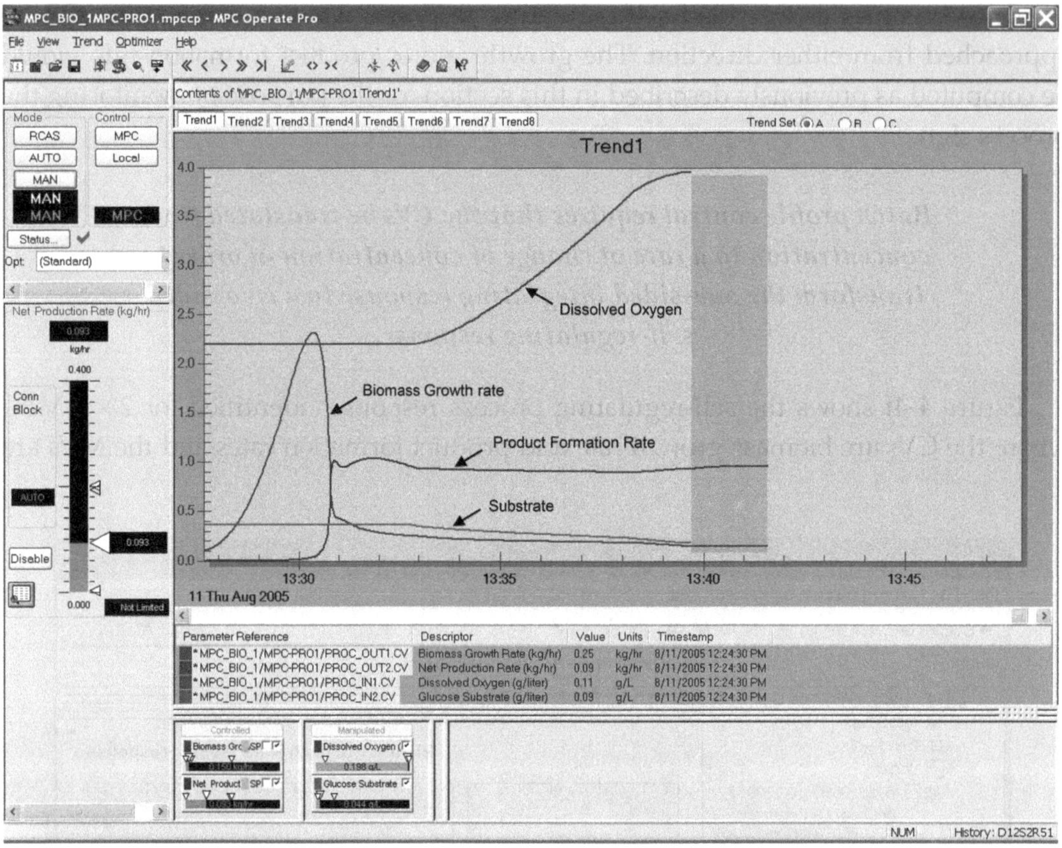

Figure 4-4g. MPC of growth rate and product formation rate by manipulating DO and the substrate.

substrate concentration and DO. MPC seeks to maintain the growth rate and product formation rate at a set point as shown in Figure 4-4g, which in turn can be a function of batch phase time. The PE is increased for the product formation rate relative to the biomass growth rate. With this setup, MPC can provide more consistent and optimized batch profiles for biomass and product concentration, which can reduce batch time and increase yield. Figures 4-4h and 4-4i show two batches for antibiotic production from a simulation running 1000 times faster than real time. In the second batch, the MPC Block defined in Figures 4-4f and 4-4g was turned on. Section 6.9 in Chapter 6 describes the kinetics and operating conditions for this optimization. If a reliable online substrate concentration measurement is not available, MPC can directly manipulate

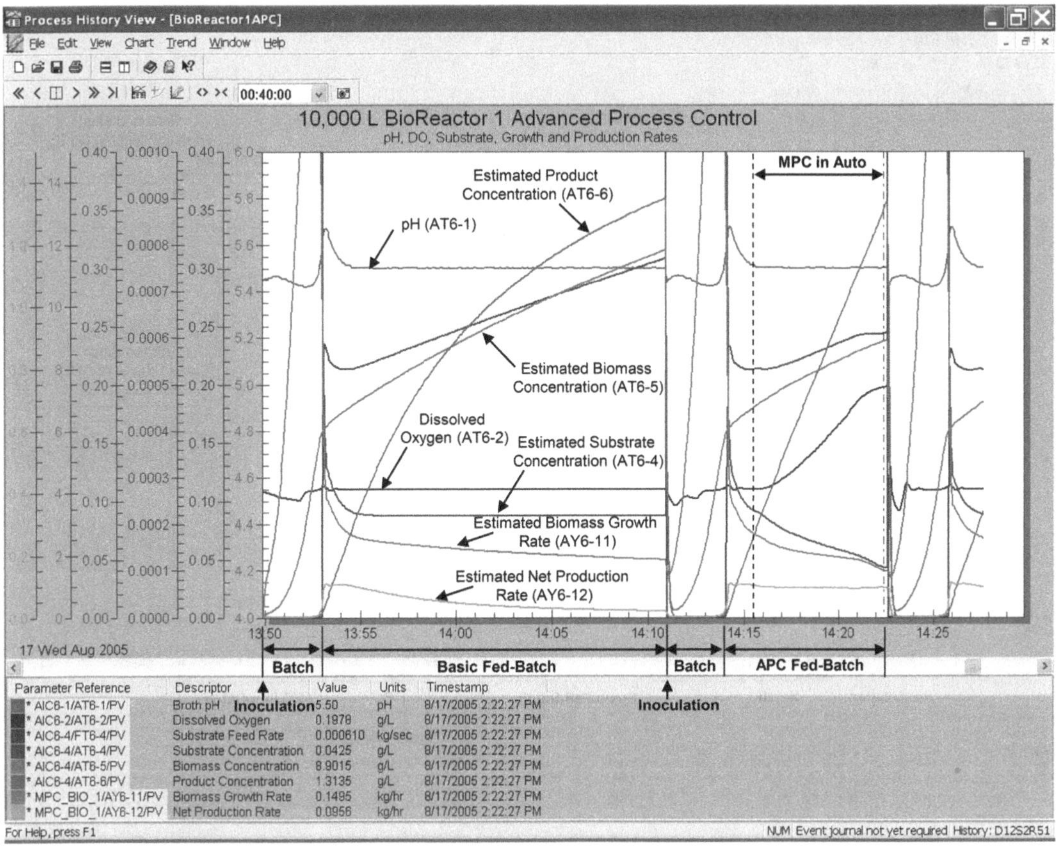

Figure 4-4h. Optimization of antibiotic batch profiles by an innovative MPC.

the substrate feed rate. The use of a digital twin running at 500 times real time for fast identification of these responses and rapid prototyping of MPC applications will be described in Chapter 5 [16]. The details, such as signal values and tag names, are not relevant. Only the general trend appearances in Figure 4.4f are important.

MPC can be set up for simultaneous biomass and product profile control.

4.5 Mammalian Cell Bioreactor Optimization

MPC can optimize mammalian cell bioreactors by controlling the cell growth rate and product formation rate at the set points as the batch progresses by the manipulation of

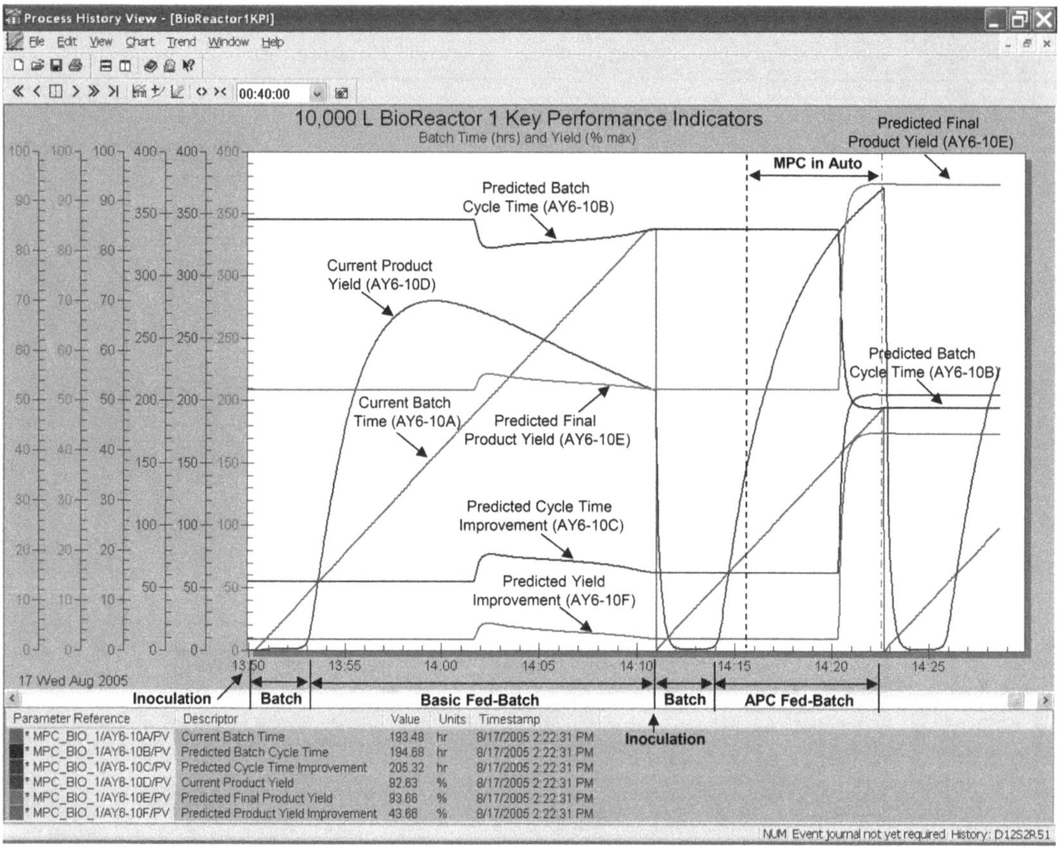

Figure 4-4i. Improvement in antibiotic key performance parameters by an innovative MPC.

glucose and glutamine flows. The growth rate is the slope of the batch-cell concentration profile, and the product formation rate is the slope of the product concentration profile. Online and at-line analyzers will be described in Chapter 7 to provide these concentrations directly or by inferential measurements or first-principle models provided by a digital twin described in Chapter 5. The relative amounts of glucose and glutamine must drastically change as cells move through their growth phases (e.g., pre-exponential, exponential, and stationary) and into product formation. An OUR computed from the analysis of off-gas oxygen content and the manipulated airflow and supplemental oxygen flow rates can provide a feedforward signal to provide some anticipation of changes in cell growth rate. Figure 4-5 shows the use of an MPC to provide this optimization [17].

Compute the slope of the batch profiles of cell and product concentrations to be the MPC CVs to manipulate glucose and glutamine flows to optimize batch profiles using a dead time block with the dead time set for the best signal-to-noise ratio.

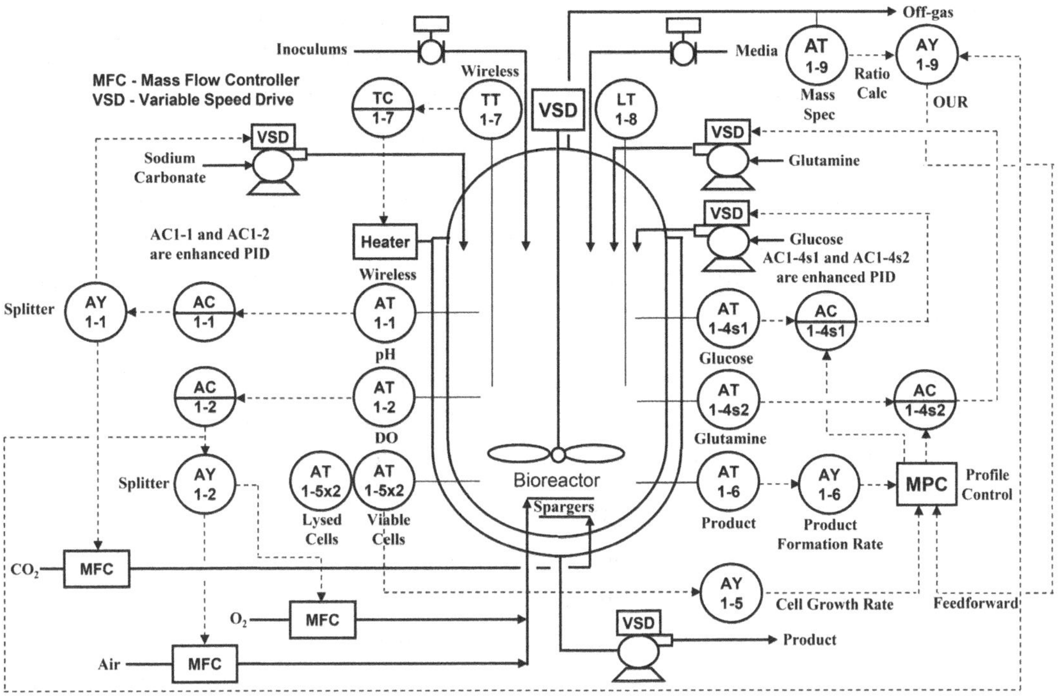

Figure 4-5. Mammalian cell bioreactors can be optimized by the manipulation of glucose and glutamine concentrations to maximize the cell growth rate and product formation rate at the best times in batch.

4.6 MPC Best Practices

Here are best practices that address every aspect of designing, testing, reporting, operating, and maintaining an MPC to maximize benefits and promote future applications [2, 18].

1. Project team members should include not only control engineers, but also process engineers and operations personnel.

2. Educating management as to the benefits of advanced control is often best done by dynamic demos using simulations in a digital twin running 500 times real time or faster as described in Chapter 9.

3. First-level support of MPC requires staff with knowledge of both the MPC and the process. Site staff must have sufficient understanding to troubleshoot and answer the questions of operations. Larger companies often have central teams for second-level support and to participate in projects.

4. Even in companies with experienced teams, it is not unusual to use outside MPC consultants. The right level of usage of outside consultants is rarely 0% or 100%.

5. Develop a functional design with input from operations, process engineering, economics staff, and instrument techs. Update the design as the project progresses, and after the project is completed to reflect the as-built MPC. Make sure the MPC is in sync and consistent with targets from planning and scheduling personnel.

6. Not all MPC variables must be determined up front in the project. Most important is identifying the MVs. The final selection of CVs and disturbance variables (DVs) can be made after plant testing, assuming data for these variables was collected.

7. A digital twin can be useful for testing a new regulatory control strategy. It can also be used to test and demonstrate an MPC, which can be quite illustrative and educational, particularly if a facility is applying for an MPC for the first time.

8. If filtering of a CV or DV for MPC is required, it must be done at the distributed control system (DCS) or programmable logic control (PLC) level. The faster scan times enable effective filtering (usually on the order of seconds) without significantly affecting the underlying dynamics of the signal. In addition, filters associated with the process variables of PID loops should be reviewed to ensure excessive filtering is not being used to mask other problems.

9. A digital twin can be useful for determining thermophysical equation parameters for PID calculated control variables (e.g., duty and MPC CVs, estimating process gains, and evaluating possible inferential predictors.

10. With most MPC products, adding MVs, CVs, and DVs is a straightforward task after models are identified. This allows starting with a smaller MPC on one part of the unit, and later increasing the scope as experience and confidence is gained.

11. Inferential measurements can be developed ahead of the plant test, which allows the model to be evaluated and adjustments made. For data driven regression based inferential measurements, one must have at least confirmed that plant and lab measurements exist that correlate with the analyzed valve. The final determination of model inputs can be made during the modeling phase.

12. Address the challenge of accurately knowing when a lab sample is collected. A technique for automating this is to install a thermocouple directly in the line of the sample point. A spike in the temperature measurement is used to detect a sample collection.

13. When implementing an inferential measurement online, it is often useful to filter inputs to the calculation to remove phantom effects such as overshoot or inverse response.

14. It may be tempting to avoid the benefit estimation and/or postaudit, especially when a company has previous successful history with MPC. But doing so carries a risk. New management may not have experience or understand the value of MPC, leading to the inevitable question: "What is MPC doing for me today?"
15. The other temptation is to forgo needed instrumentation or hardware repairs and proceed directly with an MPC project, arguing that MPC can compensate for such deficiencies. This carries the risk of not meeting expectations and MPC getting a bad reputation, which will be difficult to erase.
16. Regular reporting of relevant key performance indicators (KPIs) and benefits is seen as the best way of keeping the organization in the know and motivating MPC applications.
17. Establish a user company infrastructure to make the benefits consistent.
18. Develop corporate standards to historize and report KPI using moving averages for time periods associated with operator shift (e.g., 8 hours), batch duration (cycle time), and finally for month duration (e.g., 31 days) to be consistent with real-time accounting.
19. Screen and eliminate outliers and bad inputs and review results before reporting KPI.
20. Train operators in the use and value of the MPC for normal and abnormal operation.
21. Improve field instrumentation, valves, and basic control before MPC pretests.
22. Eliminate oscillations from overly aggressive PID tuning that excite nonlinearities.
23. Realize that changes even in PID loops not manipulated by the MPC can affect the MPC.
24. Use secondary flow loops (e.g., using Coriolis flowmeters) so that MPC manipulates a flow set point rather than a valve position to isolate valve nonlinearities from the MPC.
25. Use secondary jacket/coil temperature loops so MPC manipulates a temperature set point rather than a coolant or steam flow to isolate jacket/coil nonlinearities from MPC.
26. Avoid replacing basic control loops with MPC if the PID controller execution time must be less than 1 s or the PID gain is greater than 10 to deal with unmeasured disturbances.

27. Use inferential measurements or digital twin process variables to provide a faster, smoother, and more reliable concentration measurement, particularly for product.
28. Eliminate data historian compression and filtering to get raw data for analysis.
29. Minimize filtering so that lag is less than 10% of dead time and noise in MPC and PID output is generally less than control valve or variable speed pump deadband.
30. Conduct tests near constraint limits other than at normal operating conditions and in each distinctive important batch phase.
31. Use pretests to get step sizes and time horizons.
32. Step size should be at least five times deadband, stick-slip, resolution limit, and noise.
33. Get at least 20 data points in the shortest time horizon.
34. Use a near-integrator approximation to shorten the time horizon if optimization is not affected.
35. Get meaningful significant movement in the MVs for varying steps.
36. Make sure process gains are accurate for analysis, prediction, and optimization.
37. Use engineering knowledge and dynamic digital twin to confirm or modify process gains.
38. Reformulate the MPC to eliminate interrelationships between process input variables as seen by similar process gains in a column of the matrix.
39. Reformulate the MPC to eliminate interrelationships between process output variables as seen by process gains in one column of the matrix having the same ratio to gains in another column.
40. For small changes in dynamics, modify gains and dead times online.
41. For major changes in dynamics, retest using automated testing software.

References

1. Gregory K. McMillan and Robert Cameron, *Models Unleashed: Virtual Plant and Model Predictive Control Applications* (Research Triangle Park, NC: ISA [International Society of Automation], 2004).

2. Gregory K. McMillan, "Advanced Process Control," in *A Guide to the Automation Body of Knowledge,* 3rd ed., eds. Nicholas P. Sands and Ian Verhappen (Research Triangle Park, NC: ISA [International Society of Automation], 2018).

3. Terrence L. Blevins, Gregory K. McMillan, Willy K. Wojsznis, and Michael W. Brown, *Advanced Control Unleashed: Plant Performance Management for Optimum Benefits* (Research Triangle Park, NC: ISA [International Society of Automation], 2003).

4. Ashish Mehta, Willy Wojsznis, Dirk Thiele, and Terrence L. Blevins, "Constraints Handling in Multivariable System by Managing MPC Squared Controller" (presentation at ISA Expo, Houston, TX, October 2003).

5. Joseph S. Alford, "Bioprocess Control: Advances and Challenges" (presentation at CPC-7 to be published in *Computers and Chemical Engineering,* 2007).

6. Ken Callanan, "A Model Way to Control Fermentation," *Control* (August 2004).

7. Cleat Jerden, Tom Folger, and Matt Dee, "Advanced Control in Small Scale Biotechnology Development" (presentation at Emerson Exchange, Nashville, TN, October 2003).

8. Advanced Application Note, "MPC Implementation Methods for the Optimization of a Slow MV with Good Load Rejection by a Fast MV," Emerson (2005).

9. Advanced Application Note, "MPC Implementation Methods for the Optimization of the Response of Control Valves for Minimum Variability," Emerson (2005).

10. Gregory K. McMillan, "A Fine Time to Break Away from Old Valve Problems," *Control* (November 2005).

11. Cheng-Ching Yu and William L. Luyben, "Analysis of Valve-Position Control for Dual Input Processes," *Industrial Engineering Chemical Fundamentals* 25, no. 3 (1986).

12. James E. Bailey, and David F. Ollis, *Biochemical Engineering Fundamentals,* 2nd ed. (New York: McGraw-Hill, 1986).

13. Jens Nielsen, John Villadsen, and Gunnar Lidén, *Bioreaction Engineering Principals,* 2nd ed. (New York: Springer Science + Business Media, 2003).

14. Michael L. Shuler, Fikret Kargi, and Matthew DeLisa, *Bioprocess Engineering: Basic Concepts*, 3rd ed. (Boston: Prentice Hall, 2017).

15. Joseph S. Alford, email correspondence with author, 2006.

16. Grant Wilson, Gregory K. McMillan, and Michael Boudreau, "PAT Benefits from the Application of Modeling and Advanced Process Control of Bioreactors," *Emerson Global Users Exchange*, Nashville (October 2005).

17. Gregory K. McMillan, *Advances in Reactor Measurement and Control* (Research Triangle Park, NC: ISA [International Society of Automation], 2015).

18. Gregory K. McMillan and P. Hunter Vegas, *Process/Industrial Instruments and Controls Handbook*, 6th ed. (New York: McGraw-Hill Education, 2019).

5 Digital Twin

5.1 Introduction

The requirements for management of change are stricter than ever in the process industry. Validation requirements for pharmaceuticals involving extensive and intensive documentation of design, qualification, and verification pose an even larger motivation to minimize change. The shear complexity and effort involved limits time for creativity and discourages doing anything that changes the process as defined by research and development (R&D). The Process Analytical Technology (PAT) initiative seeks to address these concerns to promote innovation and continuous improvement.

Often the process is set based on laboratory benchtop and pilot plant runs. Even here the opportunities are narrow because batch cycle times are extremely long, especially for new biologics (e.g., 10 days for mammalian cell bioreactors), and the time to market must be minimized to make the most out of patent protection for high-value-added products.

The key to addressing this reality is the digital twin. The actual configuration, including operator interface and alarm systems, is the same as used in the lab, pilot plant, or production plant. The signals can use actual input/output (I/O) channels for communication by simulated input/outputs (SIO) or a virtual input/output module (VIM). If necessary, an actual distributed control system (DCS) can be connected with a first-principle model of the process running with communications to the DCS via open platform communications (OPC). Recent breakthroughs in kinetic modeling (detailed in the new Chapter 6) enable model parameters to be readily set so that the batch profiles in the digital twin match those in the lab and plant.

Increasingly, R&D is realizing the value of the increased capability and flexibility of a small DCS to run batches automatically, greatly reducing cycle time and nonintentional variability. The transition from lab to pilot plant to production plant of the control system and digital twin is seamless and fast, greatly reducing the validation burden.

The key to the scientific method and innovation is experimentation. Presently, experimentation is largely limited to the early phases of R&D. It is increasingly recognized that the digital twin enables rapid and free experimentation with speedup factors greater than 500 so that a 10-day batch can be completed and analyzed in 30 minutes.

The digital twin previously known as virtual plant is a relatively new concept that is easily confused with existing simulation methods for process design, configuration checkout, and operating training systems. Users may not realize that most of the existing batch process simulations are off-line and noninteractive, and that most of the real-time dynamic process simulations were originally designed for continuous processes. These real-time process simulations can develop severe numerical errors or even fail under the extreme conditions of batch operations and require interfaces for communicating I/O, controlling inventory, and coordinating with the control system in speeding up, slowing down, pausing, or resuming. The control system engineer is probably most familiar with tieback simulations because these have been predominantly used for configuration checkout and operating training systems. The process response in these tiebacks is mimicked by the trial-and-error adjustment of ramp rates. Ramps are triggered by the manual inclusion of flow path logic to include the opening and closing of valves or the turning on or off of pumps.

Section 5.2 of this chapter discusses the key features that distinguish a digital twin from existing simulation systems and itemizes the important advantages of the digital twin approach. Section 5.3 provides a spectrum of uses for digital twins. Section 5.4 concludes with the basic requirements and issues concerning implementing a digital twin.

The last two sections note the role of a benchtop system with an industrial control system in getting the most out of a digital twin. This chapter shows how a digital twin creates an environment for innovation that is the heart of the PAT initiative.

Learning Objectives

- Appreciate the significance of being able to export and import to the real plant
- Recognize the differences between different types of simulations

- Know the diverse opportunities that are essential to getting the most from PAT
- Realize the importance of a benchtop and pilot plant system with an industrial control system
- Know the basic digital twin implementation steps
- Develop the ability to nonintrusively explore, develop, and demonstrate improvements
- Understand how to develop online process metrics to automatically document benefits of improvements
- Understand how to use an MPC nonintrusively for adaptation
- Be introduced to the simpler thermodynamic requirements of bioreactors
- Recognize the importance of a charge balance for pH

5.2 Key Features

The first key feature that distinguishes a digital twin from process simulators is its ability to use the configuration, historian, displays, and advanced control tool set of the real plant without translation, emulation, special interfaces, or custom modifications. The configuration database from the real plant can be exported and then imported and downloaded into a personal computer or a control system computer just as if it were an actual hardware controller. Also, files for operator graphics, process history charts, and data history from the real plant can be copied to the digital twin so the user has the entire control system of the real plant on a computer, as shown in Figure 5-2a (intent is to convey knowledge paths and not the actual process being simulated) [1].

The use of the actual configuration, database, displays, historian, and advanced control tool set without any translation, emulation, special interfaces, or custom modification enables inherent replication of the real control system.

Most dynamic, high-fidelity process simulation software offers the ability to build a basic control strategy or sequence inside the simulation environment. However, simulation developers tend to have a process rather than a control background and focus. It is unrealistic to expect the process and batch control capability offered in simulation software to be in the same realm as the control capability of DCS software, which is the culmination of a hundred engineering years (i.e., 5 years by 20 engineers) or more of an effort by process control experts. The overall control functionality in process simulators is primitive compared to the capabilities offered in the modern DCS. The

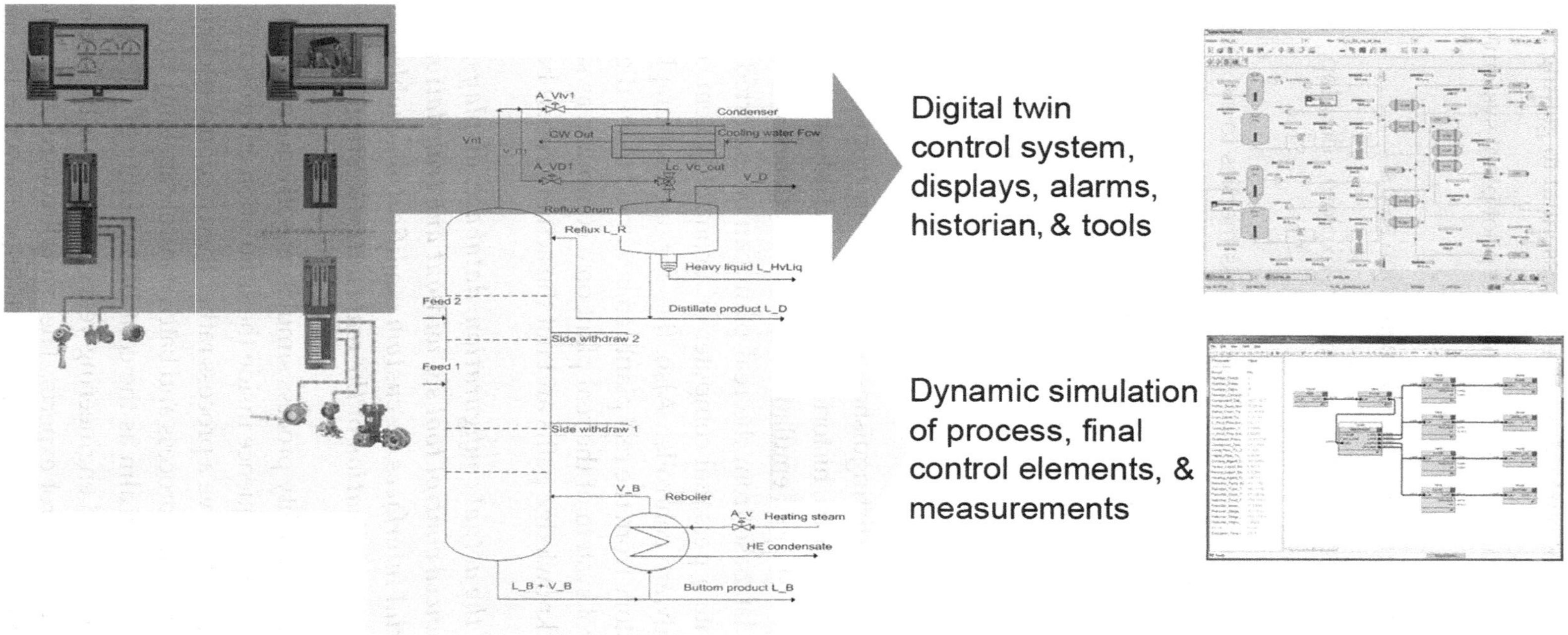

Figure 5-2a. Digital twin with actual DCS system and dynamic simulation of process and instrumentation.

DCS offers capabilities such as Sequential Function Charts and basic function blocks, the batch manager, and advanced control tools like multivariable model predictive control (MPC). These are almost nonexistent in simulators. The effort to duplicate even a simplified version of a control system in a dynamic simulation is great. At best, the user ends up with two control systems with no assurance that they will match well and no method for automatically managing changes between them.

Consequently, most simulation software now offers a standard or custom OPC interface. However, for the simulations with material balances to be able to run independently of the DCS for development and testing, the simulations must still have internal pressure and level loops set up to prevent volumes from running dry, overflowing, over pressurizing, or developing flow reversals from pressure gradient reversals, which can lead to fatal numerical errors [1]. Tables must be mapped that transfer control from these internal simulation loops to the DCS loops and initialize the proper controlled variables, set points, and manipulated variables. DCS loops that do not have a counterpart in the simulation still need to have their controller outputs initialized. The use of standard DCS blocks for split-ranged control, velocity limiting, signal characterization, and signal selection makes the proper initialization of external simulations manageable, but still difficult.

The digital twin eliminates the custom programming of control strategies into process simulations and eliminates of the significant simplification and the extra effort, verification, revision, and coordination to ensure fidelity.

The digital twin model has the material balance, component balances, energy balance, charge balance, and kinetics for a particular type of cell culture and product as detailed in Chapter 6. The user can drag blocks for process streams, mass flowmeters, pumps, fans, transmitters, control valves, and actuators into the model, and then wire them up to the bioreactor model. The process information is automatically conveyed from block to block through the "wires" that function as process and signal paths. The flow in each process path is automatically computed, as described in Chapter 6, from pressures set for streams entering or exiting the model, bioreactor pressure, and from intervening pump or fan speeds, valve positions, and pressure drops from flow resistance. It is unnecessary to duplicate level and pressure loops for inventory control and sequences for batch operation or to add interface tables for communication of I/O.

The graphical configuration of process models linked to the control modules eliminates the need to duplicate inventory and batch control or to add interface tables.

Implementing the process simulation as process models in the digital twin offers advantages beyond the elimination of the previously noted issues of duplicating loops and configuring interface tables. Foremost is the digital twin's inherent ability to run the process models at the same real-time execution time multiplier as the control modules. This ensures that the user can set a common real-time multiplier and that the simulation and control system will slow down or speed up in unison. The user can set a real-time execution multiplier of 1/30 to 30 for all modules in the digital twin. When external high-fidelity simulation packages are used, the actual real-time factor may depend on processor loading. This is because of the complexity of the calculations associated with the objectives of process or equipment design. Often during the times of greatest interest, during disturbances or failures, these simulations slow down because the integration step size has decreased for numerical stability and the control system and operator activity have increased. Even if the hook exists between the simulation and DCS real-time multipliers, the DCS is always playing catch-up.

The digital twin addresses the difficulty of ensuring that the control system is running at the same real-time multiplier as a separate process simulation during the times of greatest interest.

The speedup of the process model can be much faster than the real-time multiplier for the control system. The process model speedup is the product of speedup factors for kinetics and the material and energy balances as detailed in Chapter 6. If the kinetic speedup factor is 10 and the material and energy balance factor is 50, the process model speedup is 500. If the control system speedup is 10 that can be set via Figure 5-2b screen, the response times of the instrumentation could be reduced by a factor of 50 to keep the proportional-integral-derivative (PID) controller tuning about the same. However, the process flow ranges of final control elements and flow measurements must be increased by a factor equal to the kinetic speedup.

The process models in a digital twin can be designed to show the proper behavior for failures, start-ups, shutdowns, and batch sequences. This means that the process models can handle zero flows (closed valves and turned-off pumps or fans), empty volumes, non-equilibrium conditions, and imperfect mixing. Sophisticated external (non-digital twin) dynamic process simulations are prone to fatal numerical errors from ill-conditioned matrices used for the simultaneous solution of stream conditions by the pressure-flow solver [2]. These simulations may not only slow down but they may also shut down during extreme conditions. Consequently, the simulation of batch processes often requires specialized software that runs off-line as a single program execution. Functionally, the run conditions are set at the time of execution. These batch process simulations may appear to the user or be cited by the supplier as interactive,

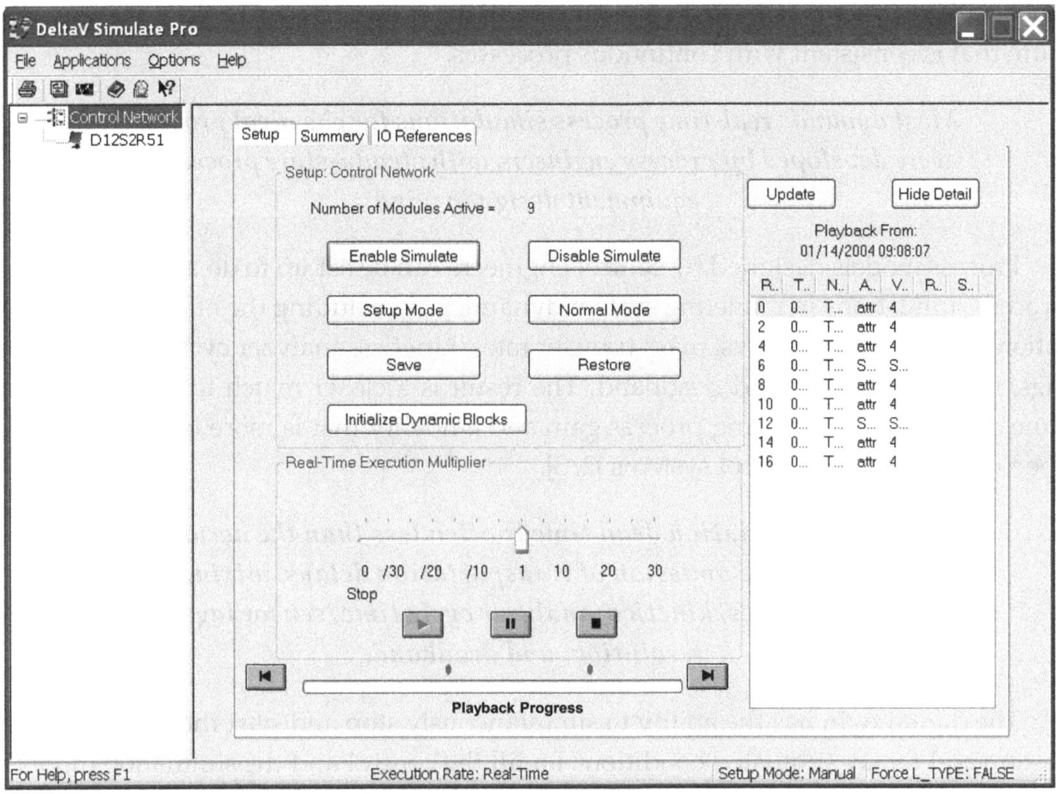

Figure 5-2b. Screen for control of digital twin speed, modes, capture, and playback.

but a change in conditions or parameters requires that the entire program and the batch process be re-executed from start to finish. The ability to include the interaction between the process and the batch and process control strategies is severely limited or impossible. Because these programs are not running in real time or a multiple of real time, it is not possible to pause, restore, or play back parts of a batch.

It is important to verify whether a dynamic simulation can handle the zero flows, empty volumes, and non-equilibrium conditions associated with batch operation.

Most external (non-digital twin) batch process simulation software offered to date is not truly interactive with either the user or the control system.

Process simulations developed by process engineers can predict self-regulating process gains relatively accurately because of the sophistication of the physical properties, thermodynamics, interactions, and equations of state needed for continuous process and equipment design. Process engineers tend to think steady state. Batch process design is noticeably absent in books and courses on chemical engineering.

Consequently, these engineers learn and think in terms of a steady state and equilibrium that is consistent with continuous processes.

Most dynamic real-time process simulations for chemical processes were developed by process engineers with steady-state process and equipment design in mind.

Process models designed by control engineers can be set up to do a better job than process simulations in modeling system dynamics by including the effect of transportation delays, mixing delays, mass transfer rates, kinetics, analyzer cycle times, sensor lags, noise, resolution, and deadband. The result is a closer match to the dead time, time constant, or integrating process gain and a fidelity that is more in tune with the spectrum of uses for control systems [2, 3].

The process simulation dead time is often less than the actual dead time because of the omission of transportation delays, mixing delays, mass transfer rates, kinetics, analyzer cycle time, sensor lags, noise, resolution, and deadband.

The digital twin has the ability to simultaneously stop and start the execution and restore and replay simulated conditions for all the control and the simulation models. Now emerging is the ability to replay actual plant data history files at high speeds to adapt and test the process models without having a connection to the actual plant.

Most simulations used for control system checkout and operator training become obsolete after start-up. The investment is lost. The digital twin offers a better chance of keeping the control system up to date by simply importing the most current configuration and enabling the simulation to better match process changes using the nonintrusive automatic adaptation described in Section 5.4.

External (non-digital twin) simulations used for control system checkout and operator training become obsolete after start-up because they lack the ability to be automatically updated and adapted.

Finally, in a digital twin everything is done in the same configuration environment that is used for the actual control system. The focus can be more on the application than learning the inevitable undocumented features and tips and techniques associated with any new simulation software and interface. The advantages offered by a digital twin are summarized as follows:

- Control system and graphics do not need to be duplicated, emulated, or translated.
- Special data interfaces, tables, and initialization issues are avoided.

- All batch, basic, and advanced control tools available in an actual control system can be readily tried out.
- Controls and simulation can run in unison at the same real-time multiplier.
- Controls and simulation scenarios can be saved, restored, and played back.
- Actual plant data can be played back at high speeds for testing and adaptation.
- Simulations can handle extreme conditions of batch operations and failures.
- Simulations can incorporate dynamics that are important for tuning and performance.
- Controls and simulation can stay up to date and have a longer life cycle.
- Engineers can work in the same environment and focus on the application.

5.3 Spectrum of Uses

The most familiar use of a digital twin is for testing and training. For the checkout of batch sequences and the training of operators, it is important to be able to repetitively and rapidly simulate batch phases. The ability to stop, start, save, restore, and replay scenarios and record operator actions is critical. For first-pass testing and familiarization of sequences and graphics, an automated tieback simulation may be sufficient. To test and learn about the interaction and performance of both control strategies and the process, the higher fidelity dynamics offered by process models is important. It opens the door to upgrading the process and control skills of technology, maintenance, and configuration engineers who support operations.

A process simulation with high dynamic fidelity is important for testing process and control system interaction and performance.

Not well recognized is that the dynamic models used often for training operators as part of an automation project have a much wider utility that today is more important than ever. There is a great opportunity to use the digital twin to maximize the synergy between operators, process control engineers, and the control systems. To start on this path, process control engineers must be given the time to learn and use a digital twin and set up online metrics for process capacity and efficiency. The digital twin offers flexible and fast exploring => discovering => prototyping => testing => justifying => deploying => testing => training => commissioning => maintaining => troubleshooting => auditing => continuous improvement showing the "before" and

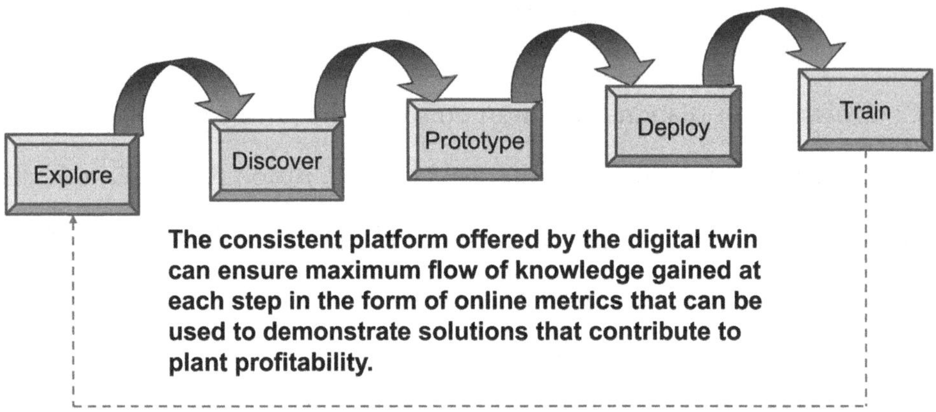

Figure 5-3a. Continuous improvement can become an inherent part of the digital twin maximizing synergy of operational, process, and control system knowledge [10].

"after" benefits of solutions from online metrics. Figure 5-3a outlines the major steps in continuous improvement maximizing innovation [10].

The capability of dynamic models for improving system performance has greatly increased, even though the use has focused mostly on training operators as an automation project nears completion. The digital twin should detail the tasks needed to address difficult situations based on the best operator practices and process knowledge and eliminate the need for special operator actions through state-based control. Advanced process control (APC) and MPC can deal with disturbances and address constraints intelligently, continually, and automatically, with great repeatability. Compare this with what operators can do in terms of constant attention, deep knowledge, and timely predictive corrections considering dead time, multivariable situations, and uncertainty in human behavior. Some operators may do well, but this is not carried over to all operators. Then, of course, an operator can have a bad day. Automation enables continuous improvement and recognition of abnormal conditions by a much more consistent operation. A better understanding by the operator of control system functionality and process performance from online metrics greatly reduces disruptions by an operator unnecessarily taking a control system out of its highest mode and/or making changes in flows. Furthermore, procedural automation can eliminate the manual operations during start-up when risk is the greatest compared to normal fed-batch or continuous operation [10].

Note that while we have singled out operators and process control engineers, the need for knowledge to attain the best performance extends to maintenance technicians, process engineers, mechanical engineers, and information technology specialists. Think of what can be realized if we were all on the same page understanding the process and operational opportunities and the value of the best measurements, valves, controllers, and software [10].

Here is a brief summary of the many possible uses of a digital twin:

- **Cause and effect relationships** – Data analytics can point to many possibilities, but the relationships identified are correlations and not necessarily cause and effect. Also, the data sets used in data analytics are limited to the range of plant operation, which by design may not show the changes possible. Process control is all about changes. We saw this from the beginning in our first course on control theory where all the variables for Laplace transforms and frequency response were deviation variables. We must see the effect of changes in process inputs on the changes in process outputs. A digital twin can make all the changes that you cannot make in a real plant. You can be creative, freeing your mind to make changes, learning the most from mistakes. We can do this all in the safe environment of the digital twin [13]. The resulting data can verify correlations, provide richer data sets for data analytics, identify the dynamic compensation needed in data analytics for the predictions by projection to latent structures also known as *partial least squares* (PLS), and provide a much leaner, faster, safer, and more focused design of experiments (DOE) for the actual plant. Root cause analysis can be greatly improved [10].
- **Interactions and resonance** – The degree of interaction and resonance depend on the dynamics, tuning, and oscillation periods. The digital twin can show all this and much more using Power Spectrum analyzers to track down the culprit if the process and automation system dead times are modeled [10].
- **Valve and sensor response** – The largest source of dead time and oscillations can be traced back in most loops to the valve and sensor response. In particular, most of the valves designed for high capacity and tight shutoff not only have a poor response time but also will cause continuous limit cycles from stiction and backlash [11]. Showing the effect of slow or noisy sensors and on-off valves posing as throttling valves can provide the justification for getting the best measurements and valves. You can show the effect of sensor and valve failures and the value of redundancy and a smooth recovery [10].
- **Process safety stewardship** – The digital twin should be used for extensive and continual testing of all the layers of protection in addition to the detailed design and implementation of the safety instrumented system (SIS). The interplay and performance of layers should be thoroughly evaluated and documented for all the important scenarios. The speed of measurements and valves is often not

thoroughly understood, and the consequences of failures should be analyzed in these scenarios [10].

- **Control system and SIS knowledge –** One of the most difficult aspect of control system and SIS design is recognizing the implications of measurement and final control element repeatability, rangeability, reliability, resolution, and response time. The digital twin enables you to explore and determine the value of the best instrumentation and best PID control and MPC, including the solutions and parameters needed. You must see how well the systems play together and help the operator in abnormal conditions. You can try all kinds of what-if scenarios to see how the system and operator perform [10].

- **Validation and regulatory support –** The validation and regulatory compliance of automation systems in pharmaceutical production requires a large expenditure of time and expertise in automation projects. The digital twin offers the verification of performance needed for confirmation and documentation [10].

- **Checkout of code –** Checking out the code and control sequences before the actual plant start-up enables the control engineer to validate that the code was written as required and designed. It allows debugging for any mistakes made during the coding and validating seen in the coordination of each sequence. The more time the control engineer spends checking out the code and exercising the sequences, the fewer issues he or she will face during the start-up and commissioning of the plant.

- **Process and equipment knowledge –** Most of our deep understanding of the implications of process and mechanical designs has come from running dynamic models. This was particularly true for the specialties of pH and compressor control. For pH control, the extraordinary sensitivity and rangeability and consequential incredible nonlinearity of a system of strong acids and bases leads to strict design requirements in terms of minimizing mixing and injection dead time and seeking better set points, and the possibility of using weak acids or weak bases and conjugate salts to help reduce the steepness and nonlinearity. Today, we have the ability to handle conjugate salts as well as a wide spectrum of weak acids and weak bases by means of an expanded charge balance equation detailed in Chapter 9 [10, 11].

- **Process equipment degradation –** The consequences of fouling of heat transfer surfaces, thermowells and electrodes, loss of catalyst activity, and plugging of filters can be studied and more optimum times for maintenance and cleaning scheduled. Better installation practices and redundancy of measurements can be justified. For pH, three electrodes with middle signal selection provide the best protection against reaction to noise, failures of any type, and fouling [10, 11].

- **Start-ups, transitions, shutdowns, and batch operation** – The design and value of procedural automation and state-based control for fed-batch and continuous processes is best achieved by trying out all conceivable scenarios in a digital twin. The digital twin is also the key for optimizing batch profiles, end points, and cycle times. If an operator claims his actions cannot be automated, there is an even a greater motivation for a digital twin to identify and test the best actions. The automated repeatable best actions reveal the other sources of changes by eliminating the variability from operator response [10, 11].
- **Optimum operating points** – Finally, you can identify and document the benefits of better set points and achieve these set points more reliably and extensively by feedforward control, override control, batch profile control, full throttle batch control, and model predictive control [10].

Before the configuration even starts in the front end of a project, the process models can be used to evaluate control strategies and advanced control tools. In the past, this was done with off-line dynamic simulations. Having ready access to an industrial tool set for both basic and advanced control and simulations that is adapted to benchtop or pilot-plant runs offers the opportunity for rapid prototyping. This can lead to control definitions that have better detail and potential performance. Benchtop or pilot-plant systems with a mini version of the industrial DCS are now available that greatly facilitate the development and scale-up of the control system [6]. Benchtop systems and pilot plants that have all the functionality of the main manufacturing systems are not yet prevalent because the development groups of these types of companies traditionally do not have the expertise (and more importantly the interest) to configure, maintain, and engineer these systems. The digital twin enables a synergy between scientists and control engineers to make the incredible capability of DCS part of the R&D process.

The digital twin can be demonstrated with the industrial batch, basic, and advanced control systems used in the benchtop or pilot-plant system. The process models can be adapted by way of a connection to a benchtop or pilot-plant system or by built-in high-speed playback of process data from experimental runs. An opportune time to take advantage of the PAT initiative is during the research and process development phase. The value of benchtop and pilot-plant systems that have an industrial DCS is significant in terms of project schedule, cost, and effectiveness because the DCS enables process control to be designed into the plant at an early stage and puts the biochemist, process technology, and configuration engineer on the same page [6]. However, the best return on investment (ROI) for PAT is realized by eventually implementing advanced process analysis and control in large-scale manufacturing processes.

In a digital twin, innovative strategies such as effective switching of the controller output can be trialed and tuned. Advanced control tools such as adaptive control, auto tuning, fuzzy logic, MPC, neural networks (NNs), principal component analysis (PCA), and PLS can be demonstrated, adjusted, and evaluated "faster than real time."

A benchtop or pilot-plant bioreactor that has an industrial control system offers significant opportunities for reducing configuration costs, improving communication, developing a digital twin, and prototyping advanced control, but the greatest ROI is realized by implementing PAT in large-scale manufacturing processes.

As a general rule of thumb, five changes are needed for each process input to develop an experimental model of a process output. For MPC, this corresponds to a minimum of five step changes in each process input, at least one of which is held long enough for the process output to reach a steady condition. For NN, PCA, and PLS it means a minimum of five batches of process input in which the respective process input differed from the normal value. Although actual plant operation is obviously the best source of data, the long batch cycle time and the desire to minimize disruptions from the introduction of perturbations severely restricts the amount of useful plant data available for the development of MPC, NN, PCA, and PLS. For example, developing a PCA with four inputs for detecting an abnormal batch would require at least 20 batches with varying inputs and at least five batches with normal inputs. If the batch cycle time is about two weeks, it would take about a year of plant production to have enough data. If you consider that you cannot deliberately optimize the spectrum of variability in the inputs, then it may take several years of production runs to have enough data.

In contrast, perturbations can be automated and introduced to a digital twin running faster than real time so that in 2 weeks there is enough data to identify models for MPC, to train NN, to develop latent variables and discriminant analysis for PCA, and to predict economic variables via PLS. The predictive ability of MPC, NN, PCA, and PLS can then be verified and evaluated by the high-speed playback of previous plant batches.

Conventional PCA assumes that all process inputs other than the ones used for the PCA are fixed. A digital twin running in real time that is synchronized with the actual plant can be used to predict the effect of variations in other process inputs by using the model-based and super model-based PCA algorithms described in Section 8.4 of Chapter 8.

The digital twin can also be used to help explore more optimal operating conditions and investigate what-if scenarios. These scenarios are important for identifying the cause of an abnormal batch. Today, PCA for batch fault detection only identifies a batch as abnormal. Logic needs to be added to diagnose the fault. Fuzzy logic rule sets have been used in conjunction with PCA to provide real-time predictive fault analysis [7]. A digital twin can help develop these rule sets off-line faster than in real time by creating scenarios and evaluating the rule sets through the high-speed playback of previous batches. An online digital twin synchronized with the real plant can be sped up and run to batch completion to predict and analyze abnormal situations based on current batch conditions. A digital twin can also be used to help create the predictive capability of the PLS "Y space" of economic variables from the PCA "X space" of process variables.

A digital twin can eliminate process testing and provides years of process data within a few weeks for advanced control, fault analysis, and performance prediction.

Additionally, a digital twin can be used to provide inferential real-time measurements of important process outputs such as nutrient, biomass, and product concentration. The built-in material balances can be used in conjunction with kinetic models to predict concentrations. These predictions are then delayed so that the values are synchronized with online or lab analysis. The concentration is shifted by a bias that is a fraction of the difference between the inferential and actual measurement, similar to what is done in the feedback correction of an NN used for property estimation [3].

A digital twin can provide faster, more reliable, and smoother measurements of concentration than online analyzers.

The rate of change of these concentrations can be used by MPC to optimize batch profiles as described in Section 4.4 of Chapter 4. A digital twin can also be run "faster than real time" to batch completion to provide an online prediction of key performance indicators (KPIs), such as batch cycle time and yield, based on current batch conditions and inferential measurements. Presently, the understanding of the value of process control in management is decreasing and the emphasis on profitability is increasing. It cannot be taken for granted that practitioners will be given the time and funding to make process control improvements. The retirement of most of the process control experts has exasperated the situation. We are at a critical point in terms of the decreasing rate of innovation and increasing tendency to copy jobs in migration projects. The best way to turn this around is by means of KPI with actual dollars correlated to process control improvements that can be revealed and tested by dynamic first-principle models in a digital twin and made available real time to operations [14].

A key insight is that the myriad improvements can be categorized as increases in process efficiency, capacity, flexibility, and safety. Increases in process efficiency show up primarily as a decrease in the ratio of the running average of raw material mass or energy used to the running average of product mass produced. Increases in process capacity show up as an increased running average of the product mass produced. Process capacity increases can be the result of higher production rates, faster start-ups, better ability to deal with abnormal operation, and greater onstream time. In all cases, the product mass must meet customer specifications. Flexibility shows up as the ability to meet different production rates or different product requirements. Safety shows up as minimizing activations of the SIS besides the obvious metric of minimizing the number of incidents, including near misses [11].

The period for metrics must be large enough to eliminate noise and inverse response and to provide the ability to make decisions based on the objective and process type. For evaluating operator and control system actions, the period is normally the batch cycle time and operator shift for batch and continuous processes, respectively. The period is a month for correlation with accounting metrics. For alerting operators as fast as possible to the consequence of actions taken (e.g., changing the controller set point or mode), the period can be reduced to be as short as six times the total loop dead time. The metrics at the end of a month, batch, and shift are historized for evaluation of plant, batch, and operator performance [11].

There is often a trade-off between process metrics. Increasing the production rate often comes at a cost of a decreasing efficiency. Similarly, changing production rates reduces both process efficiency and capacity because movement to the new process operating point takes time and the product produced in the transition may not meet specifications. Increases in yield (decrease in raw material use) can be taken as an increase in process efficiency if the raw material feed rate is accordingly decreased. There may be an accompanying decrease in the cost of recycle and waste treatment operations. Alternatively, increases in yield can be taken as an increase in process capacity by keeping the raw material feed rate constant. Prolonging a batch can improve yield and thus efficiency, but the lengthening of batch cycle time translates to less batch capacity, particularly as reaction rates or purification rates decline near end point. Time to reach a new set point can be shortened by overdriving the manipulated variable past its final resting value. For batch processes, reaching a new composition, pH, pressure, or temperature set point is often not possible without the overdrive. The process efficiency is reduced during the overdrive, but the process capacity is increased either as a reduction in batch cycle time or an increase in the continuous production rate upon reaching the set point [11].

Especially important is the translation of online metrics to the bottom-line effect on production unit profitability in the plant accounting system. This means benefits must be reported on a monthly basis and presented per accounting format and procedures. Also, obvious but often not addressed is the buy-in by the plant accounting department and plant management. This is best done by real-time accounting (RTA) [11].

The KPI in the digital and actual plant should be converted to dollars of revenue and dollars of cost with a running total for the last month, batch, and shift. Online metrics on process efficiency, often expressed as raw material mass or energy used per unit mass of product, must be multiplied by the production rate at the time and the cost per unit mass or energy used to get dollars cost per unit time. Online metrics on process capacity must have a production rate multiplied by the price of product(s) sold. This is best accomplished by getting an accounting representative to participate in the development and use of the metrics [11].

An opportunity sizing details the gaps between current and potential performance that are estimated by identifying the best performance found on cost sheets and from a DOE, which is most often done in a digital twin due to increasingly greater limitations to such experimentation in an actual plant. After completion of the opportunity sizing, a 1- or 2-day opportunity assessment was led by a process engineer with input sought and given by operations, accounting and management, marketing, maintenance, field and lab analyzer specialists, instrument and electrical engineers, and process control engineers. Marketing provided the perspective and details on how the demand and value of different products is expected to change. This knowledge was crucial for effectively estimating the benefits from increases in process flexibility and capacity. Opportunities for procedure automation and plantwide ratio control making the transition from one product to another faster and more efficient were consequently identified. Agreement was sought and often reached on the percentage of each gap that could be eliminated by potential process control improvement (PCI) proposed and discussed during the meeting. A rough estimate of the cost and time required for each PCI implementation was also listed. The ones with the least cost and time requirements were noted as *quick-wins*. To take advantage of the knowledge and enthusiasm and momentum, the quick-wins were usually started immediately after the meeting or the following week. Figure 5-3b shows this PCI opportunity assessment process [11]

An improvement in one unit's operation performance may have a negative impact on another unit's operation. For example, a decrease in steam generation from a waste furnace from more efficient operation (less waste production) decreases the steam available for power generation. Rapid changes in raw materials or energy usage may

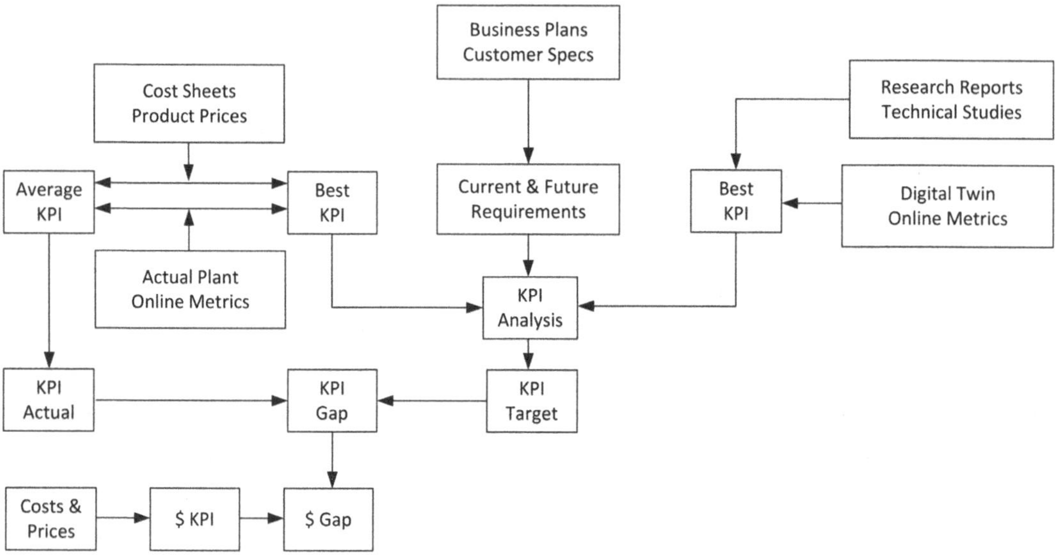

Figure 5-3b. The digital twin enables experimentation and verification to identify and achieve the best benefits from PCI [10].

upset the headers and systems that supply other systems. The online metrics should be extended to all unit operations affected to quantify the trade-offs [11].

The raw material and energy used does not have an immediate effect on product produced. Transportation and mixing delays, and time constants associated with volumes and mass transfer and heat transfer rates, prevent the start of the effect on the product and slow down the completion of the effect. Transportation delays for plug flow volumes and time constants for back mixed volumes are both inversely proportional to flow. There are dynamics and rangeability limitations associated with measurements and valves that are often not recognized and included. Also, the time to reach and settle at a new set point depends on the ability to identify a dynamic model and tune the controller to meet the best objectives [11].

The PCIs must be able to address unknowns and disturbances but also be sufficiently understood by operations to build their confidence to keep them in their highest control mode. Operators are fundamentally reluctant to relinquish manipulation of process inputs and operating points if the benefits are not seen or understood. Operators may also be initially opposed to indications and comparisons of their performance. For these reasons and many more, it is imperative that the performance metrics exhibit inappreciable noise and be consistently representative of changes in process performance. Using results in a positive way allows operators to take more ownership of process performance. The adage "if you don't measure, you can't control it" is especially important here [11].

The digital twin can speed up the benefits gained from PAT by offering users the ability to use process and end point monitoring and control, continuous improvement, and knowledge management tools in an integrated and accelerated manner. The uses of the digital twin are illustrated in the functional overview provided by Figure 1-4b in Chapter 1 and are summarized as follows:

A digital twin can accelerate the benefits of a PAT initiative.

- Testing configuration and process interactions
- Educating operators, technicians, and engineers in process and control
- Rapid prototyping of innovative control strategies and advanced controls
- Evaluating tuning settings
- Identifying MPC models
- Developing latent variables and reference trajectories for PCA
- Developing logic for fault analysis by PCA
- Predicting abnormal situations online
- Developing and testing NN and PLS models
- Making inferential real-time measurements of important concentrations
- Optimizing batch profiles
- Predicting batch KPIs, such as cycle time and yield, and PLS economic variables

Figure 5-3c shows the functional value of a digital twin, highlighting the bidirectional flows of control system and process/equipment knowledge of process control and analysis tools including online performance metrics for greater analysis, dealing with kinetics multiplicative effect in Chapter 6, and justification of improvements. The "two-way" knowledge flow is the key to improving the process/equipment and the control system besides in addition to the dynamic model and data analytics. As the fidelity of the dynamic model increases, opportunities arise for these tools to get results from the digital twin that can be used in the actual plant. The dynamic model can be run faster than real time with the tuning corrected by applying the speedup factors. New control functionality can be developed and included in the dynamic model for evaluation. If online metrics show significant improvements in control and process performance, the functionality prototyped can be added as new blocks or as improvements to existing blocks in the DCS [11].

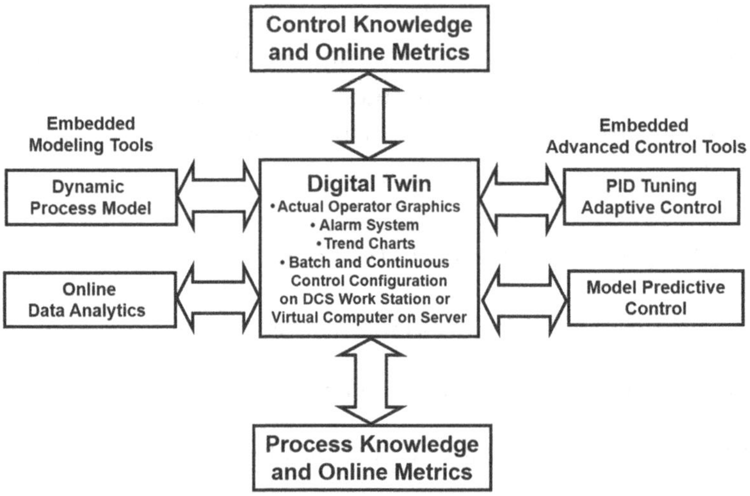

Figure 5-3c. The two-way flow of knowledge in the digital twin between tools, models, and the actual control system in the digital twin is the source of increasing synergy of knowledge between the process, control system, engineers, technicians, and data scientists [11].

5.4 Implementation

Probably the biggest obstacle to implementing a digital twin is the users' lack of knowledge of the kinetic equations that are needed to calculate the growth rates and product formation rates as a function of batch operating conditions, such as temperature, pH, DO, and substrate concentration. As detailed in Chapter 6, generalized easy-to-parametrize kinetic models have been developed that enable users to readily match historized plant batch profiles. This section discusses how an MPC can be set up to home in on the values of these parameters if they are in the ballpark. If the kinetics are unknown, NN may be able to predict kinetic parameters, terms, or rates from inferential and actual measurements. The growth rates and product formation rates are included in the net calculation of the rate of change of biomass, nutrient, and product mass in the material balance. They are then integrated to get a new accumulation or concentration of the component, as described in Chapter 6. Online calculations of the OUR and carbon dioxide production rate (CPR) as NN inputs enhance the predictive capability of the NN. The result is a hybrid first-principle and NN model [8]. In general, batch time should not be an NN input because the predictions are undesirably dependent on batch time. An NN without batch time as an input will be able to predict ethanol and cell mass [9].

Breakthroughs in kinetic models enable easily matching the batch profiles seen on trend recordings without a need to disclose proprietary R&D.

The OURs and CPRs can be used in lieu of kinetic equations, but it may be difficult to differentiate between the consumption and evolution associated with biomass growth vis-à-vis product formation rates.

If the equations are unknown, an NN may be able to predict the individual growth and production rates if there are online measurements or frequent lab measurements of biomass, product, and substrate concentration [9]. A benchtop or pilot-plant system offers the best opportunity for developing an NN. To minimize the investment, a skid of analyzers could be alternately connected to bench or plant systems to generate the test data needed for NN training and validation. This approach has proved to be successful enough for state governments to accept online NN as a permissible alternative to online analyzers for combustion emissions monitoring (CEM) systems.

Online broth or media concentration measurements and NNs can predict growth rates and product formation rates without a prior knowledge of kinetics.

Composite blocks for the bioreactor, streams, final elements, and transmitters are inserted into the digital twin process model and wired up. External input and output parameter blocks are added and connected to the final element inputs and transmitter outputs, respectively. The control system configuration is imported, and the analog input blocks in the virtual control system are switched to the simulation mode. The paths of the external inputs for the final elements are chosen via a browser to be the outputs of the analog output blocks of the virtual control system.

Similarly, the paths of the external output parameters for the transmitters are selected to be the simulation inputs of the analog input blocks of the virtual control system. Analog outputs and analog inputs of loops that are not modeled can be simulated by simple tiebacks in which the analog output is passed through signal characterizer, dead time, and filter blocks and then connected to the simulated input of the analog input block.

The concentrations, pressures, and temperatures of each inlet and outlet stream; the flow rate and pressure rise of each pump as a function of speed; the flow coefficient of each valve as a function of position; the agitator pumping rate as a function of speed; and the initial inventory, concentrations, temperature, and pressure of the bioreactor are then set. NN and PLS models that are developed from benchtop experiments or pilot-plant runs are added as necessary to enhance the kinetics described in Chapter 6. Control valve deadband and resolution are set in the *actuator* blocks, and measurement delays and lags are set in the *transmitter* blocks.

Benchtop systems offer the best opportunity for developing and validating first-principle, PLS, or NN models.

The batch sequence, basic process control system, and simulation are run in an off-line mode, and the simulation parameters are manually adjusted to match up the virtual with the real plant batch profiles of uncontrolled process outputs and manipulated process inputs. For example, consider how the model may be adjusted for better fidelity for DO control. A kinetic parameter for the oxygen limitation effect is adjusted to match up oxygen uptake rates between the virtual and actual plant. A Henry's law coefficient is then adjusted that determines the equilibrium DO (driving force for oxygen transfer) to match up pressures. Finally, an oxygen mass transfer coefficient is adjusted to match up airflows.

When representative digital twin batch profiles are obtained, the processor loading and profile alteration are checked as the real-time execution multiplier is increased, and the maximum practical speed is noted. Key model parameters are chosen as the manipulated variables, and associated process measurements are chosen as the controlled variables of the MPC for adaptation of the digital twin. An automated test sequence is then run for the MPC at the highest possible speed off-line, and the MPC models are identified and visually checked as reasonable in terms of the direction and relative magnitude of the effect [1, 4].

The digital twin is then connected to the actual plant in a read-only nonintrusive setup. The digital twin modes, set points, and batch phases come from the actual plant. The set points (targets) of the MPC's controlled variables for adaptation are externally referenced to the key respective measurements of the actual plant, as shown in the overview provided by Figure 1-4b in Chapter 1 and in the detailed signal paths outlined by Figure 5-4a. Because the relationship between model parameters and key process variables is generally nonlinear, an NN can be used to enhance the future trajectories of an MPC. Alternatively, a NN could be trained to adjust model parameters on a one-by-one basis. Because NNs are good at interpolation but not extrapolation, it is important to verify whether the inputs are within the training set range of values. The MPC offers a multivariable solution to adaptation of dynamic models [1, 4].

The use of standard tools, such as MPC, PLS, and NN, are preferred over custom and special reconciliation methods so the focus is on the application rather than the tool development and the results are more maintainable and therefore sustainable. Regardless of which technique is used, the model parameter values must be restricted to be within a practical range.

The MPC, PLS, and NN for adapting the digital twin can be developed off-line and then tested in a read-only mode.

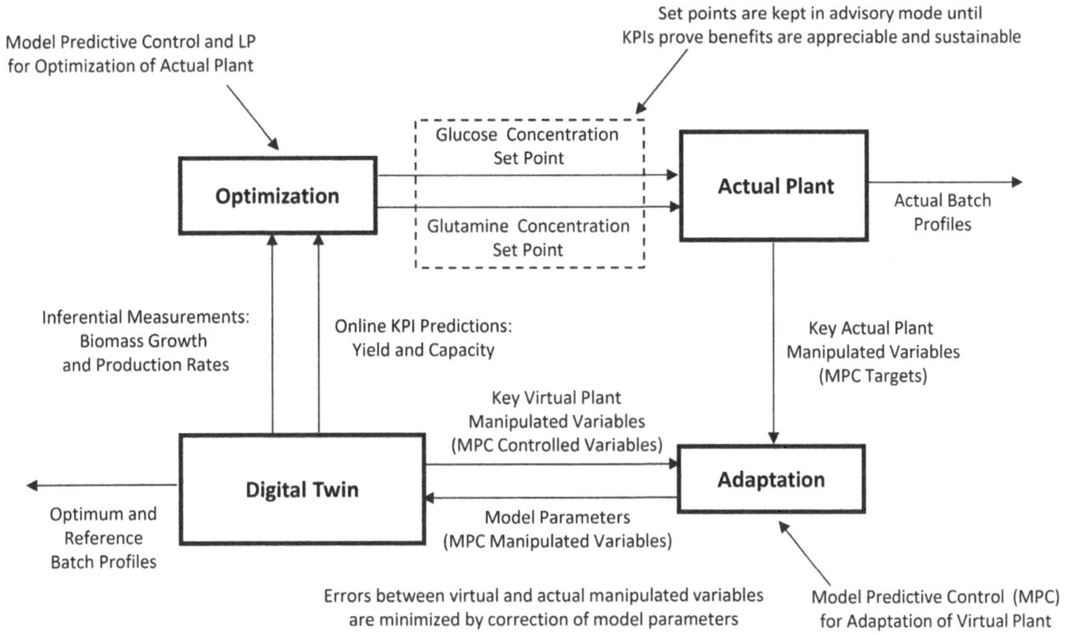

Figure 5-4a. Nonintrusive automated adaptation of model parameters to match manipulated variables with potential future optimization based on improvement in KPIs.

A mass spectrometer can be used to measure the oxygen concentration in the off-gas and a material balance can be used to measure the net difference of oxygen flow going in and coming out of the media to compute the OUR. If the DO concentration is not constant, the average rate of change of DO multiplied by the volume should be included in the calculation.

The MPC targets and controlled variables are the actual plant and digital twin manipulated variables (e.g., flows), respectively, in Figure 5-4a. That figure shows how to adapt a bioreactor model to provide more accurate inferential measurements of biomass growth and production rates that are slopes of the batch profiles for cell and product concentration. The MPC manipulated variables are model parameters (e.g., kinetic parameters) that affect these profiles [11].

The MPC can be identified and tested without the digital twin connected to the actual plant. The MPC development for adaptation is much faster than that of an actual plant MPC. This independence enables extensive adaptation and gaining associated knowledge of process relationships without any disruption to the plant [11].

The adaptation is done without affecting the actual plant because the plant's manipulated variables are being read by but not changed by the digital twin. It is critical that the digital twin have the same set points and tuning settings as the actual plant, and that the digital twin is started with controller outputs initialized to match the actual plant.

The optimized set points from MPC with inferential measurements of growth and production rate are done in an advisory mode not affecting the actual plant. Not shown in Figure 5-4a is that an MPC is run in automatic mode in another digital twin that is a duplicate of the adapted digital twin to generate and study the optimized set points. The set points are only eventually used to automatically optimize the plant if they prove more accurate, beneficial, and reliable per KPIs than an MPC with inferential measurements computed from online and at-line analyzers.

Figure 5-4b shows how an MPC was used to manipulate the Henry coefficient for oxygen to match up a change from 7700 to 7500 kPa per kmol/m^3 (intent is to convey an overview example and not the specifics of test results). This brought the controlled variable of the digital twin head pressure to its set point (target), which is the actual plant head pressure. The speed of adaptation is set by the penalty on move (PM) for the model parameter.

After adaptation, the digital twin is run off-line faster than in real time to prototype new control strategies, PCA, fault detection, and abnormal situation prediction.

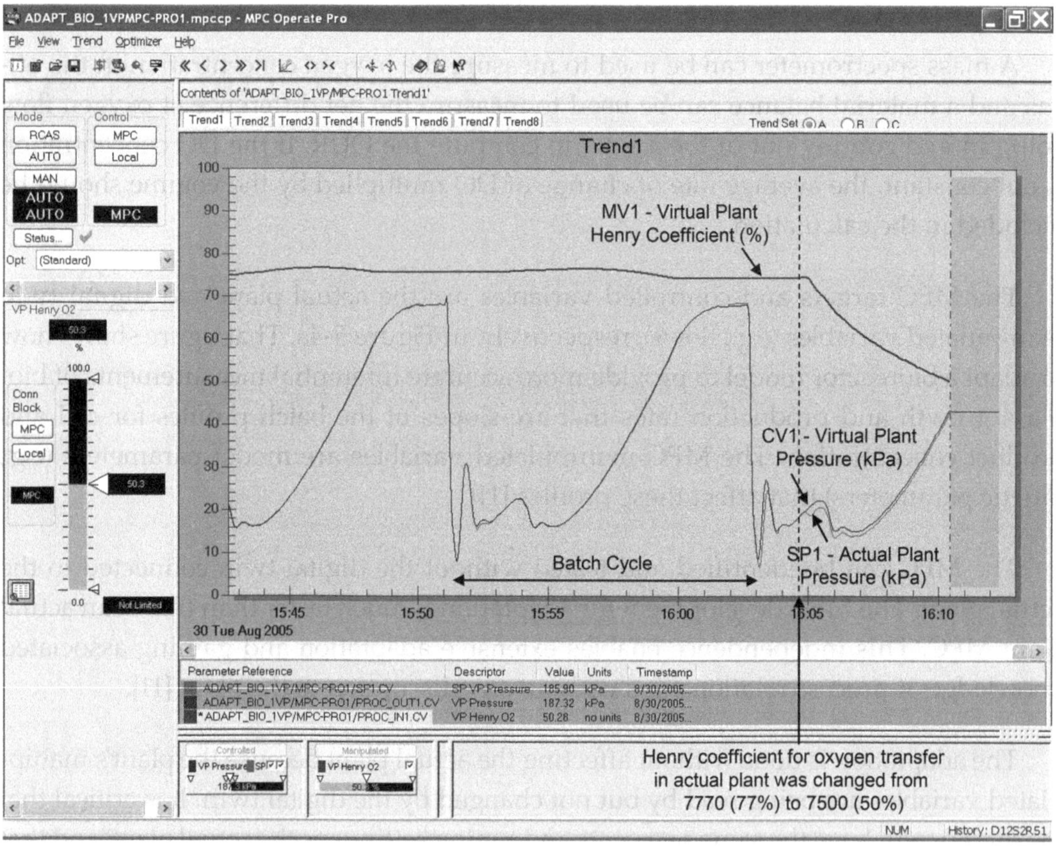

Figure 5-4b. Adaptation of Henry coefficient to match actual plant pressures.

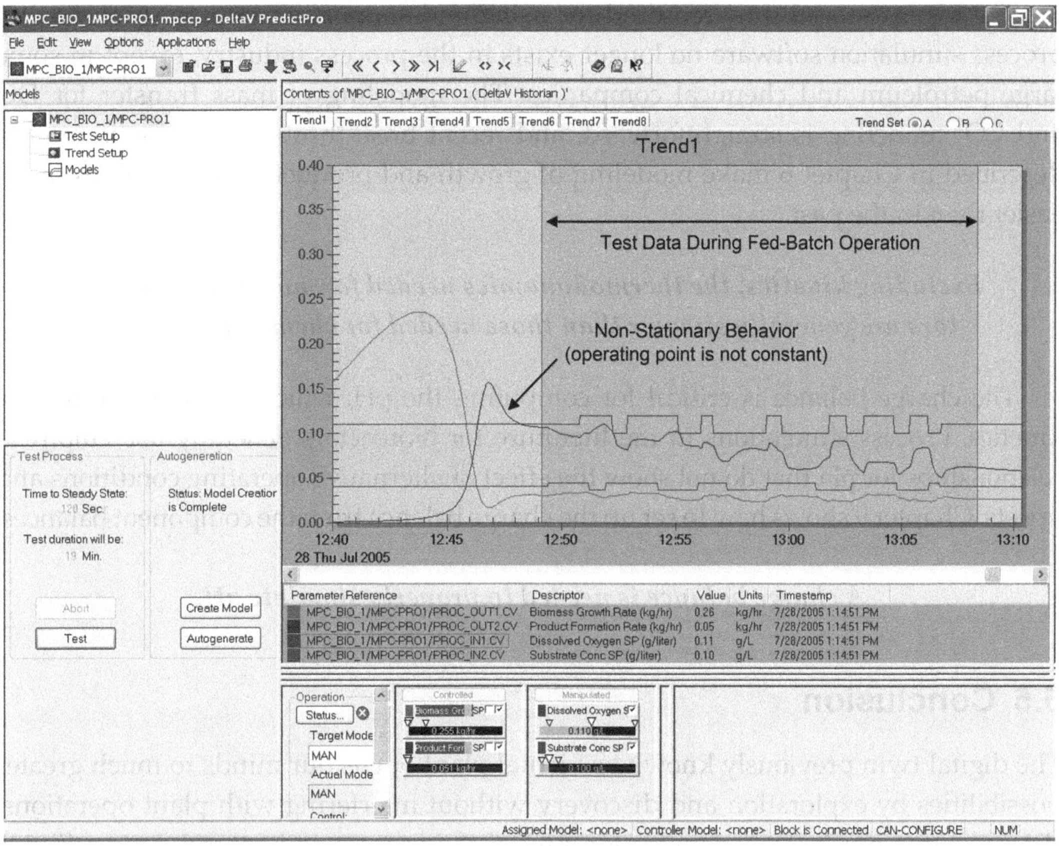

Figure 5-4c. Automated high-speed test of digital twin for model identification.

Inferential variables are computed, such as biomass growth rate and product formation rate, and the automated tests, such as those shown in Figure 5-4c (intent is to convey an overview example and not the specifics of test results), are run to identify the models of these rates versus various process inputs, such as DO and substrate concentration or feed. Even though the process operating point is moving (nonstationary) because it is a batch operation, the software is able to identify the models for optimizing growth and production rates. Inferential measurements of various concentrations are used as needed to fill in the gaps between lab measurements.

After innovative control, advanced diagnostics, and MPC applications have been prototyped and documented, they can be implemented on benchtop or pilot-plant systems. If the digital twin provides inferential measurements, it is run in real time and synchronized with these systems. After verification and validation, and accounting for scale-up factors, these applications and the digital twin can be installed for plant trials. Initially, these systems run only in the monitoring mode. When process inputs are finally manipulated, they are initially restricted to a very narrow range.

The expertise and time required for using and supporting external high-fidelity process simulation software no longer exists in the process industry, except in some large petroleum and chemical companies. The modeling of mass transfer for DO and CO_2 modeling is straightforward, and recent breakthroughs in kinetic models described in Chapter 6 make modeling of growth and product formation rates much easier than in the past.

Excluding kinetics, the thermodynamics needed for models of bioreactors are generally simpler than those needed for chemical reactors.

The charge balance is critical for computing the pH, which is important for the kinetics. Process simulations in the literature for bioreactors generally use empirical relationships for pH that do not show the effect of alternative operating conditions and upsets. Chapter 9 shows how to set up the charge balance from the component balances.

A charge balance is needed to properly simulate pH.

5.5 Conclusion

The digital twin previously known as virtual plant opens our minds to much greater possibilities by exploration and discovery without interfering with plant operations. We develop a much deeper understanding of the process and control system [5, 12]. Consequently, we can exercise much greater creativity, which translates to increasing process capacity and efficiency with benefits that are measured and documented. This leads to personal advancement and freedom for innovation. The authors of this book look forward to mentoring everyone on this path to achievement and recognition.

5.6 Digital Twin Best Practices

Here are best practices that help you use the digital twin not only for operator training but also for improving process performance and time to market.

1. Implement a digital twin that requires no changes to the DCS alarms, graphics, or logic.
2. Use first-principle process models with energy and material balances for all processes and charge balances for pH with representative physical properties and components.
3. For pH models, add a small but significant concentration of carbonic acid from exposure to atmospheric CO_2 and conjugate salts so that the titration curve

generated matches lab results, particularly in terms of the pH range and titration curve slope in the control region as detailed in Chapter 9.

4. Use an actual download of an automation system (not an imitation by emulation) that enables free exchange between the digital twin and the actual system of
   - the configuration, alarms, and APC tools (e.g., adaptive control, model predictive control, and data analytics) by exporting and importing modules and models, and
   - the historian and graphics by copying and pasting files.
5. Achieve at least medium fidelity by adjusting model parameters so that operating points sufficiently match process variables and flows on accurate process flow diagrams (PFDs).
6. Use a variable dead time block to simulate mixing and injection delays.
7. Ensure the process model can show the dynamics in start-up and transitions.
8. Have process engineers evaluate process operating conditions and response.
9. Include automation system dynamics such as the deadband and resolution limits in control valves (e.g., backlash-stiction) and in VFDs (e.g., I/O cards, speed sensor, setup), sensor transportation delays, sensing element lags (e.g., thermowell), transmitter lags (e.g., damping), wireless update rate, and analyzer cycle time and analysis time.
10. Develop, test, and document scenarios including transitions, start-up, and abnormal conditions (e.g., mechanical and automation system failures).
11. Develop and use online metrics for process efficiency and capacity.
12. Analyze and improve alarm management and the human machine interface using Center for Operator Performance and ISA standards for guidance.
13. Detail best operator responses during start-ups and to abnormal situations (e.g., equipment, valve, or sensor failures) for procedure automation (state-based control).
14. The digital twin can provide methods and practical considerations in online metrics computations, be adapted to improve fidelity by manual and automated adjustment of model parameters to match up virtual and actual flows, and be used to discover and realize process control improvement opportunities.

15. Ensure the operator training system (OTS) is continually updated and available for refreshing and enhancing operator performance.

16. Use OTS to educate technicians and engineers in maintenance and in automation, mechanical, and process technical support to improve synergy with operations. Use a multidiscipline team that includes plant operations. If the models will be deployed online, engage the operators early in the cycle. Start with your business objectives. What is the problem to be addressed, and how will you measure success in accomplishing these objectives?

17. Use a high-fidelity digital twin to develop online process metrics.

18. Synchronize process inputs with process outputs for online metrics by the individual use of dead time and filter blocks for synchronization time (e.g., total dead time and open-loop time constant).

19. Use running averages of raw materials and energy and resulting product for metrics.

20. Use different periods for running averages to show metric short- and long-term effects.

21. Improve metric signal-to-noise ratio by a filter time less than 1/5 synchronization time.

22. The shortest period for process evaluation should be greater than twice the synchronization time. Additional metrics to alert operations can have shorter periods.

23. A period equal to shift time shows the performance of operators, a period equal to the batch cycle time shows the performance of batch operations, a period equal to a month helps to predict effects on accounting cost sheets, and a period as short as six total loop dead times can alert the operator to consequences of actions taken (e.g., changing set points or modes) provided the signal-to-noise ratio is sufficient.

24. Involve accounting, plant management, and marketing in the development and use of online metrics for real-time accounting.

25. Use a digital twin for faster and more effective metrics and PCI implementation.

26. Use digital twin to reduce plant testing required for metrics and PCI.

27. Develop an MPC to adapt the digital twin without connection to the actual plant.

28. Use actual plant set points and controller tuning settings adjusted for model speedup in the digital twin.

29. To adapt the digital twin, adjust the model parameters to match the manipulated process inputs by the MPC and PID loops and noncontrolled process outputs.

30. To reduce the amount of identification needed in the actual plant for online process metrics synchronization, add injection and mixing delays and automation system dynamics to the digital twin for the development of metrics.

References

1. Gregory K. McMillan and Robert Cameron, *Models Unleashed: Virtual Plant and Model Predictive Control Applications* (Research Triangle Park, NC: ISA [International Society of Automation], 2004).

2. Walt Boyes, Dan Hebert, Larry O'Brien, and Gregory K. McMillan, "The Light at the End of the Tunnel Is a Train (Virtual Plant Reality)," *Control* (August 2005).

3. Gregory K. McMillan, "Process Modeling," in *A Guide to the Automation Body of Knowledge*, 3rd ed., Nicholas P. Sands and Ian Verhappen, eds. (Research Triangle Park, NC: ISA [International Society of Automation], 2018).

4. Terrence L. Blevins, Gregory K. McMillan, Willy K. Wojsznis, and Michael W. Brown, *Advanced Control Unleashed: Plant Performance Management for Optimum Benefits* (Research Triangle Park, NC: ISA [International Society of Automation], 2003).

5. Michael Mansy, Gregory K. McMillan, and Mark Sowell, "Step into the Virtual Plant," *Chemical Engineering Progress* (February 2002): 56–61.

6. *BioNet Bioreactor Control System*, Broadley-James Bulletin L0016B (2003).

7. Richard J. Fickelscherer, Douglas H. Lenz, and Daniel L. Chester, "Fuzzy Logic Clarifies Operations," *InTech* 52, no. 10 (October 2005).

8. R. G. Silva, A. J. G. Cruz, C. O. Hokka, R. L. C. Giordano, and R. C. Giordano, "A Hybrid Feedforward and Neural Network Model for the Cephalosporin C Production Process," *Brazilian Journal of Chemical Engineering* 7, no. 4-7 (December 2000).

9. K. Pramanik, "Use of Artificial Neural Networks for Prediction of Cell Mass and Ethanol Concentration in Batch Fermentation Using Saccharomyces cerevisiae Yeast," *IE (I) Journal – CH* 85 (September 2004).

10. Gregory K. McMillan and Martin E. Berutti, "Virtual Plant Virtuosity," *Control* (August 2017).

11. Gregory K. McMillan and P. Hunter Vegas, *Process/Industrial Instrumentation and Controls Handbook*, 6th ed. (New York: McGraw-Hill Education, 2019).

12. Christopher Stuart, Rehman Fazeem, and Gregory K. McMillan, "Improve pH Control," *Chemical Processing* (March 2017).

13. Gregory K. McMillan, and Christopher Stuart, "Simulation Breeds Innovation," *Control* (November 2019).

14. Gregory K. McMillan, "The Future Is Here," *Control* (November 2020).

6 First-Principle Models

6.1 Introduction

Dynamic first-principle modeling of bioreactors has not moved from universities to industry to any significant extent despite the reality that such models could help address the issues arising from the scarcity of process knowledge and batch data. Experimental models offered by data analytics and neural networks have been sought as a way of circumventing the need for a process-based understanding. Unfortunately, these models have significant limitations due to insufficient data as a consequence of long batch times, unknown concentrations, and a severe fundamental limitation revealed by the equations documented here for first-principle models, most notably the multiplier effect.

This chapter provides the equations and deeper understanding necessary for a digital twin described in Chapter 5 to achieve knowledge discovery and implementation detailed in Chapter 9.

Section 6.2 summarizes the challenges faced in experimental and first-principle models. This insight reveals how bioreactor batches are vulnerable and that current modeling capability, although attractive from the viewpoint of the Industrial Internet of Things (IIoT), cannot address the many unknowns [9]. Consequently, batches for new biologics that are each worth $20 million or more are at risk.

Section 6.3 offers the opportunities that a good first-principle model can provide. Section 6.4 builds on this to give an overview of the breakthroughs in equations that make these opportunities realizable. Sections 6.5 through 6.16 provide the equations including kinetics needed for the material and energy balances for the medium, sparge, and overlay. Section 6.17 summarizes all the parameters and variables including values used in a demo.

This chapter describes the significant advances in first-principle equations that enable much easier and more comprehensive models. The setup for these models is detailed and exemplified for a digital twin in Chapter 9.

Learning Objectives

- Understand how first-principle models can provide knowledge improving data analytics, process monitoring, analysis, performance metrics, and control
- Learn how new equations can readily model all types of cells and products
- Appreciate the ease with which kinetic parameters can be identified and set
- Be aware of the use of charge balance to accurately model pH
- Learn how to simply model inhibitors and inducers by setting two readily estimated parameters
- Become familiar with the material and energy balances used for all phases of cell growth and product formation
- Recognize the effect of agitation and sparge rates on the mass transfer rates

6.2 Modeling Challenges

High-fidelity dynamic models of bioreactors have not been commonly used in industry. Many practical issues limit their use mostly to academia. The largest is the setting of the kinetics that will enable the model to match laboratory, pilot plant, and production plant profiles dealing with the complexities of biological pathways. The problem stems from the perceived need to provide different equations for kinetics with parameters that are largely unknown. Research may have some ideas as to potential values, but the communication is sparse and highly proprietary. Process development largely depends on lab and pilot plant results. Because batches can take 10 days or more for the production of biologics, data on the effect of changes in batch conditions and cells is severely limited, putting batches in production worth $20 million or more at risk. Often, changes in profiles occur that cannot be diagnosed. There are often not enough representative batches for largely unknown conditions that affect cell health and productivity to develop good models using data analytics. This is particularly a problem due to batch cycle time and the changing operating conditions for new products as they move from lab to pilot plant to production. The multiplicative effect seen in the factors for the kinetics equation terms varying between 0 and 1 are problematic for any experimental method. From a fundamental view, the multiplicative effect is consistent with cell health. If the temperature is too high, cells will die regardless of whether they have enough oxygen, sugar, or nutrients. The effect of each term is diminished by each other term's factor. This complication plus the realization that data analytics offer correlations that must be analyzed as to whether they are truly cause and effect, often requiring a deeper understanding of process relationships and the limitations posed by the history of operating conditions affecting cell health.

Neural network models also suffer from these challenges, especially from an insufficient data set. The extrapolation by neural network models beyond the data set can lead to bizarre exponential predictions. Also, the slopes of batch profiles by interpolation can have unrealistic localized changes in magnitude and even direction. Overlooked in a principal objective of detecting and predicting bad batches is the necessity to have intensive and extensive data sets of abnormal operating conditions, a difficult if not impossible task unless there is a digital twin with first-principle models. Data analytics avoids the problem of extremely unrealistic changes in batch slope from interpolation and extrapolation due to insufficient data sets by data analytics' piecewise linear approach, which can handle nonlinear profiles similar to what is accomplished by a signal characterizer via its array of input and output values. Furthermore, the drill down into principal components offers considerable troubleshooting and diagnostic capabilities not seen in neural networks. Fortunately for batch analysis and predictions, dynamic compensation of inputs is not needed.

All experimental methods severely suffer from poor data sets because of missing or nonrepeatable measurements or unknown process values such as mass transfer coefficients. First-principle models can fill in the missing blanks. There is a largely untapped opportunity of synergy between first-principle models and data analytics [9, 10].

Bioreactor modeling has been extremely challenging principally because of limited data due to batch cycle time and lack of a design of experiments, historical deficiency of kinetic equations particularly in terms of ease of use and general applicability, and the many multiplicative effects.

6.3 Modeling Opportunities

Dynamic first-principle models that can readily match existing profiles of good batches could be used to greatly extend the knowledge for research and development in addition to plant operation to prevent bad batches. Additionally, inferential measurements of growth rate and production rate to match batch profiles can be developed. The prediction of important medium concentrations and the rates of change of live cells and product can be corrected by at-line and off-line analyzers. Proportional-integral-derivative (PID) control can control concentrations and optimize pH and temperature set points as detailed in Chapter 3 [8]. An enhanced PID controller that can readily deal with large and variable analytical measurement delays and failures or model predictive control can manipulate substrate concentrations as outlined in Chapter 4 to maximize growth rate and product formation rate. For example, glucose to amino acid (e.g., glutamine) ratios can be manipulated to maximize growth rate and product formation rate in the first and second half of the batches, respectively. The first-principle model can be nonintrusively and automatically adapted by an innovative but straightforward use of model predictive control (MPC) as described in Chapter 5 on how to maximize knowledge discovery using the digital twin. A key insight is that model fidelity is indicated by how well manipulated flows and uncontrolled process variables of the digital twin match those in the actual plant at the same points in the batch.

Bad batches can be prevented by first-principle models that match batch profiles enabling control of growth and product formation rates and prediction of undesirable results and the reasons why.

As described in Chapter 3, more extensive and accurate measurements of process variables and feed flows open up much greater possibilities for control. The same principle is true for modeling because accurate measurements of model inputs and outputs in the actual plant make models more accurate. The true dynamic behavior of the system can be investigated. Various scenarios can be tested, and batches that take 10 days can be completed in 40 minutes by an innovative use of kinetic and component-material-balance speedup factors. Training data and cause and effect analysis can be significantly improved, offering batch predictions. The best batch performance can be achieved by exploring, discovering, prototyping, testing, tuning, justifying, deploying, commissioning, maintaining, and continuously improving the digital twin, as described in "Virtual Plant Virtuosity" [6].

There is significant synergy between analytical measurements and first-principle models. Concentration measurements of nearly all components limiting or inhibiting growth rate or production rate are possible by at-line analyzers such as the BioProfile FLEX and near-infrared spectroscopy. The BioProfile analyzer is the most exact and comprehensive analyzer offering data on cell size and viability plus glucose, glutamine, and inhibiter concentrations. High-performance liquid chromatographs (HPLCs) can measure nutrients and products in the media. Turbidity is indicative of total cell mass. Dielectric spectroscopy can distinguish live versus dead cells. Mass spectrometers measuring oxygen in vent gas enable a computation of the oxygen uptake rate (OUR). The computation of the turbidity rate of change gives an indication of cell growth rate. The OUR combined with oxygen yield factors for cell growth and maintenance can be used to approximate the growth rate. If glucose and amino acid measurements are available, the rate of change of these concentrations with yield factors can improve the estimation of cell growth rate and product formation rates in the exponential growth phase and product formation phases, respectively. A future value block described in Chapter 9 is essential for timely rate of change computations and batch predictions with a good signal-to-noise ratio. The resulting information can provide intelligent inputs for batch data analytics and provide online metrics to diagnose batch health and, if necessary, terminate bad batches. The cost of raw materials used is insignificant with respect to loss of potential profit from wasted production time.

The significant synergy between analytical measurements and first-principle models augmented by a future value block can be used to provide intelligent inputs for batch analytics and online metrics.

6.4 Modeling Breakthroughs

Several equations enable the user to readily set parameters that provide robust high-fidelity kinetics for the effect of temperature, pH, and concentrations on cell growth rate and production rate. For temperature and pH, convenient cardinal equations are used where the optimum temperature for growth and production phases is simply the temperature or pH set point including any shifts for batch phases. The minimum and maximum temperatures complete the parameter settings. This is a tremendous advancement over the traditional uses of Arrhenius equations for temperature and Villadsen-Nielsen equations for pH that required parameters not readily available set with a precision to the sixth or seventh decimal place. For concentrations, a generalized Michaelis-Menten equation, shown to be useful for modeling intracellular dynamics, can model the limitation and inhibition effects of concentrations. The equations provide a link between macroscopic and microscopic kinetic pathways. If the limiting or inhibition effect is negligible or must be temporarily removed, the limitation and inhibition parameter is simply set to 0 g/L and 100 g/L, respectively. The biological significance and ease of setting parameters is particularly important for the equations presented [11].

These revolutionary equations enable the same generalized kinetic model to be used for all types of cells. Previously, yeast cells (e.g., ethanol production), fungal cells (e.g., antibiotic production), bacterial cells (e.g., simple proteins), and mammalian cells (e.g., complex proteins) had specialized equations developed that did not generally carry over to different cells and products.

Revolutionary simple-to-parametrize equations enable the same generalized kinetic model to be used for all types of cells.

Dynamic first-principle models in the literature were primarily created for batch processes without control system dynamics. Controllers and strategies were emulated and consequently disparate from actual capability in a modern control system. The dynamic models detailed here, along with a digital twin defined and exemplified in Chapter 9, can operate in the continuous as well as the batch mode. Measurement and final control element dynamics and the exact duplicate of controllers, historian, and operator interfaces in the lab, pilot plant, or production unit are used so that automation and process engineers and operations are working in the same automation system configuration and interface with dynamics representative of what is in the actual plant.

The equations presented are oriented toward encapsulating the main factors, which may result in a bad batch. Many models require data that encapsulate interactions that are not typically available or are difficult to quantify (e.g., covalent, ionic, and hydrogen bonding; dipole-dipole and van der Waals interactions; or specialized reaction kinetics). The equations given here facilitate high-level modeling using Michaelis-Menten and convenient cardinal equations for kinetics whose parameters are easy to adjust to match plant trend charts, avoiding such details in favor of the essentials for successful dynamic modeling and control. These essentials enable robust control strategy prototyping and predictive capabilities geared for early termination of a failed batch and process control improvements that improve batch consistency and enable optimization to improve batch performance.

6.5 General Form of Kinetics

The cell growth rate, product formation rate, and by-product formation rate (j) for process condition (i), such as temperature, pH, or component concentration, follow a general format used for biological processes where there is a maximum growth or formation rate ($\mu_{j\max}$) multiplied by a series of factors (Πr_{ji}) as shown in Equation 6-5a. This multiplicative relationship reveals an inherent vulnerability of biological processes to deterioration as a result of a single abnormal process condition. For example, even if the cells have enough sugar and nutrients, the lack of oxygen, too high of a temperature, or the wrong pH can prevent growth and product formation.

$$\mu_j = \mu_{j\max} \bullet \Pi r_{ji} \qquad (6\text{-}5a)$$

The cell growth rate, product formation rate, and by-product formation rates all follow a general format for biological processes where there is a maximum rate multiplied by a series of 0 to 1 factors for each process condition, resulting in a multiplicative effect where one bad factor ruins a batch.

The individual dimensionless rate factors (r_{ji}) (0 to 1) for the effect of media concentrations (X_i) (e.g., glucose, glutamine, precursors, inducers, and inhibitors) can be generalized by the Michaelis-Menten kinetics with an inhibition effect parameter (K_{1ji}) and limitation effect parameter (K_{2ji}) as shown in Equation 6-5b [1, 2].

$$r_{ji} = \left[\frac{K_{1ji}}{X_i + K_{1ji}}\right] \bullet \left[\frac{X_i}{X_i + K_{2ji}}\right] \qquad \text{(6-5b)}$$

The Michaelis-Menten equation can accurately model the effect of component concentrations on intracellular as well as extracellular kinetics with two simple parameters.

The individual rate factors dimensionless rate factors (r_{ji}) (0 to 1) for the effect of pH and temperature can be readily included by the convenient cardinal equations as shown in Equations 6-5c and 6-5d with readily recognizable parameters [3]. The parameters can be readily set based on set points and low and high limits and generally only require two significant figures. The Arrhenius equation for temperature kinetics and Nielsen-Villadsen equation for pH kinetics required six or more significant figures and significant research and development.

$$r_{iH^+} = \left[\frac{(pH_i - pH_{i\min}) \bullet (pH_i - pH_{i\max})}{(pH_i - pH_{i\min}) \bullet (pH_i - pH_{i\max}) - (pH_i - pH_{iopt})^2}\right] \qquad \text{(6-5c)}$$

$$r_{iT} = \left[\frac{(T - T_{\max}) \bullet (T - T_{\min})^2}{(T_{opt} - T_{\min}) \bullet [(T_{opt} - T_{\min}) \bullet (T - T_{opt}) - (T_{opt} - T_{\max}) \bullet (T_{opt} + T_{\min} - 2 \bullet T)]}\right] \qquad \text{(6-5d)}$$

The pH used in Equation 6-5c should be modeled by means of a high-fidelity charge balance that includes the significant effect of carbon dioxide and conjugate salts. *Advanced pH Measurement and Control,* Third Edition, details newly developed equations that provide this capability [4, 7].

The convenient cardinal equations are a major breakthrough because the inherent shape of the plot of growth rate and formation rate versus temperature and pH match what occurs in biological processes and the needed parameters are easily set.

6.6 Media Ordinary Differential Equations with Speedup Factors

The integration of the ordinary differential equation (ODE) for the cell concentration growing or dying or the individual *i* component mass balance concentration (X_i) being formed or utilized employs the kinetic speedup factor (k_{ks}) and the media speedup factor (k_{ms}). The computed reaction rates (e.g., cell growth and product formation rates) are multiplied by the kinetic speedup factor. The step size is the model execution time (Δt_x) multiplied by the media speedup factor. The total speedup for growth rate is the product of kinetic and media speedup factors (e.g., total speedup is 1000 for a kinetic speedup of 20 and media speedup of 50). The ODE derivative is multiplied by the step size and added to the last value (X_{io}). A small filter can be applied to each of the reaction rates to simulate an intracellular lag.

The use of a kinetic speedup factor requires significant adjustments in the digital twin. Most notably, the flow spans of measurements and capacities of final control elements must be proportionally increased. If both the flow span and flow capacity are increased by the same factor, the flow controller tuning should stay about the same, but if oscillations start, it is best to first increase the PID reset time and then decrease the PID gain. The near-integrating and true integrating response will be much faster for the primary controllers (e.g., dissolved oxygen—DO, glucose, glutamine, pH, pressure, and temperature) that manipulate these flows. The integrating process gain will be increased by the total speedup factor (product of the sparge and media speed speedup factors) due to a faster rate of change of the process variable for a given percent change in PID output (manipulated flow). In general, the PID gain must be decreased and the reset time increased. It is important that the product of the controller gain and reset time not become less than four times the inverse of the integrating process gain to prevent oscillations from violating the gain window. To ensure a smooth response, it is better to first multiply the reset time by a factor that is about ¾ the total speedup factor and then divide the PID gain by about ½ the total speedup factor. Any rate time should be divided by the total speedup factor or removed altogether. You should proportionally reduce the dead times and time constants of the automation system components (e.g., analyzer cycle time and sensor lags) by the total speedup factor. If the response is oscillatory or too slow, the PID gain should then be decreased or increased, respectively. An auto tuner should be used. Actual batch time must be computed and sequencing done based on the actual batch time rather than the model run time. Totalization of flows will be faster, but the totals should match the actual batch time except for some overshoot if the flow spans are adjusted correctly.

The actual ODE used in dynamic models is typically expressed in terms of the accumulation or depletion of the mass of each component and the total mass of all components (total media mass). The concentrations are then computed as the present mass of each component divided by the total mass. Here the equations are expressed in terms of concentrations (weight fractions) to enable the user to more readily relate to how changes in feed rate and concentration, reaction rate, and mass transfer rates affect a concentration and to approximate process gains and time constants as discussed in

Appendix F. However, some aspects can be confusing when the ODEs are expressed in concentrations as shown in Equations 6-6a through 6-6h. For example, whereas the feed concentration would always affect the accumulated mass of that respective component, for the ODE expressed in concentrations, changes in the feed rate only cause changes in the media concentration if the feed concentration is different than the media concentration as seen in the term multiplied by the feed rate being the difference in the feed and media concentrations in Equation 6-6a. To understand this concept and how we ended up with Equation 6-6a, read Appendix F, particularly Equations F-7a through F-7g, which show how the ODE is simplified by the product rule of integration for a perfusion process where the discharge flow equals the feed flow to get it into the form where the process gain and dead time can be identified. Recognize that Appendix F is working in weight fractions and this chapter is working in g/L (kg/m^3). If this is too much to absorb, you can concentrate on kinetics and mass transfer.

The equations relate the concentration of the individual components (X_l) to the rate of production or consumption of the ith component (r_i), kinetic speedup (k_{ks}), mass flow rate of the feed (F_f), mass fraction of the ith component in the feed (Z_i), volume of the holdup liquid (V_x), media speedup (k_{ms}), the model execution rate (Δt_x), hydrolysis of the product (r_h), and the initial concentration of the ith component (X_{io}).

Z_i = feed i^{th} component concentration (mass fraction)

X_i = media i^{th} component new value (g/L or kg/m^3)

F_f = feed flow rate (kg/h)

r_i = rate of production (or consumption) of the ith component (g/L per h or kg/m^3 per h)

k_{ks} = kinetic speedup factor (dimensionless)

Δt_x = model execution time (h)

k_{ms} = media speedup factor (dimensionless)

X_{io} = media i^{th} component last value (g/L or kg/m^3)

r_h = hydrolysis rate of product (g/L per h or kg/m^3 per h)

V_x = volume of the holdup liquid (L or m^3)

The ODE integration for individual i component:

$$X_i = \left\{r_i \bullet k_{ks} + \left[F_f \bullet \left(Z_i - X_i / \rho_x\right)\right] / V_x\right\} \bullet \Delta t_x \bullet k_{ms} + X_{io} \quad (6\text{-}6a)$$

The ODE integration for viable cell growth:

$$X_v = \left|(r_v - r_d) \bullet k_{ks} + [F_f \bullet (Z_v - X_v / \rho_x)] / V_x\right| \bullet \Delta t_x \bullet k_{ms} + X_{vo} \quad (6\text{-}6b)$$

The ODE integration for cell death (lysis):

$$X_d = \left|r_d \bullet k_{ks} + [F_f \bullet (Z_d - X_d / \rho_x)] / V_x\right| \bullet \Delta t_x \bullet k_{ms} + X_{do} \quad (6\text{-}6c)$$

The ODE integration for product formation:

$$X_p = \left|(r_p - r_h) \bullet k_{ks} + [F_f \bullet (Z_P - X_p / \rho_x)] / V_x\right| \bullet \Delta t_x \bullet k_{ms} + X_{po} \quad (6\text{-}6d)$$

The ODE integration for acid (e.g., lactic acid) by-product formation:

$$X_a = \left|r_a \bullet k_{ks} + [F_f \bullet (Z_a - X_a / \rho_x)] / V_x\right| \bullet \Delta t_x \bullet k_{ms} + X_{ao} \quad (6\text{-}6e)$$

The ODE integration for base (e.g., ammonia) by-product formation:

$$X_b = \left|r_b \bullet k_{ks} + [F_f \bullet (Z_b - X_b / \rho_x)] / V_x\right| \bullet \Delta t_x \bullet k_{ms} + X_{bo} \quad (6\text{-}6f)$$

The ODE integration for substrate component 1 (e.g., glucose) utilization:

$$X_{s1} = \left|r_{s1} \bullet k_{ks} + [F_f \bullet (Z_{s1} - X_{s1} / \rho_x)] / V_x\right| \bullet \Delta t_x \bullet k_{ms} + X_{s1o} \quad (6\text{-}6g)$$

The ODE for integration substrate component 2 (e.g., glutamine or precursor) utilization:

$$X_{s_2} = \left|r_{s2} \bullet k_{ks} + [F_f \bullet (Z_{s2} - X_{s2} / \rho_x)] / V_x\right| \bullet \Delta t_x \bullet k_{ms} + X_{s2o} \quad (6\text{-}6h)$$

The ODEs for the mass balances using growth and formation rates have a speedup factor for the kinetics and media whose product is the overall model speedup.

6.7 Concentration and Flow of Charges, Manipulated and Recycle Streams

The total feed rate (F_f) of individual i streams is computed per Equation 6-7a and the blended concentration of viable cells (Z_v), dead cells (Z_d), product (Z_p), glucose (Z_{s_1}), glutamine or precursor (Z_{s_2}), acid (Z_a), and base (Z_b) originating from perfusion recycle, media charge, inoculum charge, pH control, or concentration control are computed per Equations 6-7b through 6-7h:

$$F_f = \sum F_i \tag{6-7a}$$

$$Z_v = \sum(F_i \bullet Z_{vi}) \,/\, F_f \tag{6-7b}$$

$$Z_d = \sum(F_i \bullet Z_{di}) \,/\, F_i \tag{6-7c}$$

$$Z_p = \sum(F_i \bullet Z_{pi}) \,/\, F_f \tag{6-7d}$$

$$Z_{s_1} = \sum(F_i \bullet Z_{s_1 i}) \,/\, F_f \tag{6-7e}$$

$$Z_{s_2} = \sum(F_i \bullet Z_{s_2 i}) \,/\, F_f \tag{6-7f}$$

$$Z_a = \sum(F_i \bullet Z_{ai}) \,/\, F_f \tag{6-7g}$$

$$Z_b = \sum(F_i \bullet Z_{bi}) \,/\, F_f \tag{6-7h}$$

The bioreactor model can simulate perfusion processes by including recycle streams.

6.8 Specific Growth Rate Equations with Bias and Gain Terms

The viable cell growth rate (r_v) (g/L per h) is the specific growth rate (μ_v) (1/h) multiplied by the viable cell concentration (X_v) (g/L) as shown in Equation 6-8a. The specific growth rate depends on the maximum specific growth rate ($\mu_{v\max}$) multiplied by individual rate factors (r_{vi}) for process conditions as shown in Equation 6-8b.

$$r_v = \mu_v \bullet X_v \tag{6-8a}$$

$$\mu_v = \mu_{v\max} \bullet r_{vs_1} \bullet r_{vs_2} \bullet r_{va} \bullet r_{vb} \bullet r_{vO_2} \bullet r_{vH^+} \bullet r_{vT} \tag{6-8b}$$

Michaelis-Menten equations are used to model the individual rate factors for the effect of media concentrations on viable cell growth rate with bias and gain terms to provide additional flexibility as shown in Equations 6-8c through 6-8g. The bias terms provide a significant rate factor ($r_{vi} > 0$) for a low concentration. The gain term provides a greater or lesser effect at all concentrations. If the gain term can be written to, the user has flexibility to create almost any curve shape (plot of growth rate versus concentration) by making the gain term a function of concentration or other conditions. Limits are imposed that prevent the rate factors from being less than zero or greater than 1. See Figures 6-8a, 6-8b, and 6-8c for examples of the effect of DO, glucose, and a by-product (lactate) on mammalian cell growth rate.

Simple bias and gain terms applied to the kinetic equations provide additional flexibility.

$$r_{vs_1} = g_{vs_1} \bullet \left\{ \left| \frac{K_{1vs_1}}{X_{s_1} + K_{1vs_1}} \right| \bullet \left| \frac{X_{s_1}}{X_{s_1} + K_{2vs_1}} \right| + b_{1vs_1} \right\} \tag{6-8c}$$

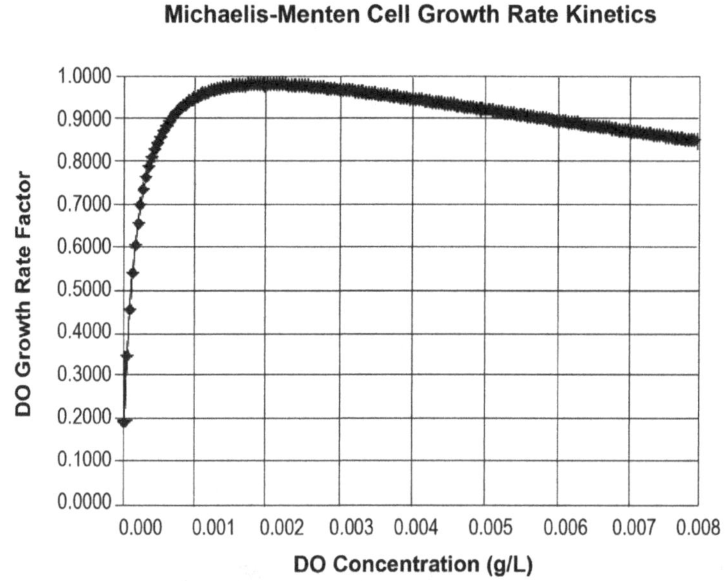

Figure 6-8a. Michaelis-Menten model for effect of DO on cell growth rate.

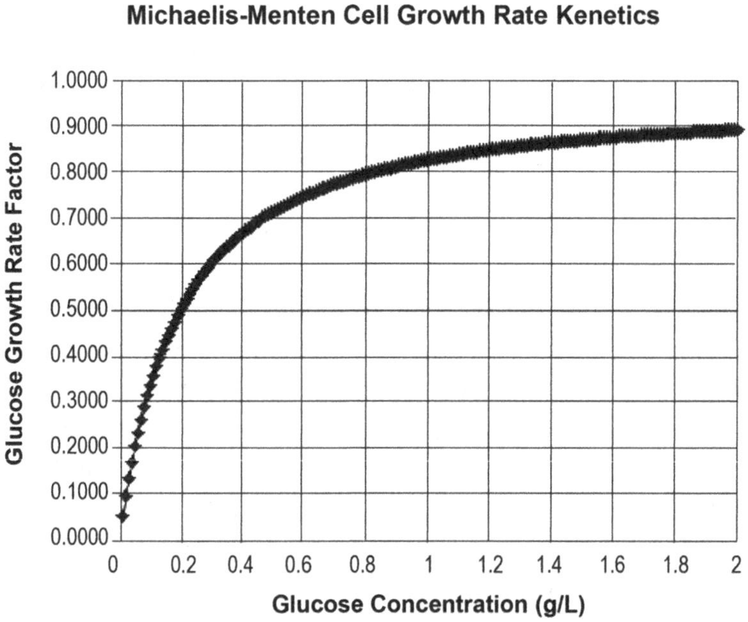

Figure 6-8b. Michaelis-Menten model for the effect of glucose concentration on cell growth rate.

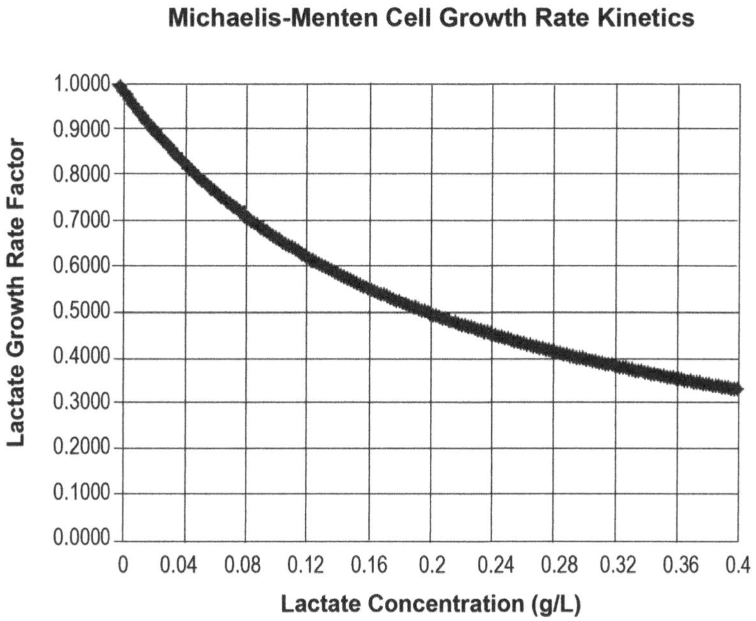

Figure 6-8c. Michaelis-Menten model for the effect of lactate concentration on cell growth rate.

$$r_{vs_2} = g_{vs_2} \bullet \left\{ \left| \frac{K_{1vs_2}}{X_{s_2} + K_{1vs_2}} \right| \bullet \left| \frac{X_{s_2}}{X_{s_2} + K_{2vs_2}} \right| + b_{1vs_2} \right\} \tag{6-8d}$$

$$r_{va} = g_{va} \bullet \left\{ \left[\frac{K_{1va}}{X_a + K_{1va}} \right] \bullet \left[\frac{X_a}{X_a + K_{2va}} \right] + b_{1va} \right\} \tag{6-8e}$$

$$r_{vb} = g_{vb} \bullet \left\{ \left[\frac{K_{1vb}}{X_b + K_{1vb}} \right] \bullet \left[\frac{X_b}{X_b + K_{2vb}} \right] + b_{1vb} \right\} \tag{6-8f}$$

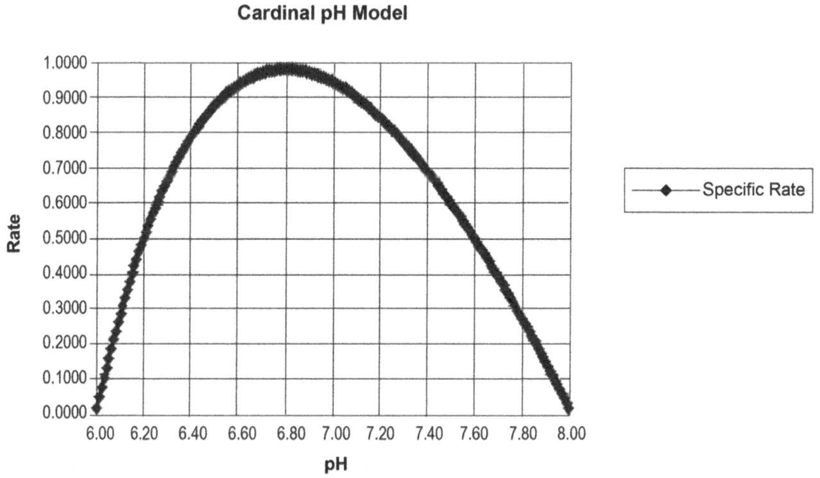

Figure 6-8d. Convenient cardinal equation model for the effect of pH on cell growth rate.

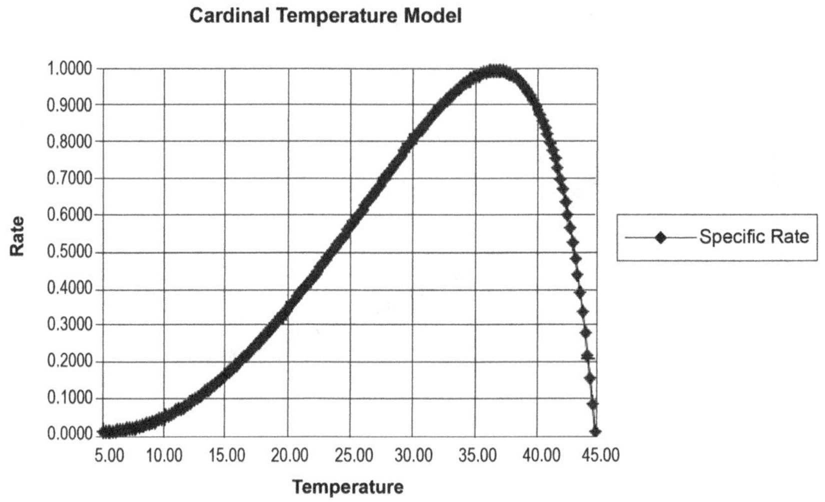

Figure 6-8e. Convenient cardinal equation model for the effect of temperature on cell growth rate.

$$r_{vO_2} = g_{vO_2} \bullet \left\{ \left[\frac{K_{1vO_2}}{X_{O_2} + K_{1vO_2}} \right] \bullet \left[\frac{X_{O_2}}{X_{O_2} + K_{2vO_2}} \right] + b_{1vO_2} \right\} \tag{6-8g}$$

Michaelis-Menten equations are used to model the effect of substrate, by-product, and DO concentrations on cell growth rate as exemplified in Figures 6-8a, 6-8b, and 6-8c.

The convenient cardinal equations enable an easily parameterized model of the rate factor for pH and temperature that is robust and made more flexible by the use of bias and gain terms. See Figures 6-8d and 6-8e for examples of the effect of pH and temperature on mammalian cell growth rate.

$$r_{vH^+} = g_{vH^+} \bullet \left\{ \left[\frac{(pH_1 - pH_{i\min}) \bullet (pH_i - pH_{i\max})}{(pH_i - pH_{i\min}) \bullet (pH_i - pH_{i\max}) - (pH_i - pH_{iopt})^2} \right] + b_{1vH^+} \right\} \tag{6-8h}$$

$$r_{vT} = g_{vT} \bullet \left\{ \left[\frac{(T - T_{\max}) \bullet (T - T_{\min})^2}{(T_{opt} - T_{\min}) \bullet [(T_{opt} - T_{\min}) \bullet (T - T_{opt}) - (T_{opt} - T_{\max}) \bullet (T_{opt} + T_{\min} - 2 \bullet T)]} \right] + b_{1vT} \right\} \tag{6-8i}$$

Convenient cardinal equations with bias and gain terms are used to model the individual rate factors for the effect of pH and temperature on cell growth rate.

6.9 Cell Death (Lysis)

The common causes of premature cell death rate, besides high temperatures and high sodium ion concentrations for mammalian cells, are high cell concentration and agitation as expressed by Equation 6-9a. The age and competition for resources and the buildup of waste products is reflected in the death rate at high cell concentrations as expressed by Equation 6-9b. The damage to cell walls and particularly to mammalian cell membranes is seen in higher death rates as agitation rates are increased as shown in Equation 6-9c. The user can include the effect of high concentrations of inhibitors such as lactic acid and ammonia as individual death rate factors via the Michaelis-Menten equations. The user can provide more complex models by setting an external calculation of death rate.

The total death rate due to lysis from impeller (r_{di}) and high concentrations (r_{dv}):

$$r_d = r_{di} + r_{dv} \tag{6-9a}$$

Cell death at high cell concentrations can be approximated by the power function:

$$r_{dv} = k_{dv} \bullet X_v^p \tag{6-9b}$$

Cell lysis (breakage) from the impeller is proportional to the agitation power (E_i) per unit volume (V_x):

$$r_{di} = k_{di} \bullet (E_i \,/\, V_x) \tag{6-9c}$$

Users can model the effect of agitation and high cell concentration on cell lysis by the simple equations provided or by externally computing and writing to the individual death rates.

6.10 Specific Product Formation Rate Equations with Bias and Gain Terms

The product formation rate (r_p) (g/L per h) is the specific product formation rate (μ_p) (1/h) multiplied by the viable cell concentration (X_v) (g/L) as shown in Equation 6-10a. The specific product formation rate depends on the maximum specific product formation rate ($\mu_{p\max}$) multiplied by individual rate factors (r_{pi}) for process conditions as shown in Equation 6-10b.

$$r_p = \mu_p \bullet X_v \tag{6-10a}$$

$$\mu_p = \mu_{p\max} \bullet r_{ps_1} \bullet r_{ps_2} \bullet r_{pO_2} \bullet r_{pH^+} \bullet r_{pT} \tag{6-10b}$$

Michaelis-Menten equations with bias and gain terms are used to model the individual rate factors for the effect of media concentrations on the product formation rate with bias and gain terms to provide additional flexibility as detailed in Equations 6-10c through 6-10g. The bias terms provide a significant rate factor ($r_{vi} > 0$) for a low concentration. The gain term provides a greater or lesser effect at all concentrations. If the gain term can be written to, the user has the flexibility to create almost any curve shape (plot of product formation rate versus concentration) by making the gain term a function of concentration or other conditions.

$$r_{ps_1} = g_{ps_1} \bullet \left\{ \left[\frac{K_{1ps_1}}{X_{s_1} + K_{1ps_1}} \right] \bullet \left[\frac{X_{s_1}}{X_{s_1} + K_{2ps_1}} \right] + b_{1ps_1} \right\} \quad \text{(6-10c)}$$

$$r_{ps_2} = g_{ps_2} \bullet \left\{ \left[\frac{K_{1ps_2}}{X_{p_2} + K_{1ps_2}} \right] \bullet \left[\frac{X_{s_2}}{X_{s_2} + K_{2ps_2}} \right] + b_{1ps_2} \right\} \quad \text{(6-10d)}$$

$$r_{pO_2} = g_{pO_2} \bullet \left\{ \left[\frac{K_{1pO_2}}{X_{O_2} + K_{1pO_2}} \right] \bullet \left[\frac{X_{O_2}}{X_{O_2} + K_{2pO_2}} \right] + b_{1pO_2} \right\} \quad \text{(6-10e)}$$

Convenient cardinal equations with bias and gain terms are used to model the individual rate factors for effect of pH and temperature on product formation rate. Often a shift in pH and temperature set points is used after sufficient cell growth to promote product formation rate over cell growth rate.

$$r_{pH^+} = g_{pH^+} \bullet \left\{ \left[\frac{(pH_i - pH_{i\min}) \bullet (pH_i - pH_{i\max})}{(pH_i - pH_{i\min}) \bullet (pH_i - pH_{i\max}) - (pH_i - pH_{iopt})^2} \right] + b_{1pH^+} \right\} \quad \text{(6-10f)}$$

$$r_{pT} = g_{pT} \bullet \left\{ \left[\frac{(T - T_{\max}) \bullet (T - T_{\min})^2}{(T_{opt} - T_{\min}) \bullet [(T_{opt} - T_{\min}) \bullet (T - T_{opt}) - (T_{opt} - T_{\max}) \bullet (T_{opt} + T_{\min} - 2 \bullet T)]} \right] + b_{1pT} \right\} \quad \text{(6-10g)}$$

The product formation rate is modeled by Michaelis-Menten and convenient cardinal equations with bias and gain terms.

6.11 Product Consumption and Degradation

Product hydrolysis (consumption) at high product formations can be approximated as shown in Equation 6-11 by a linear function:

$$r_h = k_h \bullet X_p \quad \text{(6-11)}$$

6.12 Specific By-product Formation Rate Equations with Bias and Gain Terms

The acid by-product (e.g., lactic acid) formation rate (r_a) (g/L per h) is the specific by-product formation rate (μ_a) (1/h) multiplied by the viable cell concentration (X_v) (g/L) as shown in Equation 6-12a. The specific by-product formation rate depends on the maximum formation rate ($\mu_{a\max}$) multiplied by individual rate factors (r_{ai}) for process conditions as shown in Equation 6-12b that are each computed by Michaelis-Menten equations detailed in Equations 6-12c and 6-12d. Glucose and glutamine are both utilized in the formation of lactic acid.

$$r_a = \mu_a \bullet X_v \tag{6-12a}$$

$$\mu_a = \mu_{a\max_1} \bullet r_{as_1} + \mu_{a\max_2} \bullet r_{as_2} \tag{6-12b}$$

$$r_{as_1} = g_{as_1} \bullet \left\{ \left[\frac{K_{1as_1}}{X_{s_1} + K_{1as_1}} \right] \bullet \left[\frac{X_{s_1}}{X_{s_1} + K_{2as_1}} \right] + b_{1as_1} \right\} \tag{6-12c}$$

$$r_{as_2} = g_{as_2} \bullet \left\{ \left[\frac{K_{1as_2}}{X_{s_2} + K_{1as_2}} \right] \bullet \left[\frac{X_{s_2}}{X_{s_2} + K_{2as_2}} \right] + b_{1as_2} \right\} \tag{6-12d}$$

The base by-product (e.g., ammonia) formation rate (r_b) (g/L per h) is the specific by-product formation rate (μ_b) (1/h) multiplied by the viable cell concentration (X_v) (g/L) as shown in Equation 6-12e. The specific by-product formation rate depends on the maximum formation rate ($\mu_{b\max}$) multiplied by an individual rate factor (r_{bi}) for process conditions as shown in Equation 6-12f computed by a Michaelis-Menten equation detailed in Equation 6-12g. Only glutamine is utilized in the formation of ammonia.

$$r_b = \mu_b \bullet X_v \tag{6-12e}$$

$$\mu_b = \mu_{b\max_2} \bullet r_{bs_2} \tag{6-12f}$$

$$r_{bs_2} = g_{bs_2} \bullet \left\{ \left[\frac{K_{1bs_2}}{X_{s_2} + K_{1bs_2}} \right] \bullet \left[\frac{X_{s_2}}{X_{s_2} + K_{2bs_2}} \right] + b_{1bs_2} \right\} \tag{6-12g}$$

The by-product formation rates are modeled by Michaelis-Menten equations with bias and gain terms.

6.13 Utilization Rates

The glucose utilization rate (r_{s_1}) (g/L per h) is the summation of the viable cell growth rate and the product and by-product formation rates divided by their respective yield factors ($Y_{s_1 i}$) (g cell/g glucose) plus the viable mass multiplied by a maintenance coefficient ($m_{s_1 v}$) (g glucose/g cell per h) as shown in Equation 6-13a.

$$r_{s_1} = r_v / Y_{s_1 v} + r_p / Y_{s_1 p} + r_a / Y_{s_1 a} + r_b / Y_{s_1 b} + m_{s_1 v} \bullet X_v \tag{6-13a}$$

The glutamine utilization rate (r_{s_2}) uses the same form except there is no lactic acid factor as shown in Equation 6-13b:

$$r_{s_2} = r_v / Y_{s_2 v} + r_p / Y_{s_2 p} + r_b / Y_{s_2 b} + m_{s_2 v} \bullet X_v \qquad \text{(6-13b)}$$

The utilization rates of glucose and amino acid (e.g., glutamine) are modeled by cell yield, product, and by-product maintenance factors for each substrate.

6.14 OUR and Carbon Dioxide Production Rate

The OUR (r_{O_2}) is computed similar to a utilization rate as shown in Equation 6-14a:

$$r_{O_2} = r_v / Y_{O_2 v} + r_p / Y_{O_2 p} + r_a / Y_{O_2 a} + r_b / Y_{O_2 b} + m_{O_2 v} \bullet X_v \qquad \text{(6-14a)}$$

The OUR is modeled by cell yield, product, and by-product maintenance factors.

The carbon dioxide production rate (CPR) (r_{CO_2}) can be computed by multiplying the individual OUR terms by a corresponding stoichiometric factor ($k_{CO_2 i}$) as shown in Equation 6-14b:

$$r_{CO_2} = k_{CO_2 v} \bullet r_v / Y_{O_2 v} + k_{CO_2 p} \bullet r_p / Y_{O_2 p} + k_{CO_2 a} \bullet r_a / Y_{O_2 a} + K_{CO_2 v} \bullet m_{O_2 v} \bullet X_v \qquad \text{(6-14b)}$$

The CPR is modeled by application of the stoichiometric factor to the OUR.

6.15 Agitation and Sparge

The overall mass transfer coefficient (kLa_{bO_2}) (1/h) for the oxygen transfer rate from bubbles depends on the number of impellers (n_i), the agitation power (p_a) (metric horsepower—hp), the superficial gas velocity (v_s) cm/min), the number of atmospheres (P_x/101), and an adjustable power factor (k_{pf}) per Equation 6-15a (Equation 106 in reference [6]). The power factor (e.g., 5 to 10) provides flexibility so that mass transfer rates for a given overlay pressure and concentration match those from agitation at a given speed.

$$kLa_{bO_2} = \left[[(\beta + \gamma \bullet n_i) \bullet [P_a / (V_x / 1000)]^{0.77} \bullet (v_s)^{0.67} \bullet K \bullet (P_x / 101) \right] \bullet k_{pf} \qquad \text{(6-15a)}$$

To get the conversion factor, we first must identify the factor as exemplified by following equation that converts kLz units to kg/m^3 atm^{-1} h^{-1}

K_x = mM O_2 liter^{-1} atm^{-1} h^{-1} • (32g/M • 1M/1000 mM • 1000 L/m^3 • 1 kg/1000g)
= 0.032 kg/m^3 atm^{-1} h^{-1}

The coefficient (K_x) units include a typical driving force for a steady-state condition with an air sparge and typical oxygen set point. We can back out this driving force for stated conditions to get a coefficient (K) for Equation 6-15a that enables the use of the first-principle driving force in the following equation, enabling the modeling of the dynamic response during oxygen control including the split range from air to oxygen flow (oxygen enrichment of the sparge) that greatly increases the driving force.

$$K = K_x/(X_{beO_2} - X_{dO_2}) = 0.032\ /\ 0.003 = 10.7\ \text{atm}^{-1}\ \text{h}^{-1} = 10.7\ \text{h}^{-1}\ \text{for 1 atmosphere}$$

If there is a sparge flow, the agitation power is the gassed power; otherwise, the agitation power is the impeller power as noted in Equations 6-15b and 6-15c. In both cases, the agitation power includes a kinetic speedup factor (k_{ks}) to help scale up the mass transfer coefficient with kinetics. The gassed power is computed from impeller power (P_i) (W), the impeller speed (N) (rpm), the dimensionless impeller geometric factor (m) (e.g., 0.8), impeller diameter (D_i) (m), and the volumetric sparge gas flow (F_g/ρ_g) (m/s) per Equation 6-15d (the 0.00136 factor converts from watts to metric hp). The impeller power input is computed from the dimensionless power number (N_p) (e.g., 2), media density (ρ_x) (kg/m^3), impeller speed (N) (rpm), and impeller diameter (D_i) (m) per Equation 6-15e. The superficial velocity is computed from the volumetric gas flow (F_g/ρ_g) (m/s), and the cross-sectional area (A_x) (m^2) per Equation 6-15f [5].

If sparge flow is nonzero:

$$P_a = P_g \bullet k_{pf} \tag{6-15b}$$

Otherwise:

$$P_a = P_i \bullet k_{pf} \tag{6-15c}$$

If there is a sparge flow, the agitation power is the gassed power; otherwise, the agitation power is the impeller power.

$$P_g = m \bullet [P_i^2 \bullet (N/60) \bullet D_i^3 / (F_g / \rho_g)^{0.56}]^{0.45} \bullet 0.00136 \tag{6-15d}$$

$$P_i = N_p \bullet \rho_x \bullet (N/60)^3 \bullet D_i^5 \tag{6-15e}$$

$$v_s = 10^{+6} \bullet (60 \bullet F_g / \rho_g) / 10^{+4} \bullet A_x \tag{6-15f}$$

In Equation 6-15d, the gassed power (P_g) is limited so that the ratio of gassed power to impeller power (P_g/P_i) is greater than a minimum (e.g., 0.5) and lower than a maximum ratio (1.0). The total sparge gas mass flow (F_G) and the sparge gas density (ρ_g) have low limits to prevent division by excessively small values.

The impeller power (P_i) is simply a function of impeller speed and geometry with gassed power a consequential function of impeller power and sparge flow rate.

The equilibrium DO concentration (X_{beO_2}) (kg/m^3) from bubbles is the oxygen partial pressure (P_{O_2}) (kPa) divided by the Henry coefficient for oxygen (H_{O_2}) (kPa per kmol/m^3) as shown in Equation 6-15g. The Henry coefficient units should be checked particularly because the inverse may be listed in the literature. The oxygen partial pressure is the total pressure multiplied by the oxygen mole fraction that is the arithmetic average oxygen concentration in the sparge (X_{avgO_2}). The same type of equation is used for equilibrium carbon dioxide concentration as shown in Equation 6-15h.

$$X_{beO_2} = mw_{O_2} \bullet (P_{O_2} \,/\, H_{O_2}) \tag{6-15g}$$

$$X_{beCO_2} = mw_{CO_2} \bullet (P_{CO_2} \,/\, H_{CO_2}) \tag{6-15h}$$

The equilibrium concentration of oxygen and dissolved carbon dioxide is a function of the partial pressure and the Henry coefficient, being careful not to use its inverse.

The bubble oxygen mass transfer rate (z_{bO_2}) (in kilograms per hour) is the oxygen mass transfer coefficient (kLa_{bO_2}) multiplied by the driving force, which is the difference between the bubble equilibrium DO concentration (X_{beO_2}) and the actual DO concentrations (X_{dO_2}) as shown in Equation 6-15i.

$$z_{bO_2} = kLa_{bO_2} \bullet (X_{beO_2} - X_{dO_2}) \bullet V_x \tag{6-15i}$$

The same relationship as shown in Equation 6-15j is used for mass transfer from the overlay gas in the headspace.

$$z_{hO_2} = kLa_{hO_2} \bullet (X_{heO_2} - X_{dO_2}) \bullet V_x \tag{6-15j}$$

The equations for mass transfer of carbon dioxide have a sign change because carbon dioxide is being transferred into the bubble and overlay gas in the headspace as shown in Equations 6-15k and 6-15l.

$$z_{bCO_2} = kLa_{bCO_2} \bullet (X_{beCO_2} - X_{dCO_2}) \bullet V_x \tag{6-15k}$$

$$z_{hCO_2} = kLa_{hCO_2} \bullet (X_{heCO_2} - X_{dCO_2}) \bullet V_x \tag{6-15l}$$

The driving force for mass transfer is the difference between the actual dissolved gas concentration and the equilibrium bubble or overlay concentration.

At any point in the batch profile, the overall mass transfer coefficients can be estimated from the mass transfer rates required to balance out the oxygen uptake rate and carbon dioxide production rate in Equations 6-14a and 6-14b and the concentrations in Equations 6-15k through 6-15m.

$$\sum z_i = z_{bO_2} + z_{hO_2} + z_{bCO_2} + z_{hCO_2} + z_{hH_2O} + z_{bH_2O} \tag{6-15m}$$

6.16 Dissolved Gas ODEs with Speedup Factors

The ODE for DO concentration (X_{dO_2}) includes the effect of the oxygen mass transfer rate from bubbles (z_{bi}) and headspace (z_{hi}), the oxygen uptake rate (r_{O_2}), any feed or recycle stream DO (X_{fO_2}), and the gas speedup (k_{gs}) as shown in Equation 6-16a. The effect of the change in total gas mass is included by the total mass transfer rate term (Σz_i) that includes the mass transfer of oxygen, carbon dioxide, and water vapor between bubbles, headspace, and the media. The step size (Δt_x) is the module execution time. The total speedup for the mass balance in bubbles and overlay space is the product of kinetic and gas speedup factors.

The ODE integration for DO:

$$X_{dO_2} = \left[-r_{O_2} \bullet k_{ks} + [z_{bO_2} + z_{hO_2} + F_f \bullet (X_{fO_2} - X_{dO_2}) / \rho_x - \Sigma z_i \bullet (X_{dO_2} / \rho_x)] / V_x\right] \bullet \Delta t_x \bullet k_{gs} + X_{dO_{2o}} \tag{6-16a}$$

A similar ODE can be set up and integrated for dissolved carbon dioxide concentration (X_{dCO_2}). Note the sign change for carbon dioxide evolution rate (r_{CO_2}) and the mass transfer rates (z_{bCO_2}) (z_{hCO_2}) as shown in Equation 6-16b.

The ODE integration for dissolved carbon dioxide:

$$X_{dCO_2} = \left[r_{CO_2} \bullet k_{ks} + [-z_{bCO_2} - z_{hCO_2} + F_f \bullet (X_{fCO_2} - X_{dCO_2}) / \rho_x - \Sigma z_i \bullet (X_{dCO_2} / \rho_x)] / V_x\right] \bullet \Delta t_x \bullet k_{gs} + X_{dCO_{2o}} \tag{6-16b}$$

The DO and carbon dioxide material balance depends on the OUR and CPR, respectively, the oxygen in the sparge and the overlay for DO control, the carbon dioxide in the sparge for pH control, and the mass transfer between the bubbles or overlay and the media.

Typical bubble and overlay (headspace) gas and vapor components are carbon dioxide, oxygen, nitrogen, and an inert gas such as helium.

6.17 Parameters and Variables

Values shown are for the mammalian model demo (* denotes day 1 and day 10 values shown and ** denotes kinetic parameter set to have no effect). Concentrations are expressed in gram per liter (g/L) or kilograms per cubic meter (kg/m^3) with rates per hour reflecting relatively low concentrations and the complexity of product molecular weight.

Kinetic parameters for cell growth rate:

$\mu_{v\max}$ = maximum specific viable cell growth rate (0.015 g cell/g cell per h)

K_{1vs_1} = substrate 1 (glucose) inhibition parameter for viable cell growth rate (100 g/L)**

K_{2vs_1} = substrate 1 (glucose) limitation parameter for viable cell growth rate (0.2 g/L)

K_{1vs_2} = substrate 2 (glutamine) inhibition parameter for viable cell growth rate (100 g/L)**

K_{2vs_2} = substrate 2 (glutamine) limitation parameter for viable cell growth rate (0.2 g/L)

K_{1va} = lactic acid inhibition parameter for viable cell growth rate (0.2 g/L)

K_{2va} = lactic acid limitation parameter for viable cell growth rate (0 g/L)**

K_{1vb} = ammonia inhibition parameter for viable cell growth rate (0.1 g/L)

K_{2vb} = ammonia limitation parameter for viable cell growth rate (0 g/L)**

K_{1vO_2} = oxygen inhibition parameter for viable cell growth rate (0.02 g/L)

K_{2vO_2} = oxygen limitation parameter for viable cell growth rate (0.0002 g/L)

K_{dv} = death rate factor for high concentration (0.0002 dead cell mass/live cell mass per h)

p = death rate power for high concentration (3.3)

$pH_{v\max}$ = maximum pH for viable cell growth (8 pH)

$pH_{v\min}$ = minimum pH for viable cell growth (6 pH)

pH_{vopt} = optimum pH for viable cell growth (7 pH)

$T_{v\max}$ = maximum temperature for viable cell growth (45°C)

$T_{v\min}$ = minimum temperature for viable cell growth (5°C)

T_{vopt} = optimum temperature for product formation (37°C)

Kinetic parameters for product formation rate:

$\mu_{p\max}$ = maximum specific product formation rate (0.002 g product/g cell per h)

K_{1ps_1} = substrate 1 (glucose) inhibition parameter for product formation rate (20 g/L)

K_{2ps_1} = substrate 1 (glucose) limitation parameter for product formation rate (0.2 g/L)

K_{1ps_2} = substrate 2 (glutamine) inhibition parameter for product formation rate (20 g/L)

K_{2ps_2} = substrate 2 (glutamine) limitation parameter for product formation rate (0.2 g/L)

K_{1pO_2} = oxygen inhibition parameter for product formation rate (0.02 g/L)

K_{2pO_2} = oxygen limitation parameter for product formation rate (0.0002 g/L)

k_h = hydrolysis rate parameter (0.002 g product/g product per h)

$pH_{p\max}$ = maximum pH for product formation (8 pH)

$pH_{p\min}$ = minimum pH for product formation (6 pH)

pH_{popt} = optimum pH for product formation (7 pH)

$T_{p\max}$ = maximum temperature for product formation (45°C)

$T_{p\min}$ = minimum temperature for product formation (5°C)

T_{popt} = optimum temperature for product formation (37°C)

Kinetic parameters for lactic acid and ammonia formation rates:

$\mu_{a\max_1}$ = maximum specific lactic acid formation rate for substrate 1 (0.00002 g lactate/g cell per h)

$\mu_{a\max_2}$ = maximum specific lactic acid formation rate for substrate 2 (0.00002 g lactate/g cell per h)

K_{1as_1} = substrate 1 (glucose) inhibition parameter for lactic acid formation rate (100 g/L)**

K_{2as_1} = substrate 1 (glucose) limitation parameter for lactic acid formation rate (0.2 g/L)

K_{1as_2} = substrate 2 (glutamine) inhibition parameter for lactic acid formation rate (100 g/L)**

K_{2as_2} = substrate 2 (glutamine) limitation parameter for lactic acid formation rate (0.2 g/L)

$\mu_{b\max}$ = maximum specific ammonia formation rate (0.00005 g ammonia/g cell per h)

K_{1bs_2} = substrate 2 (glutamine) inhibition parameter for ammonia formation rate (100 g/L)**

K_{2bs_2} = substrate 2 (glutamine) limitation parameter for ammonia formation rate (0.2 g/L)

Bias and gain parameters for cell growth rate:

g_{vs_1} = substrate 1 (glucose) gain for scaling viable cell growth rate (1)

b_{1vs_1} = substrate 1 (glucose) internal bias for scaling viable cell growth rate (0)

g_{vs_2} = substrate 2 (glutamine) gain for scaling viable cell growth rate (1)

b_{1vs_2} = substrate 2 (glutamine) internal bias for scaling viable cell growth rate (0)

g_{va} = lactic acid gain for scaling viable cell growth rate (1)

b_{1va} = lactic acid internal bias for scaling viable cell growth rate (0)

g_{vb} = ammonia gain for scaling viable cell growth rate (1)

b_{1vb} = ammonia internal bias for scaling viable cell growth rate (0)

g_{vO_2} = oxygen gain for scaling viable cell growth rate (1)

b_{1vO_2} = oxygen internal bias for scaling viable cell growth rate (0.15)

g_{vH^+} = pH gain for scaling viable cell growth rate (1)

b_{1vH^+} = pH internal bias for scaling viable cell growth rate (0)

g_{vT} = temperature gain for scaling viable cell growth rate (1)

b_{1vT} = temperature internal bias for scaling viable cell growth rate (0)

Bias and gain parameters for product formation rate:

g_{ps_1} = substrate 1 (glucose) gain for scaling viable cell growth rate (1)

b_{1ps_1} = substrate 1 (glucose) internal bias for scaling viable cell growth rate (0)

g_{ps_2} = substrate 2 (glutamine) gain for scaling viable cell growth rate (1)

b_{1ps_2} = substrate 2 (glutamine) internal bias for scaling viable cell growth rate (0)

g_{pO_2} = oxygen gain for scaling viable cell growth rate (1)

b_{1pO_2} = oxygen internal bias for scaling viable cell growth rate (0.15)

g_{pH^+} = pH gain for scaling viable cell growth rate (1)

b_{1pH^+} = pH internal bias for scaling viable cell growth rate (0)

g_{pT} = temperature gain for scaling viable cell growth rate (1)

b_{1pT} = temperature internal bias for scaling viable cell growth rate (0)

Bias and gain parameters for lactic acid and ammonia formation rate:

g_{as_1} = substrate 1 (glucose) gain for scaling lactic acid formation rate (1)

b_{1as_1} = substrate 1 (glucose) internal bias for scaling lactic acid formation rate (0)

g_{as_2} = substrate 2 (glutamine) gain for scaling lactic acid formation rate (1)

b_{1as_2} = substrate 2 (glutamine) internal bias for scaling lactic acid formation rate (0)

g_{bs_2} = substrate 2 (glutamine) gain for scaling ammonia formation rate (1)

b_{1bs_2} = substrate 2 (glutamine) internal bias for scaling ammonia formation rate (0)

Other parameters and variables:

F_a = agitator pumping rate (L/h)

F_f = fed-batch feed rate (kg/h of glucose and glutamine)

F_g = sparge gas flow (0.005 kg/h average during last 100 h)

F_h = headspace flow (kg/h)

F_j = flow of inlet sparge stream j (kg/h)

F_v = evaporation rate (0.0001 kg/h)

F_x = media discharge (exit) flow (e.g., drain flow for batch) (0 kg/h)

H_{CO_2} = Henry coefficient for carbon dioxide at operating conditions (kPa per kmol/m^3)

H_{O_2} = Henry coefficient for oxygen at operating conditions (kPa per kmol/m^3)

kLa_{bCO_2} = overall mass transfer coefficient of carbon dioxide from bubbles (sparge gas) (1/h)

kLa_{CO_2} = overall mass transfer coefficient of oxygen from bubbles (sparge gas) (1/h)

$m_{O_{2v}}$ = maintenance factor (0.0001 g O_2/g cell per h)

$m_{s_{1v}}$ = maintenance factor (0.0001 g glucose/g cell per h)

$m_{s_{2v}}$ = maintenance factor (0.0001 g glutamine/g cell per h)

p = death rate power for high concentration (3.3)

A_x = vessel cross sectional area (m^2)

D_i = impeller diameter (m)

N = impeller speed (rpm)

N_p = power number (2)

K = conversion factor (10)

k_{pf} = power factor (7)

n_i = number of impellers (2)

P_a = agitation power (metric hp)

P_g = gassed power (metric hp)

P_i = impeller power (W)

P_o = headspace pressure (overlay) pressure (101 kPa)

P_{CO_2} = partial pressure of carbon dioxide (5 kPa at atmospheric pressure)

P_{O_2} = partial pressure of oxygen (21 kPa at atmospheric pressure)

v_s = superficial gas velocity (cm/min)

ρ_g = sparge gas density (1 kg/m^3)

ρ_h = headspace gas density (1 kg/m^3)

ρ_m = media density (1000 kg/m^3)

r_h = hydrolysis rate (g/L per h or kg/m^3 per h)

r_{ji} = rate factor for the effect of component i on the growth or formation rate of component j

r_a = lactic acid formation rate (g/L per h or kg/m^3 per h)

r_b = ammonia formation rate (g/L per h or kg/m^3 per h)

r_d = cell death rate (g/L per h or kg/m^3 per h)

r_{O_2} = oxygen uptake rate (kg/m^3 per h)

r_{CO_2} = carbon dioxide production rate (kg/m^3 per h)

r_p = product formation rate (g/L per h or kg/m^3 per h)

r_i = formation rate of component i (g/L per h or kg/m^3 per h)

r_v = viable cell mass growth rate (g/L per h or kg/m^3 per h)

V_x = total media volume (10,000 L or 10 m^3)

X_{beCO_2} = equilibrium dissolved carbon dioxide concentration from bubbles (g/L or kg/m^3)

X_{heCO_2} = equilibrium dissolved carbon dioxide concentration from headspace (g/L or kg/m^3)

X_{beO_2} = equilibrium DO concentration from bubbles (g/L or kg/m^3)

X_{heO_2} = equilibrium DO concentration from headspace (g/L or kg/m^3)

X_{avgO_2} = arithmetic average oxygen concentration in sparge (g/L or kg/m^3)

X_{inO_2} = oxygen concentration at sparge inlet (g/L or kg/m^3)

X_{outO_2} = oxygen concentration at sparge outlet (g/L or kg/m^3)

X_{bi} = bubble component i concentration (kg/m^3)

X_i = media component i concentration (g/L or kg/m^3)

X_{hi} = headspace component i concentration (g/L or kg/m^3)

X_{dCO_2} = dissolved carbon dioxide concentration (g/L or kg/m^3)

X_{dO_2} = DO concentration (g/L or kg/m^3)

X_p = product concentration (0 – 0.8 g/L or kg/m^3)*

X_v = viable cell mass concentration (1 – 8 g/L or kg/m^3)*

X_{s_1} = glucose concentration (2 g/L or kg/m^3)

X_{s_2} = glutamine concentration (2 g/L or kg/m^3)

X_a = lactic acid concentration (0.0002 – 0.01 g/L or kg/m^3)*

X_b = ammonia base concentration (0.0004 – 0.01 g/L or kg/m^3)*

Y_{vs_1} = viable cell yield on glucose (0.4 g cell/g glucose)

Y_{ps_1} = product yield on glucose (1.1 g product/g glucose)

Y_{as_1} = lactic acid yield on glucose (1 g lactate/g glucose)

Y_{vs_2} = viable cell yield on glutamine (0.45 g cell/g glutamine)

Y_{ps_2} = product yield on glutamine (1.25 g product/g glutamine)

Y_{as_2} = lactic acid yield on glutamine (1.45 g lactate/g glutamine)

Y_{bs_2} = ammonia yield on glutamine (0.3 g ammonia/g glutamine)

Y_{vO_2} = viable cell yield on oxygen (0.5 g cell/g oxygen)

Y_{pO_2} = product yield on oxygen (0.2 g product/g oxygen)

Y_{aO_2} = lactic acid yield on oxygen (0.2 g lactate/g oxygen)

Z_v = viable cell concentration in media (mass fraction)

Z_d = dead (lysed) cell concentration in media (mass fraction)

Z_p = product concentration in media (mass fraction)

Z_a = acid by-product (e.g., lactic acid) in media (mass fraction)

Z_b = base by-product (e.g., ammonia) in media (mass fraction)

Z_i = feed i^{th} component concentration (mass fraction)

Z_{s_1} = substrate component 1 (e.g., glucose) (mass fraction)

Z_{s_2} = substrate component 2 (e.g., glutamine or precursor) (mass fraction)

Z_{ji} = gas component i concentration in inlet sparge stream j (mass fraction)

z_i = mass transfer rate of component i from bubbles and overhead (kg/h)

z_{bi} = mass transfer rate of bubble component i (kg/h)

z_{hi} = mass transfer rate of headspace component i (kg/h)

6.18 Best Practices

1. Use a digital twin with the actual configuration, graphics, and historian used in the plant.
2. Accurately model with ODE the material and energy balances for the component concentrations and temperature in the liquid phase (media) and gas phases (sparge and overlay).
3. Accurately model with ODE the material balance for the overlay pressure.
4. Accurately model pH using a charge balance that includes carbon dioxide and conjugate salts.
5. Use speedup factors for kinetics and media where model speedup is the product of both.

6. Feed rates must be scaled up to match media speedup factor.
7. Model all the kinetic rate factors, noting the ability to set individual factors equal to 1.0 to take them out of the picture until more information is available.
8. Use Michaelis-Menten equations to model the kinetic factors for the effect of the concentration of substrates, DO, and inhibitors on cell growth rate, product formation rate, and by-product formation rate.
9. For low component concentrations that do not limit growth or formation rate, set the Michaelis-Menten limitation parameter to 0 g/L.
10. For low component concentrations that do not inhibit growth or formation rate, set the Michaelis-Menten inhibition parameter to 100 g/L.
11. For inhibitors, the limitation parameter is set 0 g/L so that there is no low concentration effect.
12. For inducers, set the limitation parameter just above 0 g/L and the bias and gain so that this rate factor for product formation switches from 0 to 1 for a nonzero inducer concentration.
13. Use convenient cardinal equations to model the kinetic factors for the effect of temperature and pH on cell growth rate, product formation rate, and by-product formation rate.
14. Set the cell death rate parameter so it is approximately the cell growth rate in the stationary phase.
15. Set the optimum temperature and pH parameter for the convenient cardinal equations to be the set points that are shifted for the exponential growth and product formation phases.
16. Set the minimum and maximum temperature and pH parameters for the convenient cardinal equations to be values where effect on growth and product formation rate is catastrophic.
17. To greatly reduce the effect of temperature, the convenient cardinal minimum and maximum temperatures parameters can be set at 20°C below and 50°C above the set point, respectively.
18. To greatly reduce the effect of pH, the convenient cardinal minimum and maximum pH parameters can be set at 4 pH below and 3 pH above the set point, respectively.

19. The maximum growth and formation rate can be roughly estimated as twice the end point concentration divided by the time period of the associated growth and formation phases.
20. Model the impeller power and consequential gassed power effect on the mass transfer coefficient.
21. Compute the driving force between sparge and overlay gas concentrations and actual DO and carbon dioxide concentrations.
22. Model the mass transfer rates as the product of the mass transfer coefficient and driving force.
23. Optimize the control strategy using the gained insight and improved understanding of the process.
24. Develop inferential measurements to provide indications and early warning of problems in terms of growth, product, and by-product formation rates and death rates (e.g., cell lysis).

References

1. J. E. Haag et al., "Dynamic Modeling of Complex Biological Systems: A Link between Metabolic and Macroscopic Description," *Mathematical Biosciences* 193 (2005): 25–49.

2. J. E. Haag et al., "A General Model for Reaction Kinetics in Biological Systems," *Bioprocess and Biosystems Engineering* 27, (2005): 303–309.

3. L. Rosso et al., "Convenient Model to Describe the Combined Effects of Temperature and pH on Microbial Growth," *Applied and Environmental Microbiology* (February 1995): 610–616.

4. Gregory K. McMillan and Robert Cameron, *Advanced pH Measurement and Control*, 3rd ed. (Research Triangle Park, NC: ISA [International Society of Automation], 2005).

5. M. Moo-Young, ed., "Section 2: Transport Phenomena in Bioprocesses," in *Comprehensive Biotechnology* (Oxford, England: Pergamon Press Ltd., 1985): 30–33.

6. Gregory K. McMillan and Martin E. Berutti, "Virtual Plant Virtuosity," *Control* (August 2017).

7. Christopher Stuart, Rehman Fazeem, and Gregory K. McMillian, "Improve pH Control," *Chemical Processing* (March 2017).

8. Gregory, K. McMillan, *Advances in Reactor Measurement and Control* (Research Triangle Park, NC: ISA [International Society of Automation], 2015).

9. Gregory K. McMillan, "At the IIoT Crossroads," *Control* (February 2019).

10. Gregory K. McMillan and Christopher Stuart, "Simulation Breeds Innovation," *Control* (November 2019).

11. Gregory K. McMillan, Rehman Fazeem, and Christopher Stuart, "Bioreactor Control Breakthroughs: Biopharm industry turns to advanced measurements, simulation and modeling," *Control* (February 2020).

7 Analytical Technologies

7.1 Introduction

Advances in in-line and at-line analyzers have greatly increased their capability. The analyzers described with probes or gas samples have a fast response enabling analysis within seconds or a few minutes. Knowing how to use these analytical measurements to provide online monitoring of viable cell concentration, foreign organisms, by-products, undesirable operating conditions, cell growth rate, and product formation rate enables timely decisions for better process development and monitoring to increase production by increasing batch consistency and end point concentrations. BioProfile FLEX and chromatographs are also described that are slower but provide essential measurements of media compounds and protein concentrations, respectively. The BioProfile FLEX analyzer also provides much more reliable and accurate measurements of cell size and health than other methods. The frequency of BioProfile and chromatograph results is fast enough because changes in animal cell kinetics are slow, batch times are long (e.g., 10 days), and a long time (e.g., 1 to 4 hours) between updates is beneficial in terms of a better signal-to-noise ratio in the computation of batch slopes. Also, the enhanced proportional-integral-derivative (PID) controller described in Chapter 3 can provide smooth non-oscillatory control without retuning for large analyzer cycle times [1, 2]. The much higher cost of at-line analyzers is offset by a single analyzer's ability to be multiplexed to measure the performance of many bioreactors.

This chapter shows how to measure key aspects of cell growth, viability, and productivity using in-line analyzers with probes and at-line analyzers with sample systems.

The much greater cost of at-line analyzers, often reserved for research labs, is justified based on the much greater accuracy of results and the ability to serve many bioreactors.

Learning Objectives

- Understand how analyzers contribute to process knowledge and performance
- Learn the basic principles of operation
- Appreciate the availability of analyzers
- Be aware of limitations and calibration and basic installation considerations
- Become familiar with the latest advances in analytical measurements
- Recognize that longer cycle times of media at-line analyzers can improve signal to noise ratios
- Remember that an enhanced PID controller can provide non-oscillatory control for large cycle times
- Realize the cost of analyzers is small compared to the benefits

7.2 Overview

Dissolved oxygen (DO) electrodes are essential for closed loop DO control. Turbidity measurements traditionally have been used to provide a measurement of total cell mass. Here we address how to increase the functionality of these common measurements and look at other analytical measurements that enable the distinction of viable versus dead cells; concentration measurements of compounds promoting and inhibiting cell health; detection of sensor, totalization, or set point errors; and optimization of profiles by inferential measurements of growth rate and product formation rate.

7.3 DO

Cells quickly shut down and die from low DO content, which is modeled by the limitation parameter in the Michaelis-Menten kinetics equation. Not as well recognized is that a cell can suffer from very high DO per the inhibition parameter in the equation. Tight DO control is essential for accurate computation of the oxygen uptake rate (OUR) enabled by a mass spectrometer that is used for estimating the cell growth rate, product formation rate, and the mass transfer coefficients (KLa). Tight DO control translates changes in oxygen demand for cell growth and product formation to changes in air flow or oxygen flow that—combined with an accurate analysis of oxygen in an off-gas by a mass spectrometer—can enable a mass balance calculation of mass transfer of oxygen from sparge to *media* for cell growth and product formation.

Accurate DO measurement and tight control is essential for cell health and OUR computation.

DO is measured by an in-line electrode through the amperometric principle of an electrochemical sensor. Oxygen from the process crosses the sensor membrane. This membrane is molecule selectable, which means only oxygen can cross to the cathode. At the cathode, a fixed voltage or polarizing voltage is placed by the attached analyzer that causes the chemical reaction in Equation 7-3a to occur [3].

$$O_2 + 2H_2O + 4e^- \rightarrow 4OH^- \tag{7-3a}$$

The reaction creates a current at the cathode that travels across the electrolyte, generally a potassium chloride solution, to the anode, and the amount of current traveling across the sensor is proportional to the amount of oxygen being destroyed by the chemical reaction. An increase in the current flowing through the sensor indicates an increase in DO in the process. The whole process is governed by Ohm's law, shown in Equation 7-3b [3].

$$V = I \bullet R \tag{7-3b}$$

In this case, V is the polarizing voltage held constant by the analyzer; R is the total resistance of the process, sensor parts, and membrane; and I is the measured current that corresponds to the DO content.

The analog signal to the analyzer carries the measured current for conversion. The analyzer stores sensitivities for different oxygen measurements. This sensitivity value is used to convert the current into units used for reporting DO concentration. The use of a sensor electrical current reading to identify the concentration is why the sensor follows the amperometric principle.

A stainless-steel sensor with a surface finish designed for sterilize-in-place (SIP)/clean-in-place (CIP) processes (Figure 7-3) should be used in pharmaceutical applications. The sensor should be installed at about a 45° angle with the tip pointing down away from the spargers to reduce the possibility of bubbles at the tip. Premium DO analyzers are two times more expensive than others, but the better repeatability, easier maintenance, greater life, and advanced diagnostics (not even considering process and control performance) more than pay for the extra cost.

Use premium DO analyzers for the best performance and least life-cycle cost.

Figure 7-3. SIP and CIP DO probe.
Source: Courtesy of Emerson Rosemount Analytical.

7.4 Dissolved Carbon Dioxide

Cell metabolism results in a dissolved carbon dioxide production rate (CPR). For biologics, carbon dioxide is manipulated for pH control, lowering the media pH by an increased concentration of carbonic acid. Dissolved carbon dioxide (DCO_2) buildup that is detrimental to cell health often occurs in large bioreactors at low gas flow rates and with poor pH control often associated with unnecessary crossings of the split-range point.

DCO_2 is measured by an in-line potentiometric Severinghaus electrode. The Severinghaus electrode utilizes a bicarbonate buffer system in contact with the measured media through a selective CO_2 permeable membrane, where H_2CO_2 is carbonic acid and HCO_3^- is bicarbonate. The pH of the buffer system is measured to determine the DCO_2 concentration, as described in the equilibrium reaction in Equation 7-4a. The potentiometric Severinghaus electrode employs a glass pH sensor to measure the *pH* changes in the bicarbonate solution (i.e., electrolyte) because a glass *pH* electrode provides the highest resolution *pH* measurement [3].

$$CO_2(aq) + H_2O \overset{k_1}{\leftrightarrow} H_2CO_3 \overset{k_2}{\leftrightarrow} H^+ + HCO_3^- \qquad (7\text{-}4a)$$

By combining the equilibrium constants from Equation 7-4a for the CO_2 equilibrium reaction and using Henry's law, *pH* can be made a function of pCO_2 as shown in Equation 7-4b [3].

$$pH = -\log\left[\frac{K_{CO_2} a(p_{CO_2})}{(HCO_3^-)}\right] \qquad (7\text{-}4b)$$

The two constants, K_{CO_2} and a, are both temperature dependent. Their values are specific for CO_2 in the buffer solution [4].

An in-line sensor employs a CO_2 selectively permeable silicone membrane and a specially engineered flat membrane pH sensor. The CO_2 from the medium diffuses across the membrane until it equilibrates within the buffer solution. A change in the DCO_2 (i.e., CO_2 partial pressure) results in a pH change in the electrolyte, which is detected by the inner body pH electrode. A built-in temperature sensor is used to provide temperature compensation.

The sensor should be installed at about a 45° angle with the tip pointing down away from the spargers to reduce the possibility of bubbles at the tip.

The same principle of measurement is used off-line when scientists running a cell culture send a sample to a blood gas analyzer (BGA). This CO_2 sensor is widely used upstream for research and development, process design, and manufacturing operation.

Calibration is performed one time prior to sterilization. Process verification frequency is based on the process considered and drift observed. Accuracy is considered good for a 0% to 20% CO_2 range when the measurement precision is better than 1% (comparable to the performance of a BGA off-line).

DCO2 analyzers provide knowledge for monitoring cell health and computation of CPR and KLa.

7.5 Turbidity

Turbidity provides an inferential measurement of total cell mass including intracellular product per Beer-Lambert law detailed in Chapter 9, Section 9.10, "Instrumentation Modeling." Dielectric spectroscopy provides a measurement of viable cell mass, which subtracted from the total cell mass gives dead cell mass. The rate of change of the dead cell mass is the death or lysis rate. The rate of change of turbidity is an indication of the cell growth rate and product formation rate which, used with the OUR and oxygen, yield factors for growth, maintenance, and product formation, and the associated CPR provides an inferential measurement of the product formation rate. Before the addition of the inducer to trigger product formation, the computation for cell growth rate is simpler. Measurement of glucose and glutamine concentrations with yield factors can help improve the inferential measurements.

Turbidity measurements are based on an optical principle taking advantage of the interaction of light with undissolved (suspended) particles. Turbidity is determined by measuring the intensity of the scattered (reflected) light at a particular angle to its source, or by the loss of light intensity in transmitted light. The common name for a device that measures turbidity is *turbidimeter* [3].

All turbidimeters have two things in common:

- A light source—bulb, laser, or light-emitting diode (LED)
- One or more light detectors, typically photodiodes

The basic principle for scattered light measurement (back scattered, forward scattered, or 90° scattered) is an increase in cell concentration that increases the intensity of the scattered light and consequently the signal (or to be more specific, photo current) produced in the photodiode. For a transmitted (or optical density—OD) measurement, an increase in cell concentration increases the loss of direct light intensity and consequently decreases the current produced in the photodiode.

Process conditions such as pressure, the presence of air bubbles, particles, medium color, and medium flow velocity can interfere with turbidity readings.

Sudden pressure drops should be avoided because of the risk of air bubble formation. Any air bubbles present in the medium will scatter light and typically contribute to a higher turbidity measurement. For in-line turbidity measurement, avoid installing the sensor at a position where air may be trapped. A degasser/ air vent may be required if air bubbles are an issue.

Sample flow velocity should be maintained at a constant rate. Sudden sample velocity changes or turbulence flow could flush out particles that precipitate on the long sampling line tube and raise the turbidity value. For bioreactor online measurement, the tapping point should not be at the vessel bottom or near spargers. The sensor should be installed at about a 45° angle with the tip pointing down away from the spargers to reduce the possibility of bubbles at the tip.

A stainless-steel body and smooth ($Ra \leq 0.8$ μm) or electropolished surface are used for CIP and SIP applications. The turbidimeter can be installed in a pipeline, sample line, recirculation line, or directly in the process equipment. Process connections between the turbidimeter and process pipe/vessel can vary considerably. Typically, Tri-clamp, Neumo (aseptic), Ingold DN25, or Varivent process connections are preferred in food and beverage or hygienic applications. Other flow-through installations, including flange or wafer types of process connections, may also be available for other special applications. For high-flow, high particle load applications, sapphire sensor window construction may be required to protect turbidimeter optics [3].

Many turbidity probes in the market have an integrated electronic chip that stores the factory calibration results. This enables plug-and-measure installation for quick commissioning. Prior to commissioning, a single or multipoint check may be required for instrument validation. When necessary, a one-point or two-point process calibration may be required for medium background compensation.

Turbidimeters can provide predictions of total cell mass and growth rate—and when supplemented by measurements of viable cell concentration—can provide prediction of death rate.

7.6 Dielectric Spectroscopy

Normal, healthy animal cells have a relatively constant spherical cell size. As cells go through a natural death phase termed *apoptosis*, the shape becomes irregular and the cells shrink. Adverse seed culture and bioreactor media conditions can cause premature cell aging and apoptosis. High osmotic pressure, often due to too high of a sodium concentration from excessive use of sodium bicarbonate in pH control, causes the cells to swell and holes to appear in the membrane, resulting in the cells eventually bursting (lysis).

Dielectric spectroscopy can provide inferential measurements of cell volume, size, homogeneity, and membrane integrity as indications of apoptosis and lysis by PLS models.

Hamilton dielectric spectroscopy uses an Incyte in-line probe (shown in Figure 7-6) to make random measurements between 0.1 and 15 MHz and a nonlinear least squares fit to generate the entire frequency spectrum. The resulting peaks and beta dispersion areas in permittivity at various frequencies can be used to infer cell volume, size, and homogeneity and membrane integrity. Partial least squares (PLS), also known as *projections to latent structure,* models are generated based on laboratory test results for different possibilities [4].

The prediction of viable cell volume accuracy declines as the viability of cells significantly declines. A method has been developed to correct the prediction by applying a novel area ratio (AR) metric that is used to relate the shape of beta dispersion to a single frequency permittivity measurement to improve the accuracy of the viable cell volume throughout the batch. However, changes in batch operating conditions and resulting increases in cell apoptosis and lysis reduce the effectiveness of the correction. The use of PLS models from data sets that show most of the possibilities is advantageous [5].

Figure 7-6. In-line permittivity probe for dielectric spectroscopy.
Source: Courtesy of the Hamilton Company.

The decline in accuracy of viable cell inferential measurement decreases as the dead cell concentration increases and can be corrected by an area ratio metric for single-frequency permittivity for the cell viability.

7.7 Near-Infrared Spectroscopy

The concentration measurement of energy sources, nutrients, and toxic by-products enables more optimum scheduling of feeds in the batch sequence and even fed-batch closed-loop control of key concentrations. For example, a buildup of glucose and glutamine concentrations causes the excessive production of lactate and ammonia by-products, respectively, that are detrimental to cell health and can inhibit cell growth per inhibition parameters for these components in Michaelis-Menten equations. Lactic acid production also necessitates the addition of sodium bicarbonate for pH control that causes high cell osmotic pressure and cell lysis from a high sodium concentration. Too low of a glucose and glutamine concentration can also reduce cell growth, as seen in the limitation parameters for these components in Michaelis-Menten equations. Precursors to cell apoptosis such as metabolized amino acid and nicotinamide adenine dinucleotide + hydrogen (NAD + H) can also be measured.

Near-infrared spectroscopy (NIR) provides measurements of chemical groups from the absorption bands in the mid-infrared spectrum that are particularly strong. This can be done without sample preparation by the insertion of a fiber-optic probe (e.g., Ingold port-adapted probe with a 4 mm path length from Mettler-Toledo) into the process. Fiber-optic cables can be used to send the probe signal to a remote analyzer (e.g., Aspectrics encoded photometric NIR).

Vision chemometric software that applies mathematical and statistical methods is used to develop the PLS models. An analysis can be done as often as every 15 minutes. Calibration by withdrawal, lab analysis of reference samples, and development of updated PLS models as the batch progresses is advised every 24 hours for best accuracy. An analyzer may be used in the lab to measure glucose and ammonia content and a chromatograph used to measure amino acid content [6].

NIR can provide fast online measurements of key compounds but requires complex PLS models that should be updated by detailed analysis of lab samples every 24 hours.

7.8 Mass Spectrometers

The measurement of oxygen, carbon dioxide, and nitrogen concentrations in the vent gas flow of a bioreactor enable the computation of the OUR, CPR, and respiration quotient (RQ) and the detection of abnormal operating conditions. The OUR and CPR can be used to provide inferential measurements of cell growth rate and product formation rate and estimations of the KLa for the OUR and CPR. Changes in the KLa are indicative of poor sparge design or an error introduced in the DO and DCO_2 operating points, most likely from electrode drift but also possibly due to a set point error or poor control. The mass balance calculation of the OUR and CPR also requires an accurate, high-rangeability sparge gas flow and overlay gas flow measurements; and the concentration of oxygen and carbon dioxide entering the bioreactor.

Mass spectrometers can provide accurate, fast computations of the OUR and CPR, and an estimation of KLa when used with accurate DO and DCO2 control and sparge and overlay flow measurements.

The mass spectrum is achieved by converting chemical compounds in a sample into rapidly moving ions and computing concentrations based on a mass-to-charge ratio. The ionization process creates a family of positive ions whose mass distribution is characteristic of the parent. Mass spectrometers offer much greater inherent and sustainable concentration measurement accuracy than NIR analyzers particularly because statistical expertise and daily calibration for PLS models is not needed [7, 8].

The sample is a gas that may require some treatment to remove condensate (e.g., condensate knockout pots and heat tracing) and remove particles (e.g., filters). Sample pumps are not needed because the analyzers can operate under vacuum. All the sample lines can originate in the nonsterile parts of the bioreactor piping system. Because the bioreactor pressure is two orders of magnitude greater than the pressure in the ionization chamber, a positive flow is ensured. The analysis result is available within a few seconds.

Although the cost of mass spectrometers is an order of magnitude greater than the previously discussed in-line analyzers, a single mass spectrometer can be used for half a dozen bioreactors by multiplexing, with the main complexity being the switching and purging of sample lines.

A single mass spectrometer can be used for half a dozen bioreactors by multiplexing with the main and still provide analysis results in about a minute.

7.9 BioProfile FLEX2

The ability to accurately measure the concentrations of key components in the media, including glucose, glutamine, glutamate, lactate, ammonia, sodium, and potassium, with a resolution of 0.01 g/L plus cell density, cell diameter, cell viability, and osmolality provides nearly a complete picture of cell health. This analyzer, commonly seen in research labs, should also be used in plants for high-value-added biologics. The extremely accurate, timely, and comprehensive results and knowledge gained are well worth the investment.

The Nova BioProfile FLEX2 analyzer shown in Figures 7-9a and 7-9b has an autosampler capable of providing analysis results every 8.5 min. The sample size is extremely small (265 μL). Sample conditioning is required, and the analyzer must be installed in an environmentally controlled room. When multiplexed, the frequency of results is correspondingly reduced (e.g., 1.5 h for 10 bioreactors); but other analyzers and inferential measurements can be used to fill in the blanks in the interim, and this analyzer can be used to automatically correct computations by noting the error (synchronized computation minus the BioProfile FLEX2 result). The synchronization is achieved by the insertion of a time delay corresponding to the time between BioProfile FLEX2 analysis results for the respective bioreactor. The correction is simply a bias that is a fraction of the error added to the current computation (not a synchronized computation).

The cost of a BioProfile FLEX2 analyzer is justified by the accurate measurement of nearly all compounds and cell health factors enabling the improvement in all computations and inferential measurements.

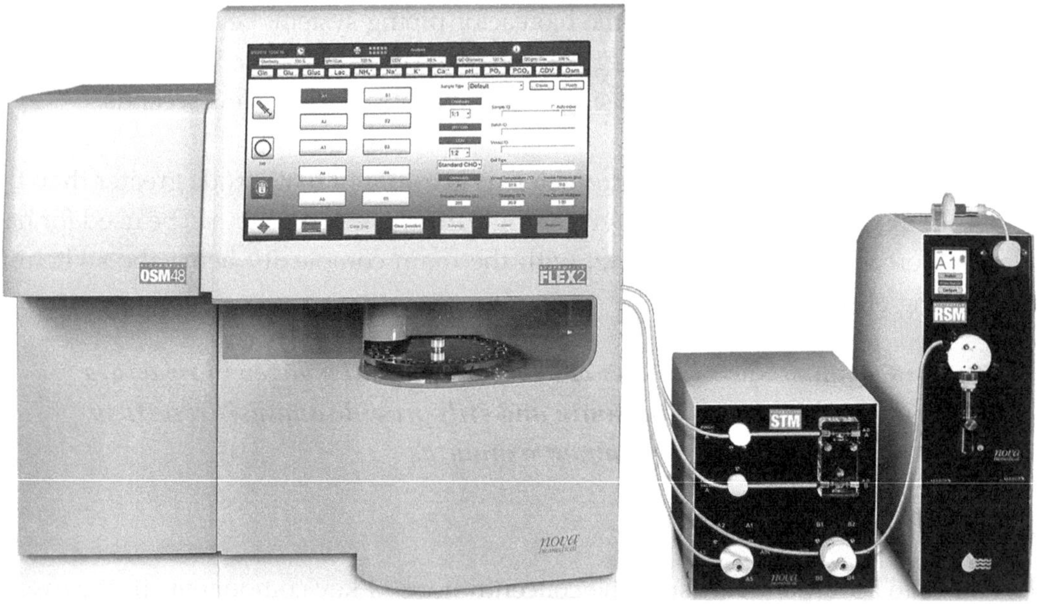

Figure 7-9a. BioProfile FLEX2 automated culture analyzer and online autosampler hardware.
Source: Courtesy of Nova Biomedical.

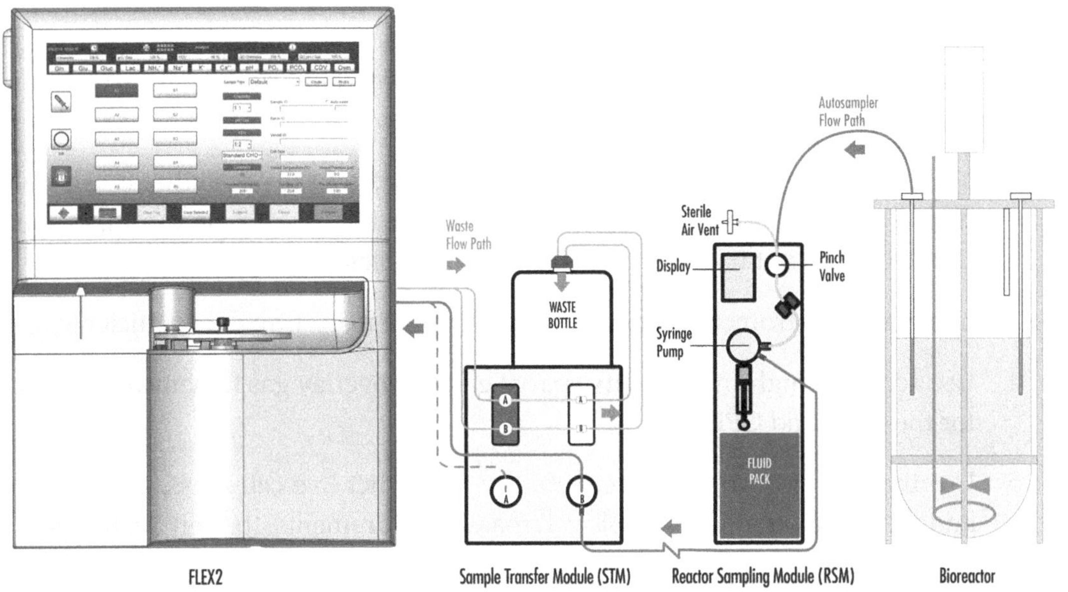

Figure 7-9b. BioProfile FLEX2 automated culture analyzer and online autosampler schematic.
Source: Courtesy of Nova Biomedical.

7.10 Liquid Chromatographs

Liquid chromatographs are used to measure protein concentration, which is ultimately what you want to know. Protein and antibody concentration measurement and purification is typically done in low-pressure, fast protein liquid chromatographs (FPLCs) due to the size and complexity of the molecule, which is unable to withstand the higher pressures and solvents in high-performance liquid chromatographs (HPLC). The protein is adsorbed in the stationary phase of agarose or polymer materials that are not as pressure stable as HPLC silica beads and are extremely sensitive to air bubbles.

FPLC is used instead of HPLC because of the sensitivity of proteins to high pressures and solvents.

The sample volume required is much larger than the column volume, which becomes a consideration as to the frequency of analysis in terms of product lost in samples. Biocompatible transparent and glass columns can be used that enable the monitoring of bubbles and condition of stationary and mobile phases during process feed and elution.

FPLC cycle time to measure product concentration may be increased to reduce product lost in samples.

7.11 Best Practices

1. Use all the latest advances in in-line and at-line analyzers to measure compounds and cells.
2. Make sure the sensor is installed correctly to reduce the possibility of gas bubbles in the media contacting the tip. Make sure the gas sample to a mass spectrometer does not have liquid droplets or particles.
3. Use mass spectrometer to detect changes in the mass transfer coefficients.
4. Use accurate, high-rangeability sparge gas and overlay gas flow measurements for the OUR and CPR.
5. Use the rate of change of live turbidity to predict live cell mass, growth rate, and, in conjunction with viable cell mass measurements, the cell death rate.
6. Use the OUR and CPR with yield factors to provide inferential measurements of cell growth rate and product formation rate.
7. Use online and at-line measurements of glucose and glutamine to optimize their concentrations.
8. Use the future value block described in Chapter 9 to compute rates of change and enhanced PID control for batch profile concentration control.
9. Use the BioProfile FLEX2 analyzer to periodically correct computations and inferential measurements.

References

1. Gregory K. McMillan, "Wireless: Overcoming Challenges in PID Control & Analyzer Applications," *InTech* 57, no. 4 (July–August 2010).

2. Gregory K. McMillan, *Tuning and Control Loop Performance*, 4th ed. (New York: Momentum Press, 2015).

3. Gregory K. McMillan and P. Hunter Vegas, *Process/Industrial Instruments and Controls Handbook*, 6th ed. (New York: McGraw-Hill Education, 2019).

4. R. A. Mattes, et al., "In Situ Monitoring of CHO Cell Culture Medium Using Near-Infrared Spectroscopy," *BioProcess International* (January 2007).

5. B. J. Downey, et al., "A Novel Approach for Using Dielectric Spectroscopy to Predict Viable Cell Volume (VCV) in Early Process Development," *Biotechnology Progress* (March 2014).

6. D. A. Skoog, *Principles of Instrumental Analysis,* 2nd ed. (Philadelphia: Saunders College, 1980).

7. C. Julian and W. Whitford, "Bioreactor Control," *BioProcess International Supplement* (January 2007).

8. Joseph S. Alford, "Bioprocess Mass Spectrometry: A PAT Application," *Journal of Process Analytical Technology* (May–June 2006).

8 Data Analytics

Yang Zhang, Research Assistant
Department of Chemical Engineering, University of Texas, Austin, Texas

Thomas F. Edgar, Professor of Chemical Engineering
Department of Chemical Engineering, University of Texas, Austin, Texas

8.1 Introduction

To meet quality, safety, efficiency, cost, and environmental targets in the process industries, it is often necessary to combine process control and monitoring methods [1]. Process control techniques were discussed in earlier chapters of this book, so this chapter focuses on monitoring techniques. Traditionally, statistical process control (SPC) charts such as Shewhart, cumulative sum (CUSUM), and exponentially weighted moving average (EWMA) charts [2, 3] have been used to monitor industrial processes for the purpose of improving product quality. Such techniques are well developed for manufacturing processes and widely used in univariate systems in which a single process variable is monitored. However, industrial process performance usually includes more than one process variable, and univariate SPC charts do not perform as well for these multivariate systems [3, 4, 5]. For example, in chemical and biological systems, hundreds of variables can be recorded regularly in a single operating unit, resulting in a large data set that is hard to analyze with univariate methods. Furthermore, because many of the variables may be correlated, they must be considered all together rather than individually. To meet these requirements, data analytics uses multivariate statistical process control (MSPC) methods that can be employed to reduce the dimension of the large raw data set while extracting useful information such as the existence of faults or abnormalities.

Since the late 1980s, industry has applied multivariate statistical projection methods to detect abnormal conditions and even diagnose the root cause of abnormal situations [6, 7]. In nearly all applications, two basic techniques are applied: principal component analysis (PCA) and partial least squares (PLS). PCA is a dimensional reduction method that identifies a subset of uncorrelated vectors (principal components) to capture the most variance in the data. PLS is a decomposition technique that maximizes the covariance between the predictor and the predicted variables for each component [4]. In some cases, PCA can be applied to a data set before using PLS to reduce the variable dimension for analysis by PLS. This step makes PLS much more efficient to use. Hoskuldsson [8], and MacGregor and Kourti [3] provide more detail on PLS.

MSPC has been applied across many industries during the past 10 years, with applications in the areas of semiconductors [9, 10], chemicals [11], mining [12], and petrochemicals. Miletic et al. [13] performed a comprehensive review of the application of multivariate statistics in the areas of steel and pulp and paper. Kourti et al. [14] showed the power of MSPC in polymer processing.

In comparison with these industries, the application of MSPC in the food, biological, and pharmaceutical industry is relatively immature. With online analyzers, plants are able to collect a tremendous amount of historical data. However, efficient tools are needed to manage these data and extract useful information. MSPC is a good candidate for this purpose as well as for achieving the manufacturing goals of high quality, efficiency, and low cost [1]. Increasing attention from both academic and industrial researchers has been focused on this promising area. In 2001, Albert and Kinley [15] from the Eli Lilly Company applied PCA to an industrial batch tylosin biosynthesis process. In 2001, Lennox et al. [16] published results on different MSPC approaches for a fed-batch fermentation system operated by Biochemie Gmbh. An MSPC application in a pharmaceutical fermentation process was studied by Lopes et al. [17] in 2002. Also, Chiang et al. [18] compared the performance of three MSPC methods in an industrial fermentation process at the San Diego biotech facility of the Dow Chemical Company.

To provide working details on applying MSPC, this chapter is organized as follows. Section 8.2 provides introductory background information on process-monitoring computational algorithms for PCA and a simple example to demonstrate how and why PCA works as a monitoring tool. In Section 8.3, the most common PCA approach for a batch process (multiway PCA) is shown, and some special features of PCA applications to batch processes are discussed. Another type of monitoring method based on nonlinear models, called *model-based PCA*, is introduced in Section 8.4. Fault detection results for a biological reactor simulation example are given in Section 8.5.

Learning Objectives

- Understand the ideas behind PCA and MSPC
- Learn how to calculate principal components from a series of correlated data

- Recognize different types of PCA and where they should be applied
- Know how to choose the number of principal components
- Understand the implications of dynamic time warping for variable batch lengths
- Learn how to perform different types of unfolding methods and the meaning behind each method
- Know which method to select among multiway PCA, model-based PCA, and super model-based PCA for a practical batch process
- Learn how to perform and analyze off-line and online multiway PCA on an industrial batch process
- Realize best practices for effective use of data analytics software

8.2 PCA Background

Short History of PCA

From 1873 to 1874, Beltrami and Jordan independently derived the singular value decomposition (SVD) method that underlies PCA [19]. The earliest work on PCA was done by Pearson in 1901 and Hotelling in 1933 [19, 20]. However, due to its heavy computational requirements, the spread of PCA was limited until personal computers became available in the 1980s. Since then, PCA has been widely used in many areas, including agriculture, biology, chemistry, climatology, demography, ecology, economics, food research, genetics, geology, meteorology, oceanography, psychology, and process control [19]. According to the *ISI Web of Knowledge* (http://isiknowledge.com), nearly 4,000 technical articles used PCA as keywords in 2004 and 2005.

Simple Example Applying PCA

A simple temperature control loop in a bioreactor can be used to illustrate how PCA works. The system consists of a fermenter, a temperature sensor, and cooling water supply. A valve manipulates the flow rate of the cooling water, and the sensor collects temperature data from the fermenter at regular time intervals. Ten pairs of normal condition data and two pairs of known abnormal runs are listed in Table 8-2a.

The raw data are plotted in Figure 8-2a. Normal operating data are depicted as "o" whereas two abnormal data points are depicted as "*". The first abnormal case represents a mismatch between the valve position and temperature, whereas the second case has a higher temperature than normal.

Table 8-2a. Raw data (* = abnormal data).

Valve Position (% max)	Temperature (°C)
10.1	25.0
9.9	25.0
10.0	25.0
9.8	25.3
9.9	25.2
9.8	25.2
10.1	24.8
10.3	24.7
10.0	25.1
10.1	24.7
10.2*	25.1*
9.9*	25.5*

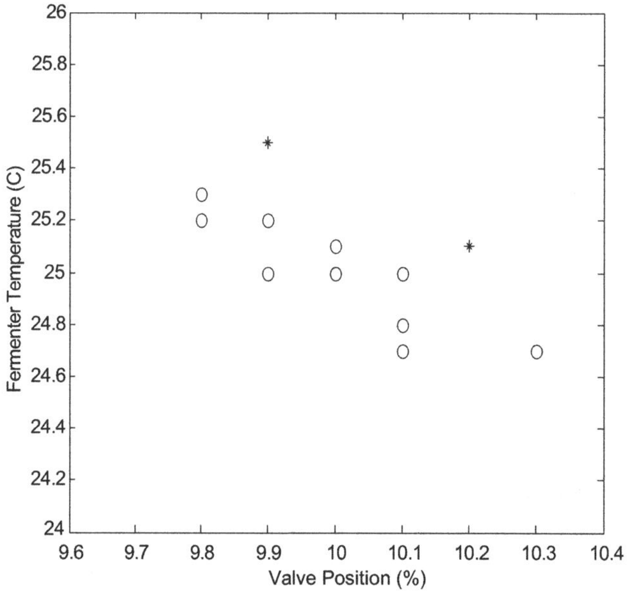

Figure 8-2a. Raw data showing normal ("o") and abnormal ("*") data points.

Control charts of valve position and temperature are shown in Figures 8-2b and 8-2c. The upper control limit (UCL) and lower control limit (LCL) are calculated by the following relationship:

$$X_{UCL,LCL} = \bar{X} \pm 2\sigma$$

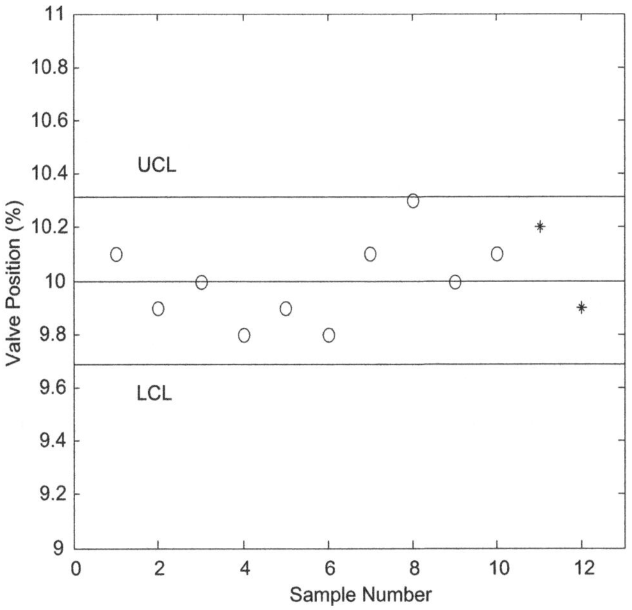

Figure 8-2b. Valve position control chart.

Note: The middle line is the average value, whereas the upper line is the UCL and the lower line is the LCL. "o" represents normal conditions and "*" indicates abnormal cases. All data lie within the UCL and LCL.

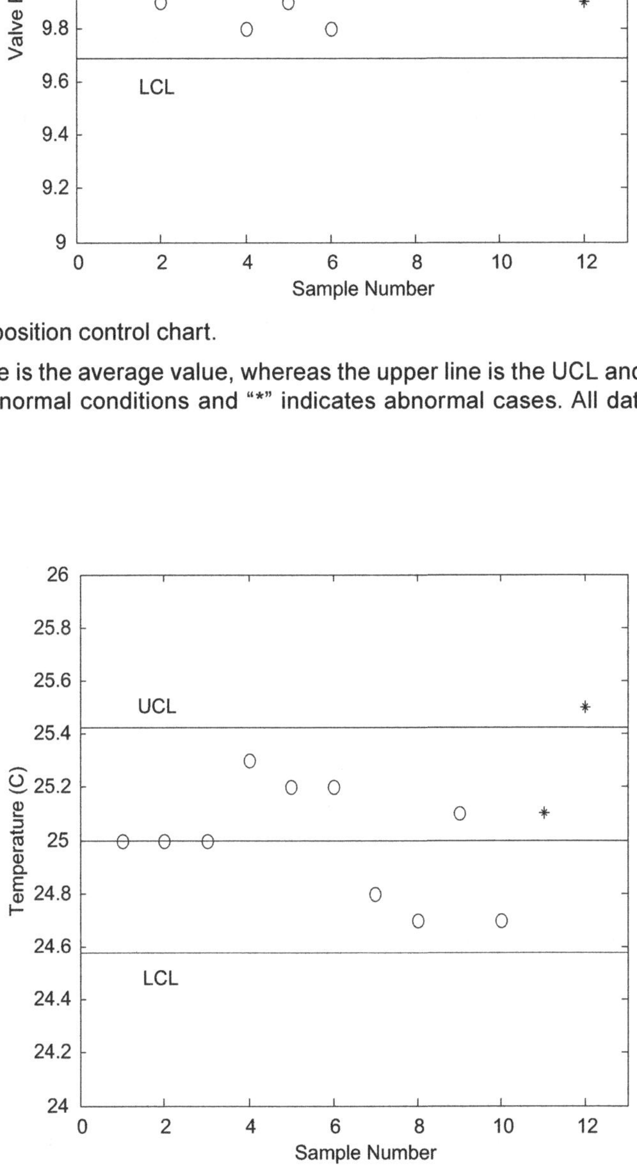

Figure 8-2c. Temperature value control chart.

Note: In this chart, "o" represents normal conditions and "*" indicates abnormal cases. Sample 12 exceeds the UCL.

$\bar{X}$ is the mean and σ is the standard deviation of the data), which indicate bounds for normal operation. Note that all valve position data fluctuate around the mean value, while temperature values also appear to be normal (*in control* using SPC terminology) except the last data point. As a result, the first abnormal reading cannot be detected by either control chart.

In industrial applications, single-variable control charts such as Figures 8-2b and 8-2c are frequently used. However, this simple example clearly illustrates that such charts have a deficiency when monitoring correlated variables. Therefore, independent variables are preferred instead of the original variables. PCA provides a way to project the raw data to the principal component subspace and obtain a series of uncorrelated variables called *principal components* (PCs). A detailed procedure is presented in the following section. Let us now walk through step-by-step sample calculations.

To carry out PCA on the raw data, first we calculate the mean, variance, and covariance (S) of the raw data (X_{raw}) after subtracting the mean:

$$\bar{\mathbf{X}}_{\mathbf{raw}} = \begin{bmatrix} \bar{X}_{1,raw} \\ \bar{X}_{2,raw} \end{bmatrix} = \begin{bmatrix} 10.0 \\ 25.0 \end{bmatrix} \tag{8-2a}$$

The mean for the valve position for normal conditions is estimated at 10.0, whereas the mean for the temperature for these same normal conditions is estimated at 25.0.

The deviations of each individual data point relative to the corresponding mean are calculated as follows:

$$\begin{aligned} X_{1i} &= X_{1i,raw} - \bar{X}_{1,raw} \qquad i = 1,2,\ldots,m \\ X_{2i} &= X_{2i,raw} - \bar{X}_{2,raw} \qquad i = 1,2,\ldots,m \end{aligned}$$

The resulting data matrix is then denoted as X (X_1 = valve position deviation and X_2 = temperature deviation), which has zero mean:

$$\mathbf{X} = \begin{bmatrix} X_{11} & X_{12} & \ldots & X_{1m} \\ X_{21} & X_{22} & \ldots & X_{2m} \end{bmatrix} \tag{8-2b}$$

where m is the number of observations. Row 1 is the mean-centered valve position data, and row 2 contains the mean-centered temperature data. From the data matrix X, one can estimate the covariance, which measures how much two variables change together. It can be calculated as follows:

$$\mathbf{S} = \mathrm{cov}(X) = \frac{X^T X}{m-1} = \begin{bmatrix} 0.0244 & -0.0289 \\ -0.0289 & 0.0444 \end{bmatrix} \tag{8-2c}$$

The covariance matrix is important because the off-diagonal terms represent the relationship between each variable (Note: **S** is symmetric). In this simple case, $s_{12} = -0.0289$, and the correlation coefficient is:

$$\rho_{12} = \frac{s_{12}}{(s_{11}s_{12})^{0.5}} = -0.88$$

This correlation coefficient indicates there is a strong linear relationship between the valve position and temperature given its absolute value is close to 1.0.

Mean centering is important when calculating the covariance matrix; therefore, the covariance matrix and PCA derive from the data variation rather than from the absolute data values.

In process monitoring, independent variables are preferred over correlated ones. Thus, we transform the correlated data to a new subspace so that the off-diagonal terms of the new covariance matrix are zero (no correlation between the new variables). Orthogonal transformation provides a way for **S** to be converted to a diagonal matrix (**Λ**) such that the off-diagonal terms are zero [7]:

$$\mathbf{U}^T\mathbf{S}\mathbf{U} = \mathbf{\Lambda} \tag{8-2d}$$

$$\mathbf{U}^T\mathbf{U} = \mathbf{I} \tag{8-2e}$$

where **I** is the identity matrix and **U** is called the *eigenvector matrix*. The rows and columns of **U** are orthonormal, so this transformation effectively forms two new variables that are linear combinations of the original variables x_1 and x_2. The diagonal terms of **Λ** are the eigenvalues, and the columns of **U** are the eigenvectors of the covariance matrix. Using available mathematical software such as MATLAB [21] (command: [u, lambda] = eig(S)), **Λ** and **U** can be calculated easily.

For **S** in Equation 8-2c,

$$\mathbf{U} = \begin{bmatrix} -0.58 & -0.81 \\ 0.81 & -0.58 \end{bmatrix}, \quad \mathbf{\Lambda} = \begin{bmatrix} 0.065 & 0 \\ 0 & 0.0039 \end{bmatrix} \text{ and:}$$

$$\mathbf{U}^T\mathbf{S}\mathbf{U} = \mathbf{\Lambda} = \begin{bmatrix} -0.58 & 0.81 \\ -0.81 & -0.58 \end{bmatrix} \begin{bmatrix} 0.0244 & -0.0289 \\ -0.0289 & 0.0444 \end{bmatrix} \begin{bmatrix} -0.58 & -0.81 \\ 0.81 & -0.58 \end{bmatrix} = \begin{bmatrix} 0.065 & 0 \\ 0 & 0.0039 \end{bmatrix} \tag{8-2f}$$

At this stage, principal axis transformation can be applied to transform the correlated raw data into independent values, which can be described as follows:

$$Y = U^T(X \quad \bar{X}) \tag{8-2g}$$

The matrix of eigenvectors (**U**) is called the *projection matrix*, and the transformed variables (**Y**) are PCs. Consider the first pair of measurements as an example:

$$\mathbf{Y} = \begin{bmatrix} Y_1 \\ Y_2 \end{bmatrix} = \mathbf{U}^T(\mathbf{X}_{raw} - \bar{\mathbf{X}}_{raw}) = \begin{bmatrix} -0.58 & 0.81 \\ -0.81 & -0.58 \end{bmatrix} \left(\begin{bmatrix} 10.1 \\ 25.0 \end{bmatrix} - \begin{bmatrix} 10.0 \\ 25.0 \end{bmatrix} \right) = \begin{bmatrix} -0.058 \\ -0.082 \end{bmatrix}$$

Λ gives the variance of transformed variables, and these two PCs are linearly independent (covariance equal to zero). To make the two PCs have the same scale or length, another way of defining PCs can be used:

$$\mathbf{Z}_i = \frac{\mathbf{U}^T(\mathbf{X}_{raw} - \bar{\mathbf{X}}_{raw})}{\lambda_{ii}^{0.5}} \tag{8-2h}$$

By dividing the normalization factor based on the diagonal elements of **Λ** (λ_{ii}), the new PCs will have a unit variance. This is quite useful because the new artificial variables (PCs) are not only independent but have unit variance as well. The values of the normalized PCs are listed in Table 8-2b, and corresponding control charts of the transformed data set $\mathbf{Z}_i$ are plotted in Figures 8-2d and 8-2e. Both

Table 8-2b. Projected data (Z1 and Z2).

PC 1 (Z_1)	PC 2 (Z_2)
–0.2275	1.3088
0.2275	1.3088
0.0000	0.0000
1.4134	–0.1782
0.8664	–0.5551
1.0939	0.7537
–0.8664	0.5551
–1.6409	–1.1306
0.3195	–0.9319
–1.1859	1.4870
–0.1355*	–3.5495*
1.8248*	–3.3509*

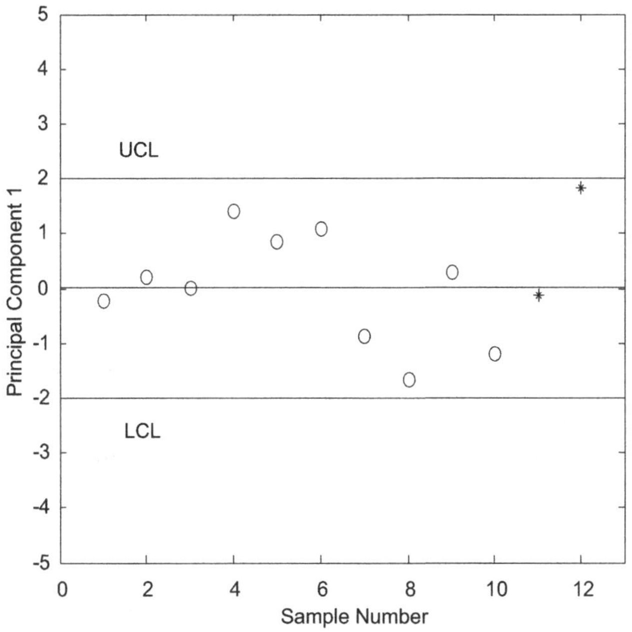

Figure 8-2d. PC 1 (Z_1) control chart.

Note: In this chart, "o" represents normal conditions, and "*" indicates abnormal cases. All data lie between the UCL and LCL.

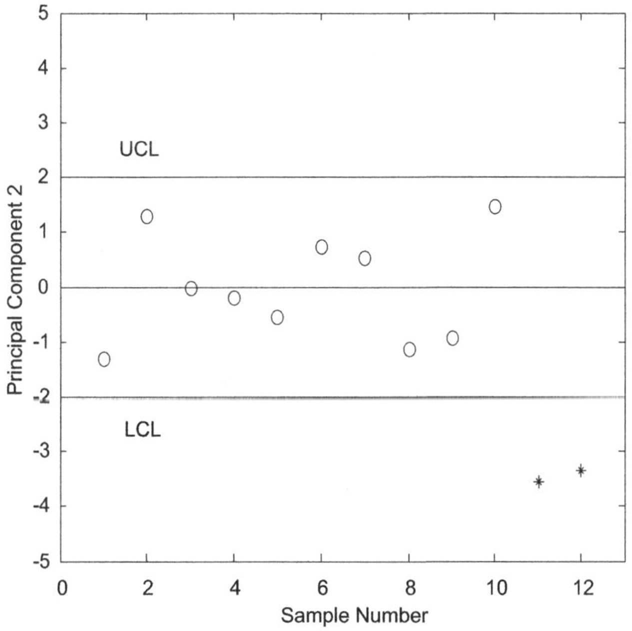

Figure 8-2e. PC 2 (Z_2) control chart.

Note: In this chart, "o" represents normal conditions, and "*" indicates abnormal cases. The last two data points are identified as abnormal on this control chart.

abnormal operation conditions are clearly identified in Figure 8-2e. In both figures, because the standard deviation of each variable is 1, the UCL and LCL are 2 and –2, respectively,

$$(Z_{UCL,\,LCL} = Z \pm 2\sigma = 0 \pm 2).$$

Both abnormal points follow the trend of the process changes (see Figure 8-2a and Table 8-2b); thus, Figure 8-2d indicates that the process is in control. However, the correlation between the two variables for the first abnormal case (valve position: 10.2; temperature: 25.1) differs from the normal behavior (i.e., the valve opens and temperature drops, which indicates a negative process gain). For that reason, Figure 8-2e shows there is a fault in the system. For the second abnormal case (valve position: 9.9; temperature: 25.5), the temperature should be higher but closer to 25. The temperature is clearly higher than 25 but not close to 25. In other words, the trend of changes between the temperature and valve position is followed but not exactly. Note that in Figure 8-2d, point 12 is close to the UCL. At the same time, the temperature change relative to the valve position is so large that it exceeds the control limit of Figure 8-2e. The role of the projection matrix **U** will be further illustrated in our analysis in the "PCA Algorithm" section later in this chapter.

In the foregoing analysis, the covariance matrix (**S**) is used to explicitly show whether the variables are correlated. However, in industrial analysis, projection can be carried out directly without explicitly computing the covariance matrix. In other words, **X** can be decomposed into the product of two matrices by:

$$\boldsymbol{X} = \boldsymbol{T}\boldsymbol{P}^T \tag{8-2i}$$

T and **P** are called *score* and *loading* matrices and have dimensions $(m \times r)$ and $(n \times r)$, respectively. Modifying Equation 8-2d, we obtain:

$$\mathbf{S} = \frac{\mathbf{X}^{\mathrm{T}}\mathbf{X}}{m-1} = \mathbf{U}\boldsymbol{\Lambda}\mathbf{U}^{\mathrm{T}} \tag{8-2j}$$

in which m is the number of observations. Substituting Equation 8-2i into Equation 8-2j, we obtain:

$$\mathbf{S} = \frac{\mathbf{X}^{\mathrm{T}}\mathbf{X}}{m-1} = \frac{\mathbf{P}\mathbf{T}^{\mathrm{T}}\mathbf{T}\mathbf{P}^{\mathrm{T}}}{m-1} = \mathbf{U}\boldsymbol{\Lambda}\mathbf{U}^{\mathrm{T}} \tag{8-2k}$$

Define $\dfrac{\mathbf{T}^{\mathrm{T}}\mathbf{T}}{m-1} = \boldsymbol{\Lambda}$

and, thus, the loading matrix (**P**) is the same as the projection matrix (**U**). The score matrix (**T**) can be calculated by post-multiplying Equation 8-2i by **P** (using $\mathbf{P}^T\mathbf{P} = \mathbf{I}$), where **P** is the eigenvector matrix of the covariance matrix):

$$\mathbf{T} = \mathbf{XP} \tag{8-2l}$$

For this simple example, the first component of the score matrix is as follows:

$$t_1 = \mathbf{X}_1\mathbf{P} = \left[(10.1 - 10.0),(25.0 - 25.0)\right]\begin{bmatrix} -0.58 & -0.81 \\ 0.81 & -0.58 \end{bmatrix} = \left[-0.058, -0.0815\right]$$

In this way, the score of all samples can be calculated. The loadings (p_j) contain the information on relationships between different rows, while the scores (t_j) describe the relationship between columns. We will discuss how score and loading matrices are used and present a more efficient calculation method later.

The control charts of independent variables (PCs) work fairly well for this simple example. However, in a real industrial application, a process unit might have hundreds of variables, so many control charts with a "false positive" error may result.

A "false positive" error describes the situation in which the system data do not contain a fault but control charts indicate there is one.

As a result of the "false positive" error problem, squared prediction error (*SPE*) and Hotelling's T^2 control charts are used to give a more definitive answer for a multivariate process. The detailed definition and calculation of *SPE* and Hotelling's T^2 for these multivariate problems are given later.

Next, we develop the values of the Hotelling's T^2 control chart. Returning to the simple example, the points in Figure 8-2f are calculated by:

$$T_j^2 = Z_j^T Z_j = Z_{1j}^2 + Z_{2j}^2 \tag{8-2m}$$

T^2 is positive for all j and thus has a minimum value of zero. The upper control limit of T^2 is related to the F distribution and can be calculated by:

$$T_\alpha^2 = \frac{k(m-1)}{m-k} F_{k,m-k;\alpha} = 10.03 \quad (T2\ \text{UCL}) \tag{8-2n}$$

where

k is the number of PCs, m is the number of normal condition data points, and α is the percentage confidence limits expressed as a fraction ($0 < \alpha < 1$). In this example, $k = 2$, $m = 10$, and $\alpha = 0.95$, indicating that there are two PCs, 10 data pairs, and 95% confidence limits.

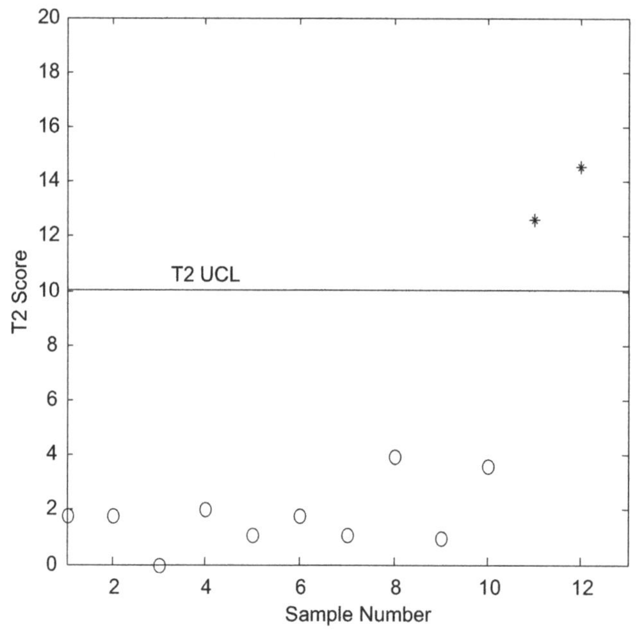

Figure 8-2f. Hotelling's T^2 control chart.

Note: In this chart, "o" represents normal conditions and "*" indicates abnormal cases.

The *SPE* control chart focuses on the magnitude of the sample projection on the residual space. For this simple case, the number of PCs is two, which is the same as original data set, so there are no remaining vectors for correlation because we started with only two variables; therefore, the residual is the null space. For more complex problems there can be many original variables, and we might only select fewer than three PCs. Therefore, we will give an example illustrating the use of the *SPE* control chart in Section 8.5, in the bioreactor example, which has a large number of process variables.

To provide a geometric interpretation of PC projection, a control ellipse is constructed for the original variables x_1 and x_2.

Combining Equations 8-2h and 8-2m, we get:

$$T^2 = (\mathbf{X}_{raw} - \bar{\mathbf{X}}_{raw})^T \frac{\mathbf{U} \quad \mathbf{U}^T}{\mathbf{\Lambda}^{0.5} \ \mathbf{\Lambda}^{0.5}} (\mathbf{X}_{raw} - \bar{\mathbf{X}}_{raw}) = (\mathbf{X}_{raw} - \bar{\mathbf{X}}_{raw})^T \frac{(\mathbf{U}^T\mathbf{U}) \bullet (\mathbf{U}^T\mathbf{U})}{\mathbf{U}^T\mathbf{\Lambda}\mathbf{U}} (\mathbf{X}_{raw} - \bar{\mathbf{X}}_{raw})$$
$$= (\mathbf{X}_{raw} - \bar{\mathbf{X}}_{raw})^T \mathbf{S}^{-1} (\mathbf{X}_{raw} - \bar{\mathbf{X}}_{raw}) \tag{8-2o}$$

Note that $\mathbf{S}^T = (\mathbf{U}^T \mathbf{\Lambda} \mathbf{U})^T = \mathbf{U}\mathbf{\Lambda}^T \mathbf{U}^T = \mathbf{U}\mathbf{\Lambda} \mathbf{U}^T = \mathbf{S}$ and $\mathbf{U}^T \mathbf{U} = \mathbf{I}$.

This leads to:

$$\frac{s_1^2 s_2^2}{s_1^2 s_2^2 - s_{12}^2} \bullet \left[\frac{(x_1 - \bar{x}_1)^2}{s_1^2} + \frac{(x_2 - \bar{x}_2)^2}{s_2^2} - \frac{2s_{12}(x_1 - \bar{x}_1)(x_2 - \bar{x}_2)}{s_1^2 s_2^2} \right] = T_{k,m,\alpha}^2 \qquad \text{(8-2p)}$$

For our example, it should be:

$$\frac{0.0244 \times 0.0444}{0.0244 \times 0.0444 - (-0.0289)^2} \bullet \left[\frac{(x_1 - 10)^2}{0.0244} + \frac{(x_2 - 25)^2}{0.0444} - \frac{2(-0.0289)(x_1 - 10)(x_2 - 25)}{0.0244 \times 0.0444} \right] = 10.03$$

which simplifies to:

$$17.83(x - 10)^2 + 9.80(x - 25)^2 + 23.22(x - 10)(x - 25) - 1 = 0$$

This ellipse is depicted in Figure 8-2g. The slopes of the major and minor axes are as follows:

$$m_{major} = \frac{\mathbf{U}(2,1)}{\mathbf{U}(1,1)} = -1.4044$$

$$m_{minor} = \frac{\mathbf{U}(2,2)}{\mathbf{U}(1,2)} = 0.7121$$

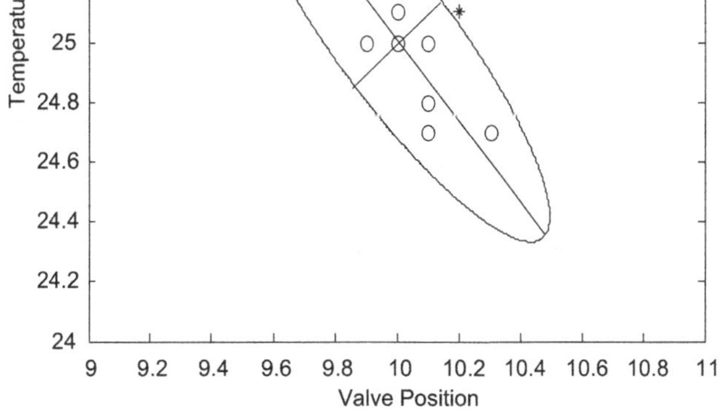

Figure 8-2g. Control ellipse for simple example.

Note: In this chart, "o" represents normal conditions and "*" indicates abnormal cases.

The lengths of the semimajor and semiminor axes are:

$$l_{major} = \sqrt{\Lambda_{11} T^2_{upper}} = \sqrt{(0.065)(10.03)} = 0.8076$$
$$l_{minor} = \sqrt{\Lambda_{22} T^2_{upper}} = \sqrt{(0.0039)(10.03)} = 0.1971$$

The center of the control ellipse is at: $x_1 = 10.0$; $x_2 = 25.0$ (the mean values of $\mathbf{X}_1$ and $\mathbf{X}_2$).

The two abnormal points lie outside of the ellipse.

For this two-variable example, the control ellipse not only shows whether the process is in control but also provides a geometric figure to explain how principal projection works. However, for high-dimensional systems, the control ellipse cannot be displayed; therefore, SPE and T2 charts are typically used in process quality monitoring.

Besides the issues mentioned in this example, other important topics in PCA include: How does one choose the number of PCs? How does one carry out PCA on continuous and batch processes? What do *SPE* and T^2 indicate? These topics will be discussed in the remaining sections of this chapter.

PCA Algorithm

Because the statistical background of PCA is described in detail by Jolliffe [19] and Jackson [20], this section will focus on the algorithm and the steps for applying PCA for both online and off-line analysis.

1. Preprocessing Data

For continuous processes, $\mathbf{X}_{\mathbf{raw}}$ is an $m \times n$ data matrix whose columns correspond to the process variables (sensors) and whose rows correspond to the samples at each time. Then, $\mathbf{X}_{\mathbf{raw}}$ is mean-centered by subtracting the mean value of each column. It is then rescaled by dividing the standard deviation before performing PCA. The data matrix after preprocessing is denoted as **X**. Centering is performed so the PCA model represents the deviation from the mean, and the scaling step is necessary so each variable has equal weight in further analysis. If different weights are needed for some variables, then a diagonal weighting function (**W**) multiplies the raw data after mean centering [22].

2. Decomposing Data

After preprocessing, **X** can be decomposed into:

$$X = TP^T + E \tag{8-2q}$$

using either SVD or the nonlinear iterative partial least squares (NIPALS) algorithm. **T** and **P** are the score matrix ($m \times r$) and loading matrix ($n \times r$) ($r = min(m,n)$), respectively. **E** is the residual matrix that results when fewer than r of the principal components are used in the model. SVD is an efficient method for identifying PCs. However, when the dimension of **X** is large (rank(**X**) > 1000), SVD can be time-consuming. NIPALS is more efficient because the PCs are calculated sequentially. The theory and algorithm of SVD can be found in mathematics and statistical modeling books (e.g., References [19] and [23]). There is an svd.m file in MATLAB [21] that provides the source code for performing SVD. In the biological fermentation system discussed in Section 8.5 of this chapter, the data matrix is very large. Thus, NIPALS is the best approach; we will describe its algorithm next [22, 24]:

1. Choose any column x_i from **X** and set it equal to t_j as the initial value.
2. $\boldsymbol{p}_j^T = \dfrac{\boldsymbol{t}_j^T \mathbf{X}}{\boldsymbol{t}_j^T \boldsymbol{t}_j}$
3. $p_{jnew}^T = \dfrac{p_{jold}^T}{\left\| p_{jold}^T \right\|}$
4. $\boldsymbol{t}_j = \dfrac{\mathbf{X}_{pj}}{\boldsymbol{p}_j^T \boldsymbol{p}_j}$
5. If t_j converges, go to step 6; otherwise, return to step 2.
6. $\mathbf{X} = \mathbf{X} - \boldsymbol{t} \bullet \boldsymbol{p}^T$
7. Go to step 1 to calculate the next PC.

Note that in step 6, the data residual is calculated based on using the current PC. The calculated $\boldsymbol{p}_j$ vectors correspond to the eigenvectors of the covariance matrix of **X**, which is defined as follows:

$$\mathbf{S} = \text{cov}(\mathbf{X}) = \frac{\mathbf{X}^{\mathrm{T}}\mathbf{X}}{m-1} \tag{8-2r}$$

So, for each $\boldsymbol{p}_j$:

$$\mathbf{cov}(\mathbf{x})\boldsymbol{p}_j = \lambda_j \boldsymbol{p}_j \tag{8-2s}$$

where λ_j is the jth eigenvalue of the covariance matrix of X and the solution of NIPALS is in descending order ($\lambda_1 > \lambda_2 > \ldots > \lambda_r$). As pointed out earlier, the loadings p_j contain the information on relationships between variables, while the scores t_j describe the relationship between samples.

The magnitude of λ_j describes the amount of information captured by the corresponding t_j and p_j. Therefore, the set of t_1 and p_1 (corresponding to λ_1) contains the most variation in **X**. In general, the first few pairs of t_j and p_j capture most of the variance.

The p_j's are orthonormal:

$$p_i^T p_j = \begin{Bmatrix} 0, i \neq j \\ 1, i = j \end{Bmatrix}$$

And the t_j's are orthogonal, namely:

$$t_i^T t_j = 0, i \neq j, t_i^T t_j = \lambda_i, i = j$$

Furthermore, by NIPALS the magnitude of λ_j is already in descending order.

3. Choosing the Number of PCs

The first k pairs of loadings and scores are stored in matrices T and P, and these pairs form an optimal model of the system. Therefore, choosing the minimum number of PCs (k) is an important decision. The key idea in this step is to select the number of PCs that will accurately reflect the system behavior and filter process noise. Typically, for a complex industrial system, more than one method is used and compared to reach a final conclusion. For example, in [25], a simulation example was analyzed by different methods; the results are given in Table 8-2c. It shows that different methods

Table 8-2c. Number of PCs determined from each analyzed method (Table 1 in reference [25]).

Method	Reactor	Boiler	Incinerator
Cumulative percentage variance	3	1	15
Residual percentage variance	Ambiguous	1	Ambiguous
Average eigenvalue	1	1	7
Parallel analysis	1	1	7
Embedded error function	2	No solution	No solution
Autocorrelation	2	No solution	2
PRESS	3	1	15
R ratio	2	2	5
Akaike information criterion	2	No solution	No solution
Minimum description length	2	No solution	No solution
VRE_{cov}	2	1	2
VRE_{cor}	2	1	2

reach different conclusions regarding the number of PCs needed for three different unit operations. Furthermore, some methods show ambiguous results or no solution at all. Finally, note that usually two or three PCs are sufficient to explain most of the variance in the system.

The choice of how many PCs to use is somewhat subjective and depends on the user's experience with the process. Some selection methods—for example, VRE (variation of reconstruction error), parallel analysis, and cross-validation— may show better reliability and effectiveness.

4. Projecting Data

Based on the information thus far, a sample x ($n \times 1$) is projected to the PC space ($\hat{x}$) and residual space ($\tilde{x}$) by:

$$\hat{x} = \mathbf{P}\mathbf{P}^{\mathrm{T}} x \tag{8-2t}$$

$$\tilde{x} = \left(I - PP^{T}\right)x \tag{8-2u}$$

5. Fault Detection Indices

Once a PCA model is developed by analyzing the historical data under normal operating conditions, new observation samples can be projected to the PC subspace to obtain their scores and loadings and compare them further with the control limits. Typically, the squared prediction error (*SPE*, also referred to as the *Q statistic*) and Hotelling's T^2 statistic are used to calculate control limits for multivariate cases. See [7] and [26] for discussions that interpret the role of *SPE* and T^2 in fault detection. If a new data set is collected, *SPE* measures the magnitude of the sample projection on the residual space (Equation 8-2v), while Hotelling's T^2 statistic considers the variations in the PC space (Equation 8-2w) [7].

SPE is preferred over Hotelling's T^2 in fault detection applications because of their nature as described in the simple example presented earlier. In other words, use residual space to detect faults that are not explained by the model.

$$SPE = \left\|(\mathbf{I} - \mathbf{P}\mathbf{P}^{\mathrm{T}})x\right\|^2 \tag{8-2v}$$

$$T^2 = x^T \mathbf{P}\mathbf{\Lambda}^{-1}\mathbf{P}^{\mathrm{T}} x \tag{8-2w}$$

where

$\mathbf{\Lambda}$ is a diagonal matrix: $\mathbf{\Lambda} = \mathrm{diag}(\lambda_1, \lambda_2, \ldots, \lambda_k)$.

The upper control limit of *SPE* is given by Jackson and Mudholkar [27] as follows:

$$SPE_{\alpha} = \theta_1 \left[\frac{c_{\alpha}\sqrt{2\theta_2 h_0^2}}{\theta_1} + 1 + \frac{\theta_2 h_0 (h_0 - 1)}{\theta_1^2} \right]^{\frac{1}{h_0}} \tag{8-2x}$$

where

$$\theta_i = \sum_{j=k+1}^{n} \lambda_j^i, i = 1,2,3$$

and

$$h_0 = 1 - \frac{2\theta_1\theta_3}{3\theta_2^2} \tag{8-2y}$$

k is the number of PCs retained, and α is the significance level, while c_{α} is the standard normal deviation corresponding to the upper 1-α percentile.

The UCLs of Hotelling's T^2 statistic can be calculated by using the F-distribution [20]:

$$T_{\alpha}^2 = \frac{k(m-1)}{m-k} F_{k,m-k;\alpha} \tag{8-2z}$$

where F is the F-distribution and m is the number of observations. As discussed earlier, the two indices measure different variations of the sample. *SPE* is preferred over Hotelling's T^2 in fault detection calculations. Besides the two general indices, several other metrics are proposed in the literature, including combined indices [28], global Mahalanobis distance [7], and Hawkins's statistic [20].

Different Types of PCA and Applications

Section 8.2 introduced the core idea and basic algorithms of PCA. Because PCA has been applied broadly, many variations have been developed to meet specific requirements. Table 8-2d lists some of the variations of PCA and their usage in chemical and biological applications.

8.3 Multiway PCA

Section 8.2 includes some important definitions and algorithms for applying PCA to continuous processes. However, for batch and semibatch processes, there is no steady

Table 8-2d. PCA variations and their applications in chemical and biological process monitoring.

Processes	Method	Articles	Method Features
Time Invariant Continuous Process	PCA	Kresta et al. [29]	Reduce the system dimension.
Time Variant Continuous Process	Recursive PCA	Li et al. [30]	Monitor the adaptive model.
Dynamic Continuous Process	Dynamic PCA	Ku et al. [31], Russell et al. [32]	Consider the time correlation between variables.
Nonlinear Continuous Process	Nonlinear PCA	Dong and McAvoy [33]	Consider the nonlinearity in the data.
Large-Scale Processes	Multiblock PCA	MacGregor et al. [34]	Decompose a large matrix into smaller blocks.
Multiscale Continuous Process	Multiscale PCA	Bakshi [35]	Deal with signals using different scales.
Continuous and Batch Processes	Model-Based PCA	Rotem et al. [11]	Use the first-principle model to simulate the process.
Batch Process	Multiway PCA	Nomikos and McGregor [22, 36, 37]	Unfold three-dimensional batch data to two dimensions.

state and the historical trajectories usually contain considerable nonlinearity, so no effective online monitoring technique existed before 1995. Nomikos and MacGregor [22, 36, 37] originally introduced the basic ideas of multiway PCA and PLS methods to monitor batch processes in real time. In this section, we provide the general ideas and computational steps for performing PCA on batch processes and introduce one variation of PCA (multiway PCA).

Reference Trajectory

Batch processes usually exhibit some batch-to-batch variation because the process variables deviate from their specified trajectories, errors occur during the charging of the materials, and disturbances arise from variations in impurities [22]. Besides the variations between batches, the trajectories have general similarities because of the kinetics and transport limits in the system. Fault detection extracts the system's intrinsic dynamics as a reference trajectory and then measures a sample trajectory's deviation with respect to the reference. If the deviation is out of the control limits, an alarm is triggered.

In some cases, the reference trajectory is used directly in fault detection and identification without performing the PCA step [38, 39, 40].

PCA of a Batch Process

The multiway PCA procedure for monitoring and analyzing batch processes can be summarized in terms of the following two procedures:

Procedure A: Building the PCA Model Steps

1. Synchronizing batch trajectory (optional; see Section 8.3).
2. Unfolding the normal operating condition's historical data $X(I \times J \times K)$ into a two-dimensional array (Figures 8-3a to 8-3c). I is the number of batches, J is the variable number, and K is the sampling time.
3. Normalizing the data.
4. Calculating the PCs and extracting score and loading matrices.
5. Obtaining the UCLs of *SPE* and Hotelling's T^2.

Procedure A: Detecting Faults Tests

- **Off-line test** – Calculate the score of a new batch at the end of a run, and compare it with the UCLs. This procedure is usually used to check product quality.
- **Online test** – Calculate the score of a new batch at regular time intervals during the batch, and compare the results with the UCLs. An online test is used to perform real-time monitoring. If an alarm is triggered, the contribution plot, which shows the contribution of each variable to the scores of *SPE* and Hotelling's T^2

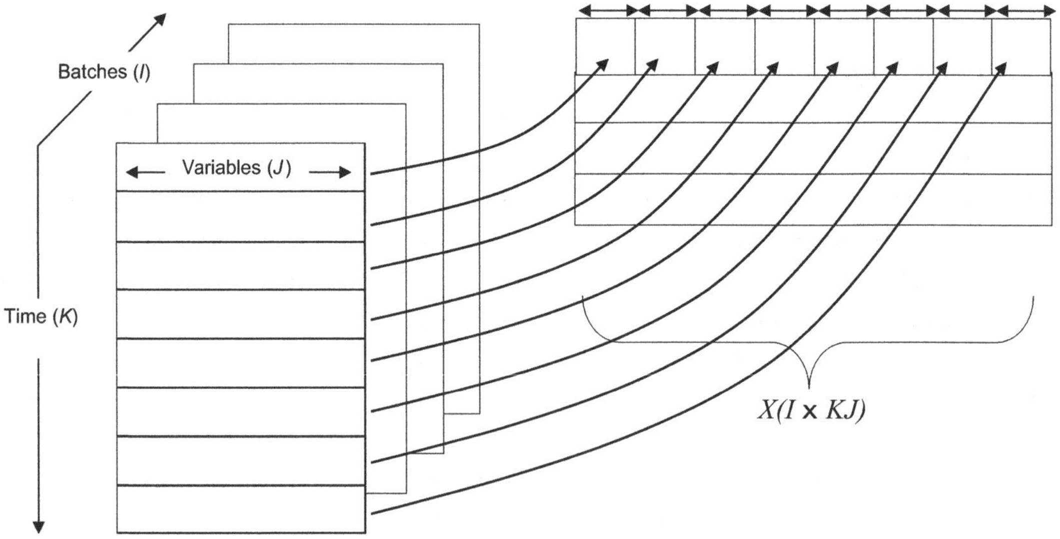

Figure 8-3a. Unfolding data matrix according to batch-wise $X \rightarrow X(I \times KJ)$.

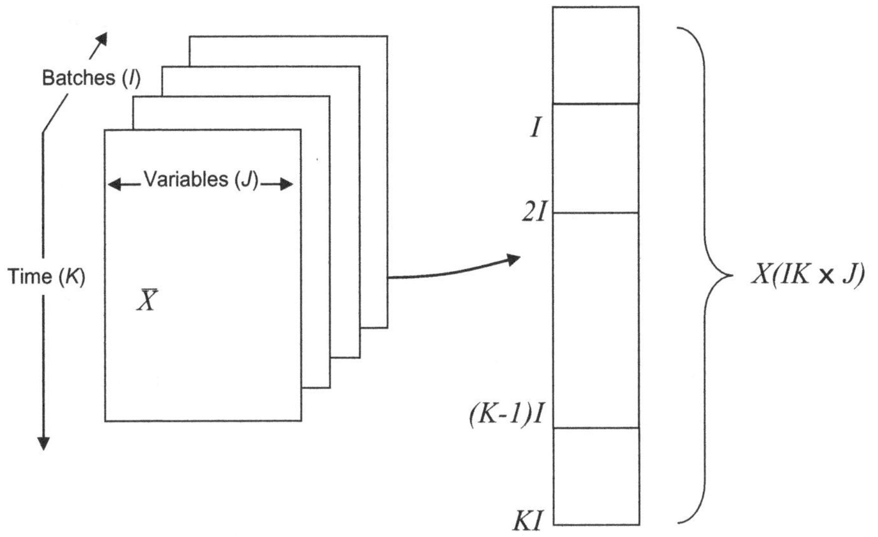

Figure 8-3b. Unfolding data matrix according to variable-wise $\bar{X} \to X(IK \times J)$.

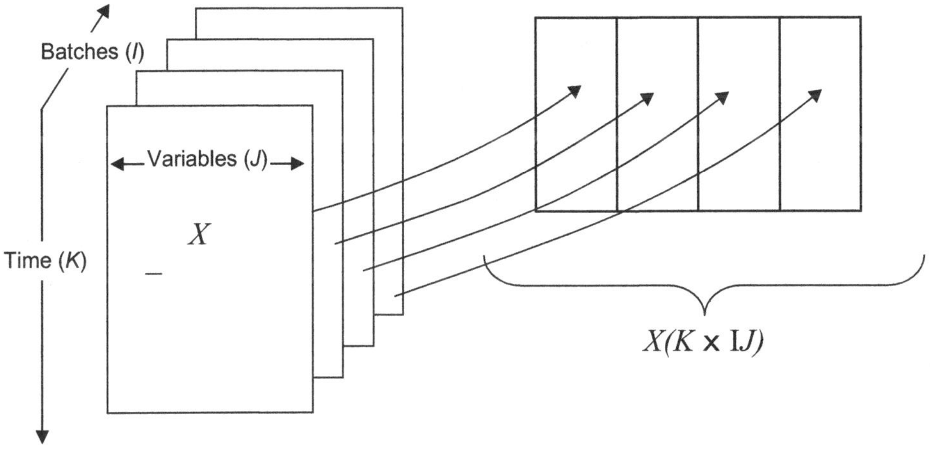

Figure 8-3c. Unfolding data matrix according to time-wise $X(K \times IJ) \to \bar{X}$.

at a specific time, is used to diagnose faults. The algorithm for the online test depends on the specific unfolding method used. A detailed PCA algorithm for batch-wise unfolding is given in the next step on historical data unfolding.

To build the model, pay special attention to the following specific steps.

Historical Data Unfolding (Step A.2) and Normalization (Step A.3)

The main difference between continuous process and batch process historical data is that the former is stored in a two-dimensional matrix, whereas batch data is usually stored in a three-dimensional matrix. Therefore, to perform PCA, the data set should be unfolded to two dimensions. See Figures 8-3a, 8-3b, and 8-3c.

In step A.2, all the trajectories are assumed to have equal length. Usually, the historical data are stored in $X(I \times J \times K)$, where I is the number of batches, J is the variable number, and K is the sampling time (observations).

Recently, Kourti [41] pointed out that the measurements may not be equally spaced and that some variables do not exist throughout the whole processes. These situations are variations of the general case (K is the same for each batch), but in this chapter J and K are assumed constant.

Van Sprang et al. [42] provided a critical evaluation for online batch processing and discussed different data-unfolding methods. There are three ways to unfold the data: batch-wise, variable-wise, and time-wise (Figures 8-3a, 8-3b, and 8-3c). Besides van Sprang et al., Lee et al. [43], Westerhuis et al. [44], and Kourti [41] have also compared the three different unfolding methods.

For batch process monitoring, it is recommended that you include batch time as a variable when performing PCA.

The unfolding method leads directly to the variation information that PCA extracts. Batch-wise unfolding focuses on analyzing the differences among batches, variable-wise unfolding attempts to discover the variability between variables, and time-wise unfolding is used to extract the correlation among samples (at different times). The first two methods are widely used in batch monitoring, and we will discuss them in detail here. After unfolding, the mean-centering and rescaling steps (step A.3) are carried out as described in Section 8.2 (data preprocessing).

When batch-wise unfolding is used, the mean trajectory reflects the dynamics of the system. After processing, the residual is the deviation of the specific trajectory from the average dynamic profile of the process. The dynamic behavior of the batch process is removed by this method, which is a big advantage, but the row vector data will not be complete until the end of a batch (Figure 8-3a). In PCA online monitoring, the score and loading calculations need a complete data set. Thus, one has to predict the future values for the whole batch [22, 36], which is time-consuming and can add uncertainty, especially during a batch's initial period. The variable-wise method discovered by Wold et al. [45] does not have this problem because only the current time data matrix ($I \times K$) is needed for each time (a small data matrix shown in Figure 8-3b). The shortcoming of the variable-wise approach is that the system dynamics are still included in the data set after preprocessing. Thus, the calculated loadings will contain the correlations between variables, with time dependency included [41, 43]. Furthermore, Westerhuis et al. [44] found that variable-wise analysis offers little benefit to monitoring, because it focuses on the wrong source of variations in the data.

Recently, Lee et al. [43] combined the two batch-wise and variable-wise methods, which we called *hybrid-wise unfolding,* and applied it successfully to a bioreactor fault detection case. At first, the data set is unfolded batch-wise, and the mean-centering and scaling steps are performed. After that, the data is rearranged variable-wise. In this way, the time dependency is canceled and future data prediction is also avoided.

Besides data unfolding methods, several other methods, such as parallel factor analysis (PARAFAC) [46, 47] and Tucker models [18, 48], analyze three-dimensional data directly. Westerhuis et al. [44] also found that unfolding is preferred to other direct three-dimensional analysis methods. However, Chiang et al. [18] suggested the unfolding method gives similar results to the three-dimensional analysis method. After all, after unfolding, PCA is able to capture most of the variation of the system and fits the data best.

Procedure B: PCA Algorithm Steps

The data preprocessing (first three) tasks in procedure B steps are the same as those in procedure A steps.

After data preprocessing, the data set is similar to what we have in steady-state continuous processes, so the following steps in procedure B are the same as in continuous process. The UCLs are calculated by Equations 8-2x to 8-2z. In procedure B, the off-line monitoring is composed of data preprocessing (as in procedure A) and data projection (as in a continuous process [Section 8.2]). The online monitoring approach is different. Using batch-wise unfolding as an example, the main steps are as follows:

1. Perform the normalization steps in step A.3 and analyze the new incomplete batch data ($k < K$).
2. Calculate the new score of the batch using the "PCA projection" method (the third method for anticipating the future data is given in [22]), which also takes the missing data into account:

$$\boldsymbol{t}_k = (\mathbf{P}_k^T \mathbf{P}_k)^{-1} \mathbf{P}_k^T xnew_{,k} \tag{8-3a}$$

3. The projection of x on principal and residual spaces is:

$$\hat{x}_{new} = t_{new}\mathbf{P}^T \tag{8-3b}$$

$$\tilde{x} = x_{new} - \hat{x}_{new} \tag{8-3c}$$

   Equations 8-3b and 8-3c are similar to Equations 8-2t and 8-2u. **P** is the calculation result of loading the matrix from step A.

4. Calculate the scores of Hotelling's T^2 and SPE by:

$$SPE_{new}(k) = \sum_{j=1}^{J} \tilde{x}_{new,jk}(k)^2 \tag{8-3d}$$

$$T^2_{new} = t_{new}(\mathbf{S})^{-1} t^T_{new} \tag{8-3e}$$

   where **S** is the covariance matrix of the score calculated in step A. These two equations are variations of Equations 8-2v and 8-2w. Step B.4 is repeated for every new sample.

   For batch-wise unfolding, future data anticipation is needed, while for variable-wise unfolding, this step is skipped, as discussed previously. Lee et al. [43] give a detailed algorithm scheme of variable-wise PCA algorithm.

5. Compare the results of SPE and T^2 with the upper control limits.

   The steps after data preprocessing are nearly universal for all types of PCA in batch processes. Thus, in later sections, the preprocessing steps are discussed in detail.

Batch Data Synchronization

Even for successful batches, certain batch operating steps may require different durations from batch to batch. If PCA is used directly without synchronizing trajectories, the result can be misleading. For a bioreactor, the oxygen concentrations, fermenter temperature, and other variables may vary around their normal profiles. Unsteady behavior will affect the duration of each stage and result in different batch lengths. Furthermore, synchronization between different batches is a necessary assumption to perform batch data monitoring [22, 37].

Techniques such as dynamic time warping (DTW) [49], correlation optimized warping [50], indicator variable matching [36, 51], and data interpolation have been introduced to synchronize batch profiles. In some cases, the indicator variable approach does not work [41]. Simple interpolation within the data can sometimes be employed, but it does not guarantee synchronization (the batches may still reach completion at different times). Dynamic time warping is more complicated, but it captures the dynamics, matches patterns for batches of different lengths, and has broad application. DTW, first used in speech recognition, was introduced for batch data analysis by Kassidas et al. in 1998 [49]. After that, the DTW method was successfully applied to process monitoring in industrial emulsion polymerization [49], spectroscopic profile analysis [50], semiconductor production monitoring [9], chromatographic data alignment [52], and other applications. Therefore, DTW is used in this chapter.

The core idea behind DTW is to align batches of different durations by determining the minimum accumulated path length between two trajectories.

Ramaker et al. [53] and Pravdova et al. [50] give clear explanations of how DTW works. We will use a simple example to demonstrate the algorithm, and later the method will be used on a bioreactor fermenter case. Suppose there are two trajectories $\mathbf{X}_1$ and $\mathbf{X}_2$ of different lengths and each trajectory has two variables:

$$\mathbf{X}_1 = \begin{bmatrix} 0 & 0.5 & 2 & 1.9 & 4 \\ 2 & 3 & 4 & 3.9 & 5 \end{bmatrix}$$

$$\mathbf{X}_2 = \begin{bmatrix} 0 & 0.1 & 0.5 & 1.9 & 4 & 4.1 \\ 1.9 & 2 & 3.1 & 4 & 5.5 & 5 \end{bmatrix}$$

$\mathbf{X}_1$ is a 2×5 matrix while $\mathbf{X}_2$ is 2×6; each row represents a process variable. Figure 8-3d shows the optimal warping path between the two trajectories. The line represents the

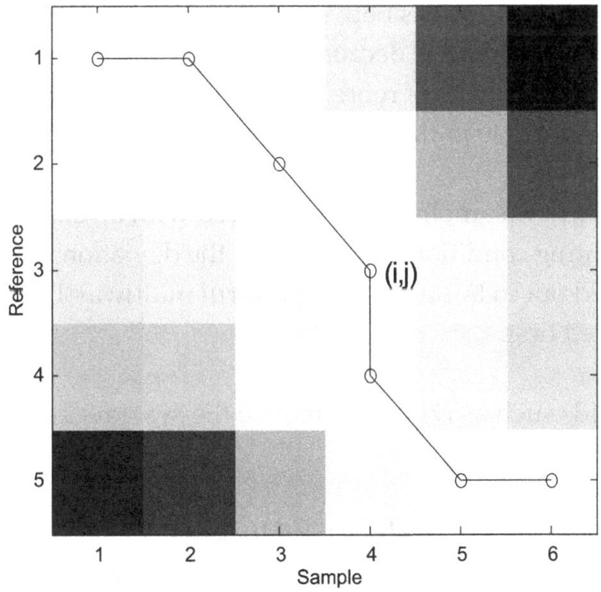

Figure 8-3d. Warping path by DTW.

optimal path, and the coordinates of small circles represent the synchronized point between two trajectories. For example, i = 4 and j = 3 means that the fourth point in the sample trajectory ($\mathbf{X}_2$ [1.9; 4]) should be synchronized with the third point in the reference trajectory ($\mathbf{X}_1$ [1.9; 3.9]), which appears to make sense in Figure 8-3d.

For implementing PCA off-line, DTW can be carried out as described earlier because all data is available. However, for online cases, no complete trajectory is available, so we must decide how to align current batch time with the reference trajectory. Kassidas et al. [49] suggest a way to do this aligning that is similar to the off-line case. This step is carried out whenever a new observation is taken; thus, the computational load could be higher. However, with the development of personal computers, the alignment is calculated very quickly compared to the interval between observations. For a detailed online algorithm of DTW, see reference [49].

8.4 Model-Based PCA

Model-based PCA (MB-PCA) was first developed by Wachs and Lewin [54] in 1998 with the aim of removing the process's dynamic behavior and nonlinearity without performing data synchronization. This method has been successfully applied to monitor many processes, including a batch reactor with cooling jacket [54], the Tennessee Eastman simulation of a plant-wide process [54], a three-stage ethylene compressor [11], and an N-type metal oxide semiconductor (NMOS) fabrication process [10].

Algorithm of MB-PCA

Though there are many similarities between multiway PCA and MB-PCA, in this section we focus on their differences. Because biological processes are generally run in a batch mode and a batch process is more complex than a continuous process, we discuss a batch process example in this section.

In multiway PCA, new sample data is compared with an existing data set that consists of normal operating condition trajectories. If the deviation is too large, an alarm is triggered. As pointed out in Section 8.3, to perform multiway PCA, all the trajectories must be synchronized first.

However, methods such as DTW will reduce the system's ability to detect faults as noted earlier.

In MB-PCA, a first-principle model is used to describe the nonlinearity and dynamics of normal operating conditions. The sample data is compared with the calculated data using a first-principle model, and PCA is performed on the residual between the

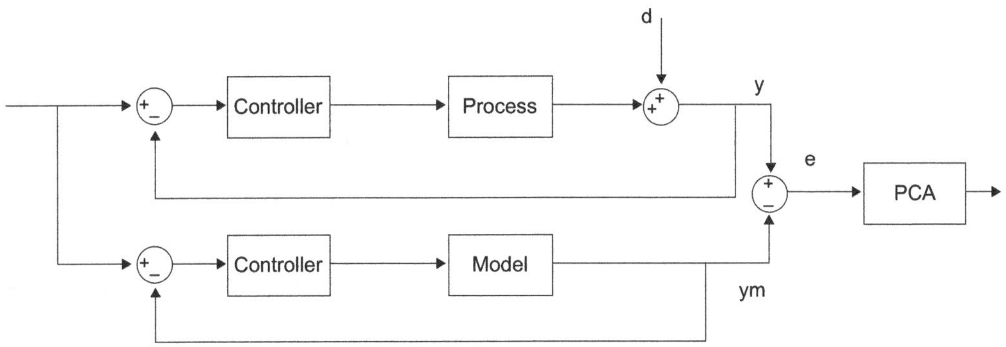

Figure 8-4a. MB-PCA flow scheme.

model predictions and the data (Figure 8-4a). If the first-principle model is perfect, no dynamics will be left in the residual data set.

As a result, the residual vectors of every batch can be adjusted to have the same length without having to perform complex calculations. Then, the residual data matrix is unfolded—for example, $E(I \times KJ)$—and scaled according to:

$$\tilde{e}_{ij} = (e_{ij} - e_{j,mean}) / \sigma(e_j) \tag{8-4}$$

Finally, PCA is performed on the unfolded data, as described in Section 8.3.

However, if the first-principle model is imperfect, the residuals will still include dynamic effects and can lead to incorrect monitoring results.

In summary, MB-PCA can be used when a first-principle model is accurate enough to describe the dynamics of a process. It can effectively enhance the performance of PCA [10, 11, 54]. However, for large-scale systems in which the dynamics may not be well known, the MB-PCA can lead to wrong decisions.

Super Model-Based PCA

As mentioned in Section 8.4, MB-PCA is useful when the process can be accurately described by a first-principle model. However, in an industrial environment this assumption is generally not realistic.

McPherson et al. [55] added a residual modeling stage to modify MB-PCA so the structural dynamics in the residuals would be removed.

Figure 8-4b is a schematic of super model-based PCA (SMB-PCA). The residual data (E matrix in MB-PCA) and output from the plant are considered in the error model.

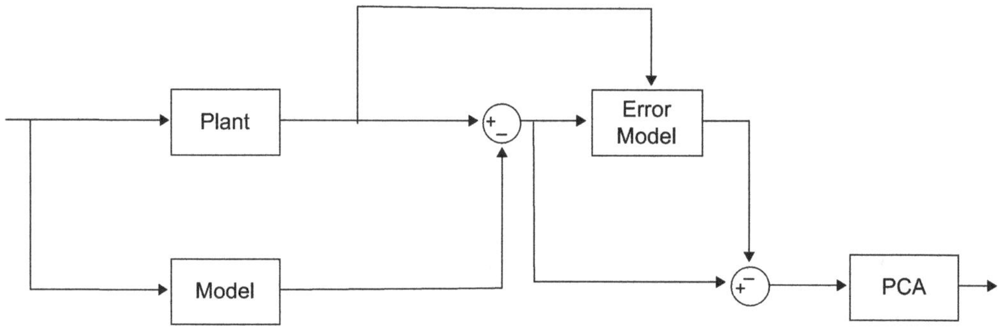

Figure 8-4b. SMB-PCA flow scheme.

The error model is used to predict the remaining structure in *E*, and the structural part is subtracted again from the residual matrix (*E*). Then PCA or other monitoring techniques are utilized. So, when using a batch process as an example, the main steps in SMB-PCA are as follows:

Procedure B: Building the PCA Model Steps

1. The values predicted by the model are subtracted from the raw data, and data preprocessing is performed, resulting in a residual matrix (*E*).
2. The structural part in the residual matrix is estimated using the error model.
3. Estimated error values are subtracted from the residual matrix to obtain an unstructured residual matrix ($\hat{E}$).
4. PCA is applied to extract information from $\hat{E}$.
5. UCLs of *SPE* and Hotelling's T^2 are obtained.

Procedure B: Monitoring

The data preprocessing tasks are the same as those in step A and in data projection. Score calculations are the same as those in multiway PCA.

Residual model building is a key step in SMB-PCA. McPherson et al. [26] tested five ways to model the structured residuals: partial least squares (PLS), auto-regressive with exogenous input (ARX), dynamic PLS, dynamic nonlinear PLS, and dynamic canonical correlation analysis (CCA). In McPherson et al.'s work, an exothermic batch reactor process (the same example that Wachs and Lewin used to develop MB-PCA) was monitored. The results suggested that all the dynamic error models are excellent at detecting faults, whereas the PLS with SMB-PCA does not perform well because the variables in the residual matrix mainly have dynamic nonlinear correlation.

Table 8-2d and the foregoing discussion show that PCA has many variations, and most of them have specific applications. Multiway PCA and MB-PCA are attractive for batch processes because they can be applied easily and offer good monitoring capability. Figure 8-4c shows a flowchart for choosing a monitoring technique for a batch system. Table 8-4 compares the shortcomings and advantages of each method and gives guidance on the choice of methods. For a simple system in which the physical model is exactly known, MB-PCA is recommended because the synchronization and unfolding steps can be skipped. For a large system for which physical information is sufficient but limited, SMB-PCA may perform better, offering a good balance between computational loading and fault detection ability.

Multiway PCA is a more universal method than the MB-PCA and SMB-PCA because it does not require a physical model but uses normal operation data; therefore, it can be implemented with relative ease and speed.

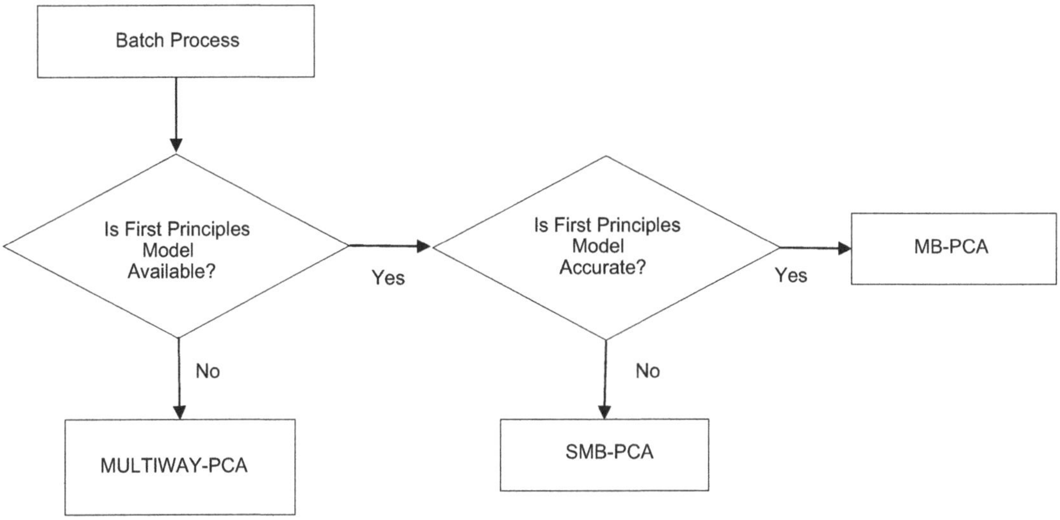

Figure 8-4c. Flowchart of PCA variation for batch process.

Table 8-4. Multiway PCA, MBPCA, and SMBPCA comparison.

	Multiway PCA	MB-PCA	SMB-PCA
Physical Model	Not needed	Accurate physical model information	Basic physical model information
Synchronization (such as DTW)	Suggested	Not needed	Not needed
Different Unfolding Method	Further treatment (multiple options)	Does not matter	Does not matter
Applicable System	Does not matter	Simple system	Does not matter

8.5 Fault Detection

Bioreactor Simulation

In this section, we use a detailed nonlinear dynamic model to simulate control of a batch penicillin fermentation process. The high-fidelity fermentation model used here is proposed by Birol et al. [56]. The simulated bioreactor volume is 1000 L, and the time scale is one thousand times faster than real time. Figure 8-5a shows one process

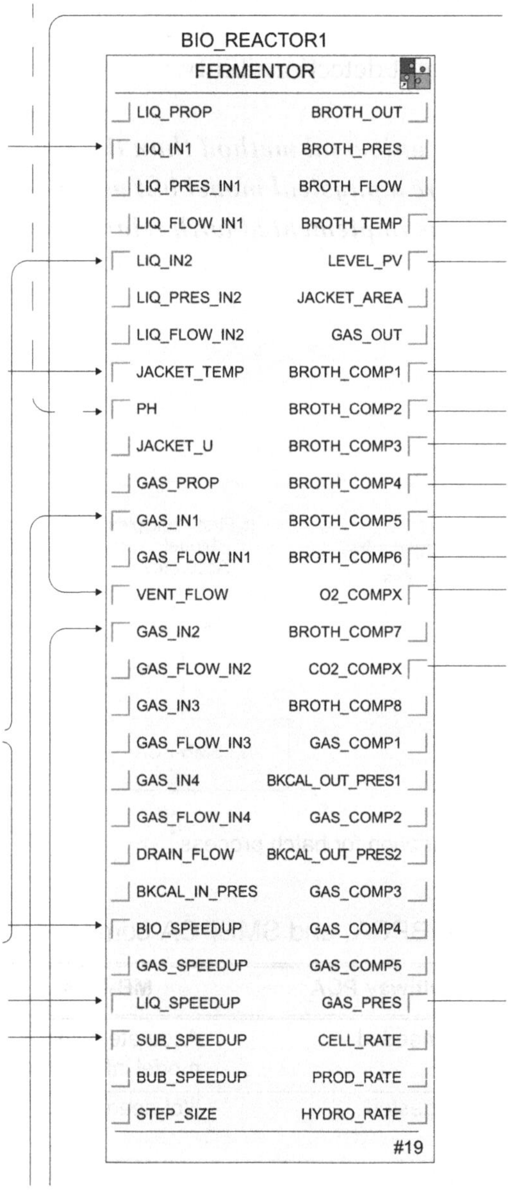

Figure 8-5a. Process module of bioreactor.

module of the system. Small variations are added to the simulation input variables to mimic real industrial plants. In this way, all the normal batches will fluctuate around a mean trajectory.

The simulated data can be directly read from the open platform communication (OPC) server, which is convenient for online monitoring and control. There is also a historical data server that can be used to save historical data for future use. Figure 8-5b shows the trends of some important process variables that change during the run. PID controllers are used to control dissolved oxygen (CO), substrate concentration, and broth pH.

PCA Application

Nine important process variables are used in PCA. These include process inputs such as base reagent flow (kg/s) and vent flow (kg/s); the process output variables include head pressure (kPa), biomass concentration (g/L), product concentration (g/L), biomass growth rate (kg/h), current product yield (% max), and batch time (h).

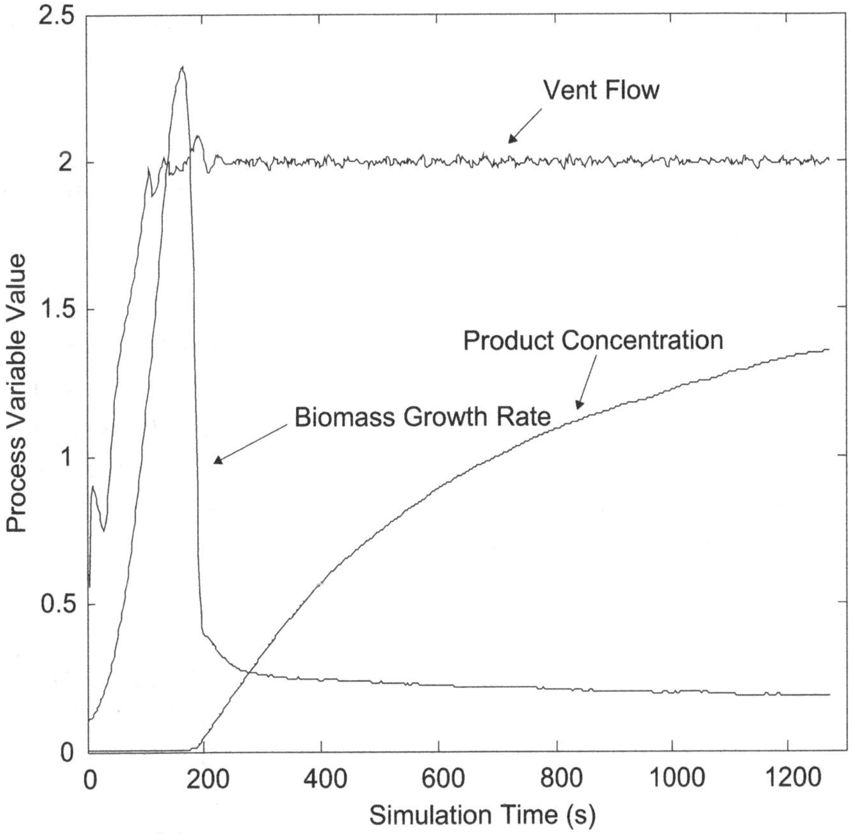

Figure 8-5b. Process variable value from process history view.

Which monitored process variables are selected for PCA is an important decision; the choice will depend on the user's experience in this field. You must monitor the right variables to identify abnormal behavior.

Twenty normal batch profiles generated by the simulator described earlier are used to build a PCA model and to calculate UCLs ($\alpha = 0.99$) of *SPE* and T^2 control charts. The multiway PCA calculation results are shown in this section, and both online and off-line tests are applied. For the off-line test, the batch-wise unfolding method is used. For the online case, both variable-wise and hybrid-wise unfolding methods are applied. Related algorithms are listed in Section 8.3 of this chapter [43]. DTW is used to synchronize different batch lengths [51]. For a similar process, four PCs were retained in the PCA model [43]. In our case, for an off-line test, six PCs are retained, capturing 54% of the variation. Two PCs are used for online testing with variable-wise unfolding, which retains 84.5% of the total variation. For the online test with the hybrid-wise unfolding case, two PCs capture 78.9% of the information.

For our simulation, if the batches are relatively well synchronized, or if changes in batch completion time are a symptom to diagnose batch problems, the DTW step can be bypassed.

Two types of faults are tested by multiway PCA: (1) pH sensor failure at time zero and (2) initial biomass seed overload (from 1 kg to 1.5 kg). To compare and test the PCA approach, a normal batch was also tested.

With respect to diagnosing abnormal batches, Table 8-5 lists the off-line test results, and Figure 8-5c shows the off-line *SPE* control chart of normal and abnormal cases. They illustrate that PCA treats the normal batch data as a good batch, and both faults are detected successfully. As mentioned in previous sections, *SPE* is focusing on the residual space, which measures the deviation between good batches (those used to build PCA model) and tested ones. The results show that with six PCs, the residual space can correctly distinguish a faulty batch from good ones.

Table 8-5. PCA off-line test results.

	T^2	SPE	T^2 99% UCL	SPE 99% UCL
Normal batch	0.09	0.67	23.8	0.74
pH sensor failure	1.63	3.08*	23.8	0.74
Initial biomass seed overload	0.40	1.34*	23.8	0.74

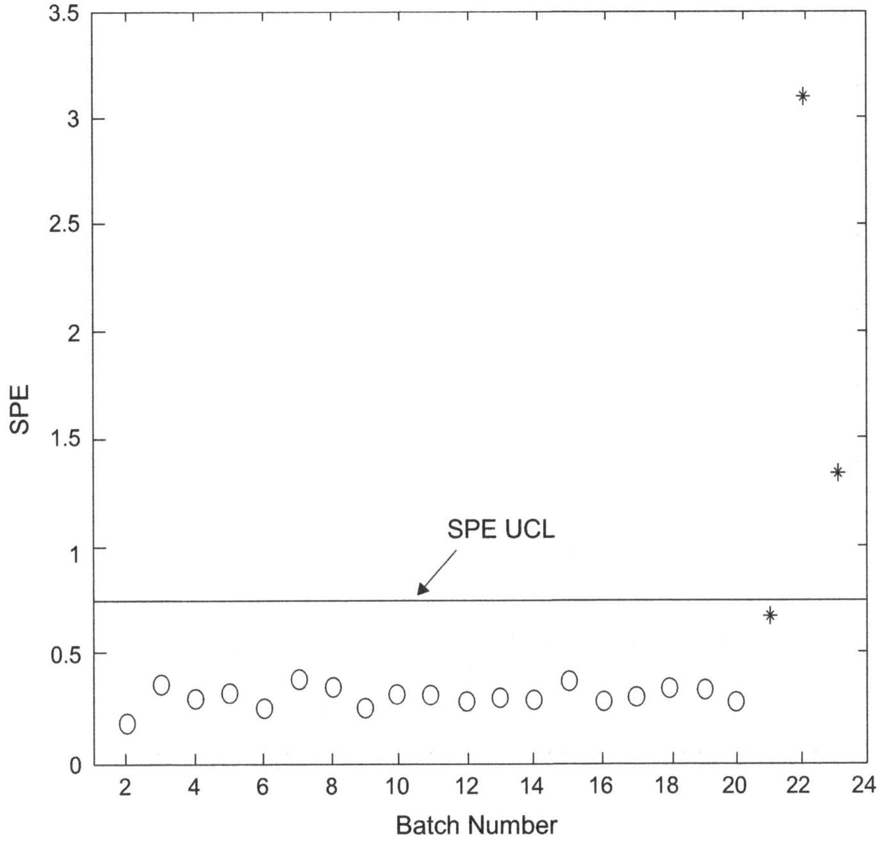

Figure 8-5c. *SPE* off-line control chart.

Note: In this chart, the open symbols represent batches used in the PCA model, and the asterisks (*) represent batches tested for abnormality.

As mentioned in Section 8.3, off-line PCA is typically used to check the quality of the final product and is performed at the end of every batch. To achieve the purpose of monitoring, online PCA should be applied. In this section, variable-wise and hybrid-wise unfolding methods [43] are used. An important feature of these methods is that they do not need to estimate unfinished batch profiles because only current data is used (Section 8.3).

Figures 8-5d to 8-5i show the control charts for the online monitoring of T^2. Both approaches classify normal batches correctly. In the pH sensor failure test, for both methods, the T^2 scores exceed the UCL at an early stage, which means that the fault is detected rapidly. However, for the case of initial biomass overload, the two methods give different answers (Figures 8-5f and 8-5i). The variable-wise unfolding results

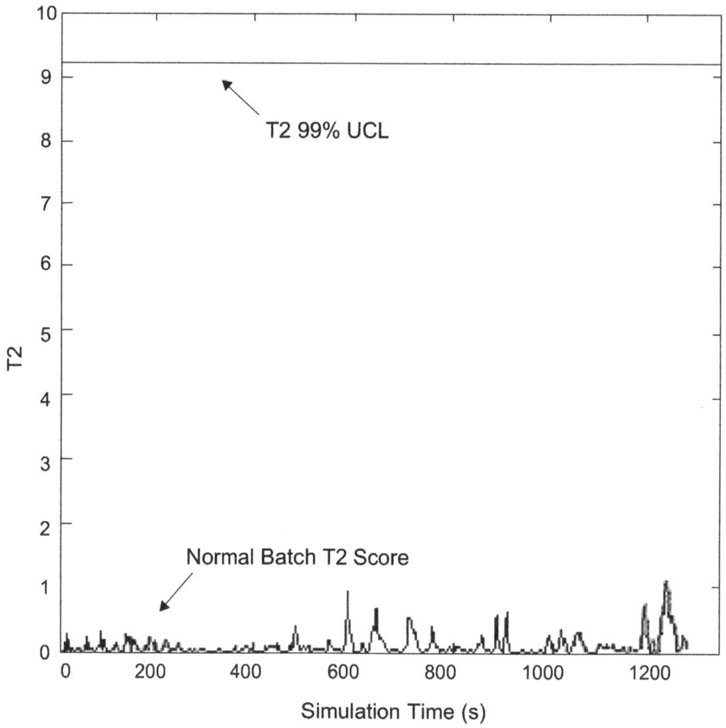

Figure 8-5d. T^2 score of another normal batch with variable-wise unfolding.

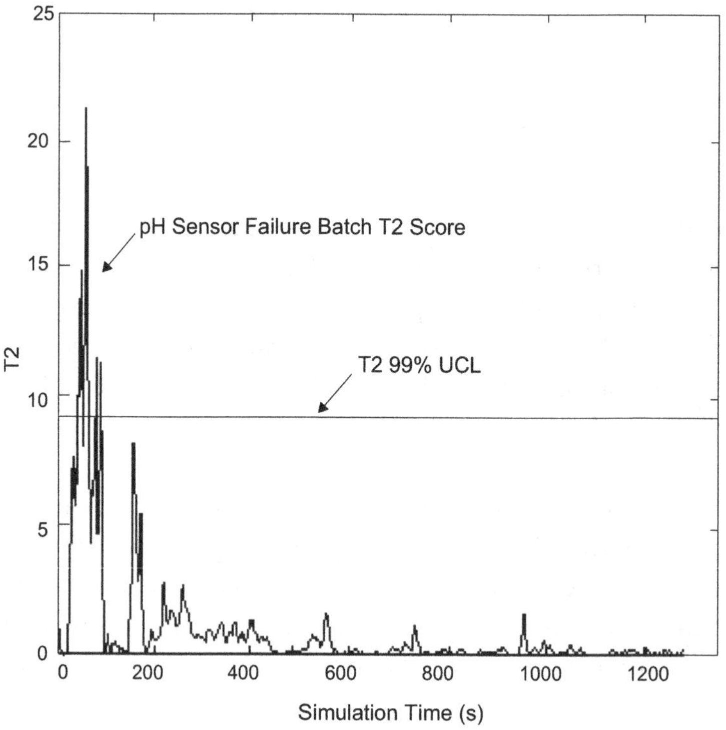

Figure 8-5e. T^2 score of pH sensor failure batch with variable-wise unfolding.

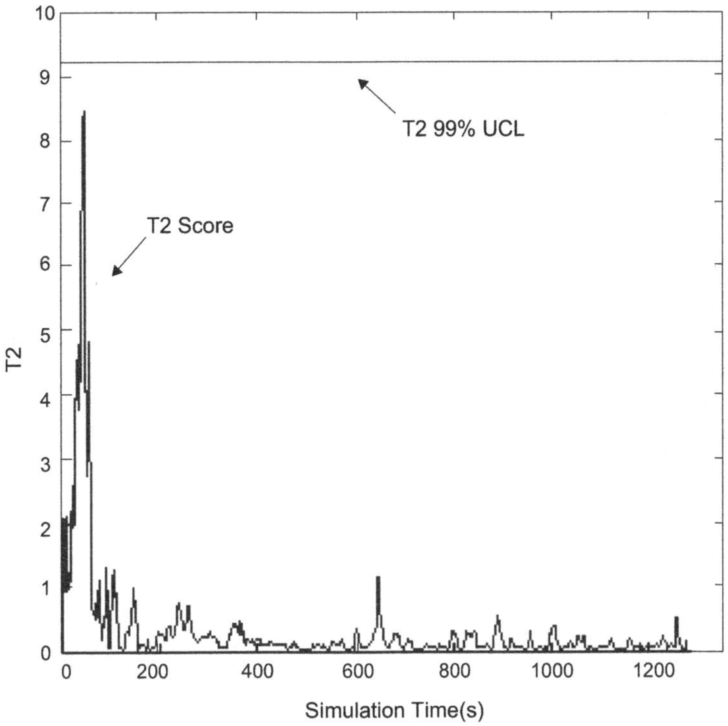

Figure 8-5f. T^2 score of initial biomass overload batch with variable-wise unfolding.

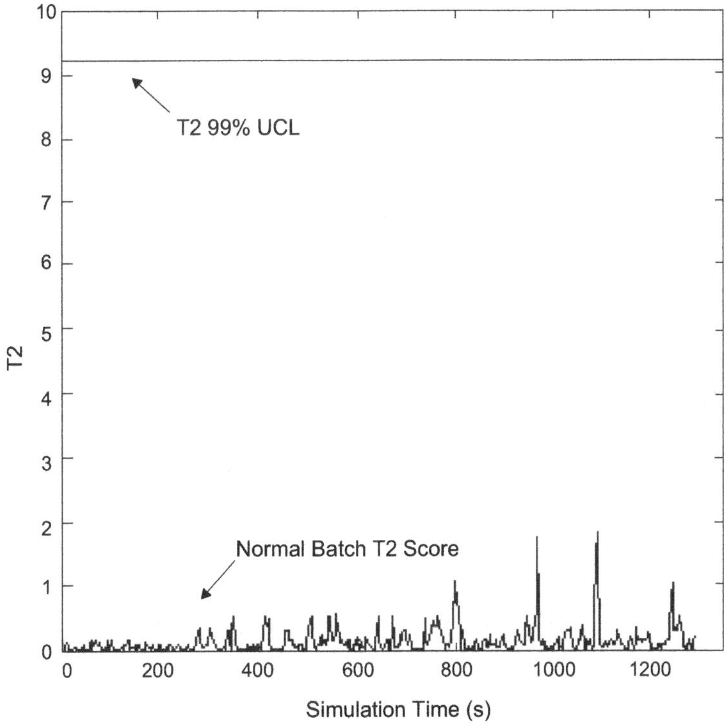

Figure 8-5g. T^2 score of another normal batch with hybrid unfolding.

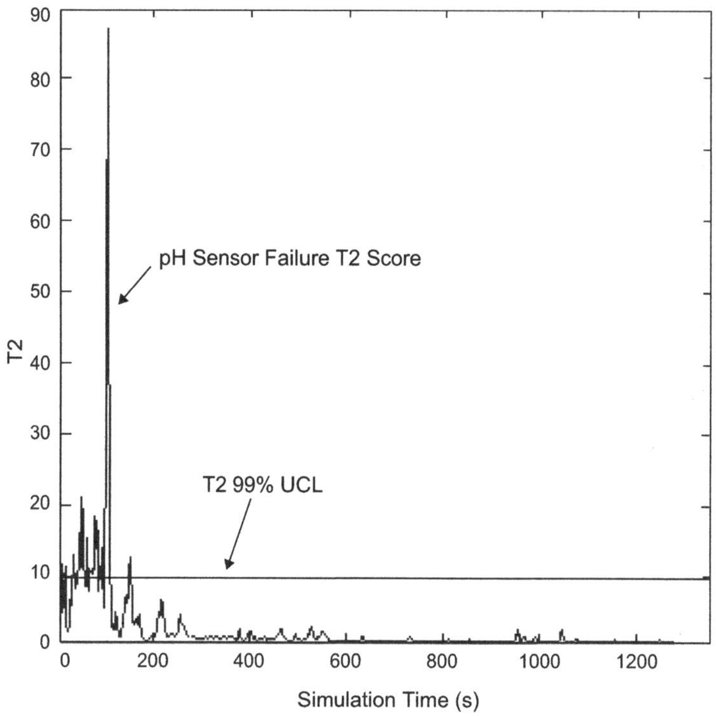

Figure 8-5h. T^2 score of pH sensor failure batch with hybrid unfolding.

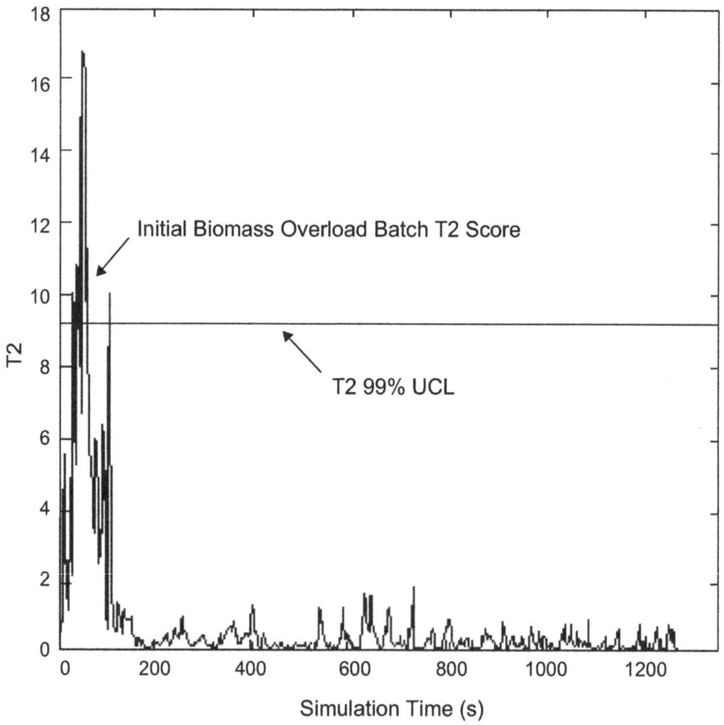

Figure 8-5i. T^2 score of initial biomass overload batch with hybrid unfolding.

indicate that the batch is good, but the hybrid-wise unfolding shows that the batch has abnormality at the beginning. In Section 8.3, we pointed out that variable-wise unfolding cannot completely remove system dynamics from a raw data set, and this fact will influence the performance of PCA, as the example demonstrates.

Hybrid-wise unfolding works better than variable-wise unfolding because when performing mean-centering, the process dynamic is subtracted.

In off-line monitoring, only one value is calculated for every batch (Figure 8-5c), which indicates the quality of the product. In real applications, a UCL is calculated to determine when the product should be discarded. For online monitoring, current process variable values are measured at regular time intervals and analyzed by PCA or other MSPC methods. If the computed results exceed the corresponding UCL, then further fault diagnosis and analysis should be performed to isolate and correct the process problem. Online PCA can detect abnormalities early in the batch process, normally giving you enough time to take remedial action that is critical as a bad batch, especially for new biologics, can mean a loss of millions of dollars and weeks of production.

8.6 Data Analytics Best Practices

The success of data analytic projects, as with any engineering endeavor, depends greatly on the steps taken in applying the analytic technology. To address this application, a multidiscipline team should be formed that includes the toolset provider, as well as experts from plant operations, plant maintenance, information technology (e.g., data scientists), and process and automation system engineering. Best practices are summarized below [57]:

1. Use a multidiscipline team that includes plant operations. If the models will be deployed online, engage the operators early in the cycle.
2. Work with lead operators and engineers to select key inputs and outputs but be open to including any inputs (especially utilities associated with a unit operation) to help identify unsuspected causes of poor process performance.
3. Enter feedstock and recycle stream data if material properties are required.
4. Collect lab data on key compositions and quality parameters.
5. Survey instrumentation to improve rangeability and repeatability and eliminate deadband and stiction. Identify loop dynamics and tune controllers to meet objectives.

6. Ensure lab results, that are time-stamped as to when the sample was taken, are inputs to the online analytic toolset. Address the common situation where lab results are communicated to the operator by phone or through a lab terminal that is not tied to the control system.

7. Realize changes in the incoming raw material properties may directly impact batch operation and quality parameters and must be inputs to the online analytic toolset. Be aware that properties associated with each material shipment provided by the supplier may only be available as data from supplier data provided to purchasing.

8. Include controller outputs in the data. Tight control by process loops transfers most of the variability from the controlled variable to the manipulated variable (e.g., PID output).

9. Work with lead operators and engineers to understand the dynamics in the process and instrumentation system and the direction key parameters should take in response to process changes and conditions.

10. For continuous processes, align data by dynamic compensation of process inputs. For plug flow (e.g., pipelines, extruders, sheet lines, and many gas unit operations), include a time delay on each process input that matches the response time of a key process output. For back mixed liquid volumes, set a time delay equal to the total dead time and a time lag equal to the process time constant on each input.

11. Apply data analytics to each important unit operation using a cascade approach of one unit operation's model outputs affecting a downstream unit operation's model inputs to see the effect on the whole process. Avoid trying to develop a single model for inputs near the beginning of a process and outputs near the end of a process.

12. Limit the number of PCs to two to three if possible based on the percentage of variation that each PC contributes. Select those with the higher contribution and whose accumulated contribution adds up 75% or more of the total variation. Add more intelligent measurements if necessary. Do not exceed five PCs.

13. Start with a simple first pass at the model and test it off-line. If the model does not capture 70% of the variation, then additional parameters may be required. It may also be possible that the data analytics will not work for the selected problem.

14. Work with the tools before engaging the team. It is important to understand the limitations of the tools and the support available for the tools.

15. Drill down into the PCs, eliminating those with little contribution and evaluating those with large contribution to understand and verify cause and effect.

16. Once models are developed that appear to work, run them in the engineering lab or some other location that does not directly impact the operators.

17. Look for false indications and missed indications. Ensure changes in predicted outputs are smooth with proper sign. The models may have to be tuned before deploying them.

18. Include a way to detune or disable the models online. If the process changes significantly, measurements change, or material properties change substantially, then the models may have to be regenerated.

19. Disable compression in the historian on key parameters used in the data analytics.

20. Evaluate gradual feedback batch end point predictions by a bias that is a fraction of the difference between the prediction and the at-line or lab analyzer result for the batch.

21. Use a digital twin to conduct a design of experiments to expand the range of conditions that can be used as inputs, and to test and improve the ability to diagnose and predict key continuous and batch compositions for a wide range of operation.

22. For biological processes, restrict tests to focus on the effect of a particular process variable (e.g., glucose concentration) while the other process variables are held constant at their optimum to deal with the pervasive multiplicative effect of cell kinetics [58].

References

1. Theodora Kourti, "The Process Analytical Technology Initiative and Multivariate Process Analysis, Monitoring and Control," *Analytical and Bioanalytical Chemistry* 384 (2006): 1043–48.

2. Leo H. Chiang, Evan L. Russell, and Richard D. Braatz, *Fault Detection and Diagnosis in Industrial Systems* (New York: Springer-Verlag, 2001).

3. John F. MacGregor and Theodora Kourti, "Statistical Process Control of Multivariate Processes," in *Control Engineering Practice* 3 (1995): 403–14.

4. Jong-Min Lee, "Statistical Process Monitoring Based on Independent Component Analysis and Multivariate Statistical Methods" (PhD diss., Department of Chemical Engineering, Pohang University of Science and Technology, Pohang, South Korea, 2004).

5. Robert L. Mason and John C. Young, *Multivariate Statistical Process Control with Industrial Applications* (Philadelphia: ASA-SIAM, 2002).

6. Sergio Valle-Cervantes, "Plant-Wide Monitoring of Processes Under Closed-Loop Control" (PhD diss., Chemical Engineering Department, University of Texas at Austin, Austin, TX, 2001).

7. S. Joe Qin, "Statistical Process Monitoring: Basics and Beyond," *Journal of Chemometrics* 17, no. 8–9 (2003): 480–502.

8. Agnar Hokuldsson, "PLS Regression Methods," *Journal of Chemometrics* 2, no. 3 (1988): 211–28.

9. Gregory A. Cherry and S. Joe Qin, "Batch Synchronization and Monitoring Using Data Interpolation and Dynamic Time Warping with Principal Component Analysis" (presentation at TWMCC meeting, Austin, TX, 2005).

10. Sivan Lachman-Shalem, Nir Haimovitch, Eitan N. Shauly, and Daniel R. Lewin, "MBPCA Application for Fault Detection in NMOS Fabrication," *IEEE Transactions on Semiconductor Manufacturing* 15, no. 1 (2002): 60–70.

11. Y. Rotem, A. Wachs, and D. R. Lewin., "Ethylene Compressor Monitoring Using Model-Based PCA," *AIChE Journal* 46, no. 9 (2000): 1825–36.

12. Theodora Kourti, "Monitor," *Chemometrics and Intelligent Laboratory Systems* 76 (2005): 215–20.

13. Ivan Miletic, Shannon Quinn, Michael Dudzic, Vit Vaculik, and Marc Champagne, "An Industrial Perspective on Implementing Online Applications of Multivariate Statistics," *Journal of Process Control* 14, no. 8 (2004): 821–36.

14. Theodora Kourti, Jennifer Lee, and John F. MacGregor, "Experiences with Industrial Applications of Projection Methods for Multivariates Statistical Process Control," *Computers and Chemical Engineering* 20, supplement 1 (1996): S745–S750.

15. Sarolta Albert and Robert D. Kinley, "Multivariate Statistical Monitoring of Batch Processes: An Industrial Case Study of Fermentation Supervision," *Trends in Biotechnology* 19, no. 2 (2001): 53–62.

16. B. Lennox, G. A. Montague, H. G. Hiden, G. Kornfeld, and P. R. Goulding, "Process Monitoring of an Industrial Fed-Batch Fermentation," *Biotechnology and Bioengineering* 74, no. 2 (2001): 125–35.

17. J. A. Lopes, J. C. Menezes, J. A. Westerhuis, and A. K. Smilde, "Multiblock PLS Analysis of an Industrial Pharmaceutical Process," *Biotechnology and Bioengineering* 80, no. 4 (2002): 419–27.

18. Leo H. Chiang, Riccardo Leardi, Randy J. Pell, and Mary Beth Seasholtz, "Industrial Experiences with Multivariate Statistical Analysis of Batch Process Data," *Chemometrics and Intelligent Laboratory Systems* 81, no. 2 (2005): 109–19.

19. Ian T. Jolliffe, *Principal Component Analysis*, 2nd ed. (New York: Springer-Verlag, 2002).

20. J. Edward Jackson, *A User's Guide to Principal Components* (New York: John Wiley, 2005): 8–21.

21. Paul Nomikos and John F. MacGregor, "Multivariate SPC Charts for Monitoring Batch Processes," *Technometrics* 37, no. 1 (1995): 41–59.

22. Gilbert Strang, *Linear Algebra and Its Applications*, 2nd ed. (New York: Academic Press, 1980).

23. Paul Geladi and Bruce R. Kowalski, "Partial Least Squares Regression: A Tutorial," *Analytica Chimica Acta* 185 (1986): 1–17.

24. Sergio Valle, Weihua Li, and S. Joe Qin, "Selection of the Number of Principal Components: The Variance of the Reconstruction Error Criterion with a Comparison to Other Methods," *Industrial & Engineering Chemistry Research* 38, no. 11 (1999): 4389–4401.

25. Barry M. Wise, and Neal B. Gallagher, "The Process Chemometrics Approach to Process Monitoring and Fault Detection," *Journal of Process Control* 6, no. 6 (1996): 329–348.

26. J. Edward Jackson and Govind S. Mudholkar, "Control Procedures for Residuals Associated with Principal Component Analysis," *Technometrics* 21, no. 3 (1979): 341–49.

27. H. Henry Yue, and S. Joe Qin, "Reconstruction-Based Fault Identification Using a Combined Index," *Industrial & Engineering Chemistry Research* 40, no. 20 (2001): 4403–4414.

28. J. V. Kresta, J. F. MacGregor, and T. E. Marlin, "Multivariate Statistical Monitoring of Process Operating Performance," *Canadian Journal of Chemical Engineering* 69, no. 1 (1991): 35–47.

29. Weihua Li, H. Henry Yue, Sergio Valle-Cervantes, and S. Joe Qin, "Recursive PCA for Adaptive Process Monitoring," *Journal of Process Control* 10, no. 5 (2000): 471–86.

30. Wenfu Ku, Robert H. Storer, and Christos Georgakis, "Disturbance Detection and Isolation by Dynamic Principal Component Analysis," *Chemometrics and Intelligent Laboratory Systems* 30, no. 1 (1995): 179–96.

31. Evan L. Russell, Leo H. Chiang, and Richard D. Braatz, "Fault Detection in Industrial Processes Using Canonical Variate Analysis and Dynamic Principal Component Analysis," *Chemometrics and Intelligent Laboratory Systems* 51, no. 1 (2000): 81–93.

32. D. Dong and Thomas J. McAvoy, "Nonlinear Principal Component Analysis—Based on Principal Curves and Neural Networks," *Computers and Chemical Engineering* 20, no. 1 (1996): 65–78.

33. John F. MacGregor, Christiane Jaeckle, Costas Kiparissides, and M. Koutoudi, "Process Monitoring and Diagnosis by Multiblock PLS Methods," *AIChE Journal* 40, no. 5 (1994): 826–38.

34. Bhavik R. Bakshi, "Multiscale PCA with Application to Multivariate Statistical Process Monitoring," *AIChE Journal* 44, no. 7 (1998): 1596–1610.

35. Paul Nomikos and John F. MacGregor, "Monitoring of Batch Processes Using Multiway Principal Component Analysis," *AIChE Journal* 40, no. 8 (1994): 1361–75.

36. Paul Nomikos and John F. MacGregor, "Multi-way Partial Least Squares in Monitoring Batch Processes," *Chemometrics and Intelligent Laboratory Systems* 30, no. 1 (1995): 97–108.

37. Yang Zhang and Thomas F. Edgar, "Application of MB-PCA and Golden Batch to Bioreactor Monitoring and Control" (presentation at TWMCC meeting, Austin, TX, 2006).

38. Michael A. Boudreau, Gregory K. McMillan, and Grant Wilson, "Maximizing PAT Benefits from Bioprocess Modeling and Control," *Pharmaceutical Technology* (November 2006).

39. David Emerson, "Production Performance Ratings Measure Effectiveness of Batches" (presented at ISA's online Technical Information and Communities, 2004).

40. Theodora Kourti, "Multivariate Dynamic Data Modeling for Analysis and Statistical Process Control of Batch Processes, Start-ups and Grade Transitions," *Journal of Chemometrics* 17, no. 1 (2003): 93–109.

41. Eric N. M. van Sprang, Henk-Jan Ramaker, Johan A. Westerhuis, Stephen P. Gurden, and Age K. Smilde, "Critical Evaluation of Approaches for On-Line Batch Process Monitoring," *Chemical Engineering Science* 57, no. 18 (2002): 3979–91.

42. Jong-min Lee, Chang Kyoo Yoo, and In-Beum Lee, "Enhanced Process Monitoring of Fed-Batch Penicillin Cultivation Using Time-Varying and Multivariate Statistical Analysis," *Journal of Biotechnology* 110, no. 2 (2004): 119–36.

43. Johan A. Westerhuis, Theodora Kourti, and John F. MacGregor, "Comparing Alternative Approaches for Multivariate Statistical Analysis of Batch Process Data," *Journal of Chemometrics* 13, no. 3–4 (1999): 397–413.

44. Svante Wold, Nouna Kettaneh, Hakan Friden, and Andrea Holmberg, "Modelling and Diagnostics of Batch Processes and Analogous Kinetic Experiments," *Chemometrics and Intelligent Laboratory Systems* 44, no. 1–2 (1998): 331–40.

45. Age K. Smilde and Durk A. Doornbos, "Three-Way Methods for the Calibration of Chromatographic Systems: Comparing PARAFAC and Three-Way PLS," *Journal of Chemometrics* 5, no. 4 (1991): 345–60.

46. Rasmus Bro, "PARAFAC, Tutorial and Applications," *Chemometrics and Intelligent Laboratory System* 38 (1997): 149–71.

47. Age K. Smilde, "Three-Way Analysis: Problems and Prospects," *Chemometrics and Intelligent Laboratory Systems* 15, no. 2–3 (1992): 143–57.

48. Athanassios Kassidas, John F. MacGregor, and Paul A. Taylor, "Synchronization of Batch Trajectories Using Dynamic Time Warping," *AIChE Journal* 44, no. 4 (1998): 864–75.

49. V. Pravdova, B. Walczak, and D. L. Massart, "A Comparison of Two Algorithms for Warping of Analytical Signals," *Analytica Chimica Acta* 456 (2002): 77–92.

50. Theodora Kourti, Jennifer Lee, and John F. Macgregor, "Experiences with Industrial Applications of Projection Methods for Multivariate Statistical Process Control," *Computers and Chemical Engineering* 20, supplement 1 (1996): 745–50.

51. Giorgio Tomasi, Frans van den Berg, and Claus Andersson, "Correlation Optimized Warping and Dynamic Time Warping as Preprocessing Methods for Chromatographic Data," *Journal of Chemometrics* 18, no. 5 (2004): 231–41.

52. Henk-Jan Ramaker, Eric N. M. van Sprang, John A. Westerhuis, and Age K. Smilde, "Dynamic Time Warping of Spectroscopic BATCH Data," *Analytica Chimica Acta* 498, no. 1–2 (2003): 133–53.

53. A. Wachs and D. R. Lewin, "Process Monitoring Using Model-Based PCA," in *Proceedings of IFAC Symposium on Dynamics and Control of Process Systems* Corfu, Greece (1998): 86–92.

54. Lindesay McPherson, Julian Morris, and Elaine Martin, "Super Model-Based Techniques for Batch Performance Monitoring" (presented at the 12th European Symposium on Computer Aided Process Engineering, 2002).

55. Gulnur Birol, Cenk Undey, and Ali Cinar, "A Modular Simulation Package for Fed-Batch Fermentation: Penicillin Production," *Computers and Chemical Engineering* 26, no. 11 (2002): 1553–65.

56. Gregory K. McMillan and P. Hunter Vegas, *Industrial/Process Instruments and Controls Handbook,* 6th ed. (New York: McGraw-Hill, 2019).

57. Gregory K. McMillan, "At the IIoT Crossroads," *Control* (February 2019).

9 Models to Improve Operator, Automation, and Process Performance

9.1 Introduction

Dynamic simulators became the key to success of digital systems starting more than 30 years ago in the first migrations from analog control. These simulators simplify checkout and development of increasingly complicated displays, alarms, and control systems. Moreover, a working simulator created in lockstep with control development affords an efficient transition to operator training and provides future opportunities for experimentation to improve process efficiency. Large process manufacturers have realized the benefit of off-line dynamic simulations to improve processes and control systems. However, these off-line simulations required the writing of code for the functional duplication of controls, severely limiting the scope and reality of the plant and the automation system being investigated.

Today, first-principle dynamic simulations for gaining process knowledge and improving product quality, process capacity, and efficiency can become one with the simulations for operator training and system checkout in the digital twin. Advancements in packaged commercial simulation software enable users to focus on the heart of the simulation rather than writing and rewriting code for model [10, 19].

First-principle dynamic simulation has become more accessible than ever and offers deep insight into process behavior and exciting avenues for building the understanding of operators and control engineers.

Problems with traditional integrated control and safety systems (ICSS) testing and commissioning are all too familiar. Plant operators commonly face delays in start-up and normal operation due to ICSS configuration and graphics issues, lack of trained operators, and last-minute changes required in control strategies and operating procedures. A digital twin solution with project milestones aligned to that of plant commissioning and start-up, especially in the case of greenfield projects, enables one to leverage the tremendous benefits of dynamic simulation during commissioning, start-up, and across the plant life cycle. It is important to perform adequate planning on system maintenance considerations in the beginning of a digital twin project. The maintenance focuses on equalizing the off-line control system and plant model with any significant changes in the real plant.

A properly maintained digital twin provides an off-line, safe, and reliable system to support a variety of phases of the plant life cycle (including conducting ICSS factory acceptance test (FAT), enabling virtual commissioning, reviewing operating procedures, and training operators well ahead of plant commissioning and start-up).

A true digital twin uses imports of control modules and exact copies of the operator interface files, providing a seamless experience for trainees and engineers.

The digital twin simulation can start out by providing the pressure, temperature, and level response by traditional first-principle models. The opportunity exists to match oxygen, reagent, glucose, and nutrient demand and cell growth and product formation by investment in a model that includes a charge balance, kinetics, and mass transfer dynamics as detailed in Chapter 6. The model can be used for batch and continuous (e.g., perfusion and purification) processes. Recent advances enable the builder of the model to match profiles without the use of proprietary process research and development (R&D) knowledge. Plus, if R&D becomes involved in the use of digital twin, the knowledge gained can be used to improve its kinetics securely. Furthermore, the model can be used for the development of data analytics projection to latent structures (PLS) models and inferential measurements as noted in Chapter 5.

The additional investment in models with the ability to match profiles can be less than expected using the equations in Chapter 6 and the techniques detailed in this chapter. The benefits can be greater than expected by leveraging the synergy of everyone in R&D, process, operations, maintenance, automation, and information technology, enabling a deeper understanding of what is important to improve profitability [6].

This chapter shows how to achieve the synergy gained by expanding operator training systems to match plant profiles without the need for proprietary R&D data.

Learning Objectives

- Understand the possible scope and possible benefits of simulations
- Learn to identify data and specify dynamic models in bioprocesses
- Appreciate the power of simulation as a tool for generating deep insight
- Be aware of the many factors impacting performance in bioprocesses
- Become familiar with models available for creating a digital twin
- Recognize the aspects of the modeling process, equipment, and instrumentation
- Understand how to realize performance improvements through application of a digital twin

9.2 Overview

Real-time dynamic simulator modeling provides a wellspring for creativity as well as unparalleled benefits in the realization of robust control strategies and operator training.

A true digital twin simulates a close approximation of the real response of physical plants by providing ab initio—first principles—modeling.

Although there are many sophisticated models (encompassing the wide world of diverse thermodynamic and physical effects), a simple model often provides a workbench from which to not only craft elegant control strategies but also create impactful understanding. Dynamic simulators enable off-line (but real) control systems to interact transparently with a close approximation of the real world. This requires not only a dynamic simulation but also a dynamic simulation that can run in real time (supporting update rates of a minimum of 1 s and preferably 500 ms). This is critical for leveraging this potential. With such a dynamic and real-time model in hand, engineers may test control strategies that could impact millions of dollars' worth of process materials and utilities digitally, proactively testing strategies in an environment where the only limitation is time. Such models also enable fledgling operators to react to uncommon conditions—such as start-up and shutdown—in a low-risk training environment without risking damage to actual productivity or equipment.

9.3 Bountiful Boundaries

The model's fidelity and effectiveness correlate almost directly to the amount of data the user may provide as well as the thoughtfulness with which boundaries are drawn. The very minimum data needed to get started with a model is the set points and operating limits for pH, temperature, and dissolved oxygen (DO) and the profiles of the reagent, oxygen, utility, and glucose flows [1–3, 5]. Assuming good control, model fidelity is seen in the match of the model to process manipulated flows throughout the batch. Turbidity, mass spectrometers, and dielectric spectroscopy measurements can enable much better modeling and add the ability to infer cell lysis and product formation rates [14–18].

Often, the simulation of a single unit operation generates significant insight before doing simulations for large swaths of the physical plant.

A bioreactor might serve as a powerful case for a simple unit operation study, whereas optimization of utility steam consumption may require simulation of a complex network of units and pipes. The key insight is always one of boundaries.

To gain deeper knowledge of complex utilities and coupled unit operations, simulations beyond each unit operation in isolation are expanded to reveal the interplay of operations together.

A boundary (as in a differential equation) marks the edge of where the interesting dynamics begin and where disturbances occur. Contrary to many expectations, the inputs to a process are often where disturbances (upsets) enter the dynamics of a system. The first step in constructing a digital twin is thoughtful identification of the boundaries.

In the example of a bioreactor, the boundaries could be drawn at the reagent feeds. A bioreactor running in a batch mode's state (a vector composed of variables such as turbidity, product production, and pH) enables simple modeling by drawing boundaries at the feeds and draws of the system. In the case of pH neutralization, three tanks in series provide agreeable dynamics for control. Trying to construct a model with boundaries drawn between the tanks (isolating them from one another) gives an incomplete picture of the true process dynamics. This short-circuits insight at best and provides an unrealistic simulation at worst. The key is to draw the boundaries around elements that are both coupled and consequential.

9.4 Data Drives Dynamics

Boundaries alone color a large portion of a simulation's response; however, even the most thoughtfully chosen boundaries give woefully inadequate results if they are constructed without good data.

The quality of the model depends directly on the quality of the data used in the configuration of the model.

Using good data as input leads to a better model. After selecting boundaries, simulation engineers must collect the data that govern the physics of the problem. In a perfect world, key operation parameters are historized and are directly relevant for simulation. Although that is not always the case, these parameters show only the response. To understand the generation of that response, the physical characteristics of the system must be taken into consideration. The model configurations and tuning parameters follow plant design documents such as piping and instrumentation diagrams (P&IDs), heat and material balance (H&MB) diagrams, piping isometrics, and equipment and device process data sheets. In the absence of this data, models may be based on plant operating data such as snapshots of control system graphics and historians. This is often the case if there is significant deviation in plant operation from design or non-steady state operation seen in batch processes. We will consider the physical parameters for vessels, valves, compressors, and pumps. To begin, consider the simple case of a vessel.

Most of the dynamics of a process system are the interplay of lag and dead time. In the case of a vessel, these are generally functions of agitation and—more importantly—physical dimensions: height, radius, and dimensions of the bottom and top of the vessel. Many simulators do not natively consider residence time distributions or imperfect mixing. However, the flow rates from the tanks are impacted by the hydrostatic head, which directly correlates to vessel dimensions.

A critical facet of system dynamics that often is overlooked resides within the dynamics of the valves in the system.

Valve dynamics depend chiefly on resolution, backlash, and slew rate. Resolution dictates the coarseness of the valve response. The backlash accounts for the deadband of the actuator. The slew rate governs how fast a control valve can respond. (More details on modeling this response are given in Section 9.10.) Frequently, these effects are inaccurately labeled *hysteresis*. Hysteresis fails to accurately encompass the dynamics of the valve. A frictionless valve that can instantly respond and reach the exact actual position is an ideal, never a reality. Innumerable valves with their own dynamic characteristics play together to exert a profound impact on the response of the field. Accounting for these characteristics is critical for showing the proper system response.

Pumps are the prime movers of the digital plant. The most critical piece of information for simulating a pump's response is the pump curve. The pump curve's intersection with the resistance curve of the piping shows the operation point. The actual pump curve will provide the response at a variety of operating conditions and respond appropriately. The pump curve's response is inextricably tied to the piping specification and liquid properties. A perfect pump curve paired with an inaccurate capture of the resistance will provide skewed results. However, in the absence of a curve, a single operation point can be used. It is important to keep in mind that for a system of flow equations to be tractable, the curves must be monotonic (ever increasing or decreasing) in the first quadrant of the coordinate system.

Compressors serve as movers for gas phase streams. Configuration of a compressor requires entry of the compressor map and updates of the current compressor speed. Compressor behavior can become complex by comparison with the introduction of surge, and simulating surge requires identification of data to the left of the surge line [20].

9.5 Digitizing the Plant for Process Performance

The simulation process requires a methodical inventory of the unit operations in the area to be digitized. The first step is to identify all the equipment and the boundaries. The data for each unit operation and the advantages to boundary identification were explained in the previous sections. In this section, we focus on showing the configuration process for each unit operation. The screenshots shown here are from one simulation platform; however, the data required between platforms is generally about the same.

Each unit operation has a set of parameters that fix the dynamics of the system.

The primary settings required to configure a vessel as shown in Figure 9-5a are the physical dimensions as well as the conditions that fix the state of the model at start-up (such as temperature and pressure).

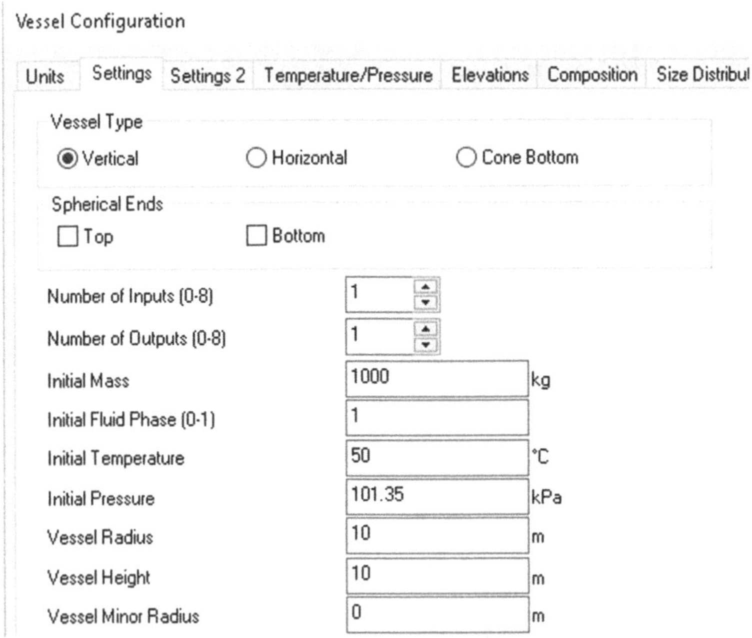

Figure 9-5a. Example vessel configuration.

When modeling a valve, there are three primary values to encapsulate: the valve inherent flow characteristic curve, the flow coefficient (C_v), and the response of the actuator. Modeling the response of the actuator is discussed in Section 9.10 around instrument modeling. The valve conductance should be obtained from vendor literature. The valve curve may be quickly configured according to its type as shown in Figure 9-5b. If the valve has unique characteristics, a custom curve may be provided.

To fully test batch processing performance, it is important to model all equipment around the bioreactor as well as upstream and downstream process units. The interplay of these units affects the performance of the unit and the fidelity of the model as a whole. To fully examine the implications of these dependencies, pipes should be modeled with all valves, pumps, filters, and other equipment to fully test flow paths.

Filtration is a standard process in the bioprocessing industry. Depending on the model fidelity, filtration can be modeled based on a particle size at a component level or using a more rigorous size distribution. A model can track a particle size distribution throughout a simulation if necessary.

A chromatography column can be modeled using two holdups: one holdup for the mobile phase and another for the stationary phase. Calculations within the model implement resin-ligand binding to distribute components of the inlet stream to the

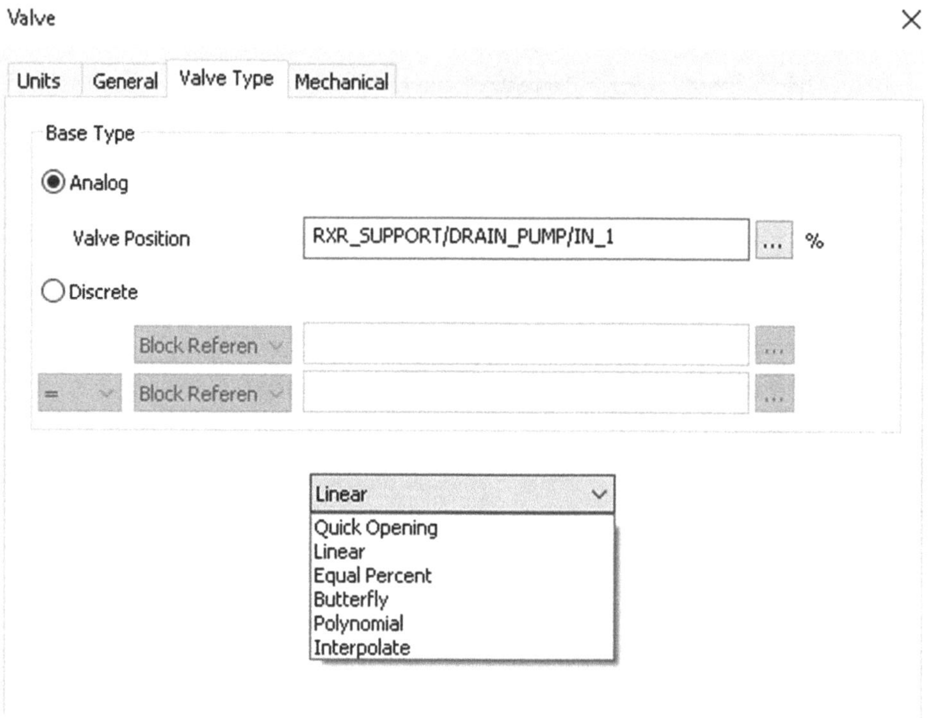

Figure 9-5b. Valve configuration.

appropriate phase. Separation during the elution phase can be simplified where data is not available to match existing plant data using a curve fitting approach. Optical density can be calculated empirically as a function of component concentration or simply simulated based on an integration of elution volume through the column. The stationary phase components will be moved to the mobile phase when the expected curve-fitted optical density is achieved. The actual operation of a chromatography column is complex, and more information is given in Section 9.7, "Equipment Modeling."

9.6 Titration Curve Modeling

Bioreactor performance critically depends on pH. The species growing inside the bioreactor (be it mammalian or bacterial) will produce some acidic by-products during its growth (such as lactic acid). During the course of the batch, the accumulation of this waste product lowers the pH of the broth. In turn, this adversely impacts cell growth. To prototype control strategies using the digital twin, modeling the electrolytes involved in the running system provides the basis for pH simulation. The topic of electrolyte modeling is a multifaceted and complex topic that could easily fill an entire book. However, the basic concepts needed to prepare a simulation are straightforward: electroneutrality, dissociation equilibrium for acids, and species balances [4, 7].

Electroneutrality provides a simple kernel for the computation of pH after identifying which species are present in the system.

The overall expression for electroneutrality is a simple one. This expression assumes that the sum of the concentrations of species multiplied by their charges is zero as shown in Equation 9-6a. That is:

$$0=\sum_{i=1}^{n}\pm\varepsilon_i \bullet [X_i^{\pm\varepsilon_i}] \qquad (9\text{-}6a)$$

Where i is an index over the species (X_i) in the system of charge ε_i. The next concept required to model pH is the concept of the dissociation equilibrium constant. The equilibrium constants for aqueous components are generally given by *pKa* values. The *pKa* values express the ratio of products to reactants of the dissociation reaction, such as for an acid as exemplified in Equation 9-6b:

$$H_xA^\theta \leftrightarrow H^+ + H_{x-1}A^{\theta-1} \qquad (9\text{-}6b)$$

The equilibrium expression is then given by Equation 9-6c:

$$K_{a1}=\frac{[H^{+1}][H_{x-1}A^{\theta-1}]}{[H_xA^\theta]} \qquad (9\text{-}6c)$$

Note that θ is the value of the net charge on the species on the left-hand side of the equilibrium expression.

The *pKa* is simply the base ten logarithm of K_{a1} as defined in Equation 9-6d:

$$pK_{a1}=-\log(K_{a1}) \qquad (9\text{-}6d)$$

The auto-ionization of the solvent also must be considered in any system, as shown for water in Equation 9-6e:

$$pK_w=[H_3O^+][OH^{-1}] \qquad (9\text{-}6e)$$

A species balance simply accounts for the amount of material added to the system. In a system where only a monovalent acid (*HA*) is added to a constant volume as exemplified by Equation 9-6f:

$$X_a=[HA]+[A^-] \qquad (9\text{-}6f)$$

where X_a is the initial concentration of the acid.

These concepts combined enable formulation of a simple charge balance that allows for reasonable simulation of pH for systems of low ionic strength and where salts are assumed to dissociate completely. Regardless of the ionic strength or solubility of the salts in the system, the first important step in such simulation is the identification of the species involved. The simulation developer should list all salts, acids, and bases used within the system. For each salt involved, the developer must also list the conjugate acid-base pair (e.g., sodium carbonate corresponds to carbonic acid). After identification of the species involved, the pKas for each species (and at each dissociation step) must also be identified. For immensely strong acids and bases, the first dissociative step is often so great as to be unmeasurable. Note that for any aerobic system under study, carbon dioxide will contribute to the charge balance as carbonic acid. For an in-depth derivation of the charge balance and an example of its application, see Appendix H, "Charge Balance to Model pH." A first-principle pH model can be created based on the knowledge of reagents added plus carbon dioxide and by-products produced. When using a commercial dynamic simulator, the user may configure the components and the dissociation data and enable the simulator to simply compute the pH value automatically.

For many cases, this simple charge balance provides a reasonable method for computing the pH in bioprocess systems.

The pH calculation is complicated by the effects of ionic species activity coefficients in higher ionic strength solutions (typically above 0.1M). Rigorous calculations can be incorporated in the pH algorithm to consider ionic species activity values. However, the return on the additional computation involved should be evaluated thoroughly. In most cases, with the available correlations, ionic species activities cannot be determined precisely. Also, most dissociation constants are known at 1% accuracy and industrial pH measurement is rarely known at much higher precision.

Whenever the user has collected the data for simulation, it is critical to ensure the modeling is accurate by obtaining the titration curve for the system under study.

Repeating this experiment in the digital twin enables the component and assumption set to be verified. In most bioreactors, some form of buffering is added to aid in controlling pH.

Any system will have some buffering effects that depend on the species present in the system. Even when no buffer is added, carbonic acid and carbonate will provide noticeable buffering.

In the titration curve in Figure 9-6, the additions of a strong base to an acidic system show relatively small amounts of change. This stems from the buffering provided

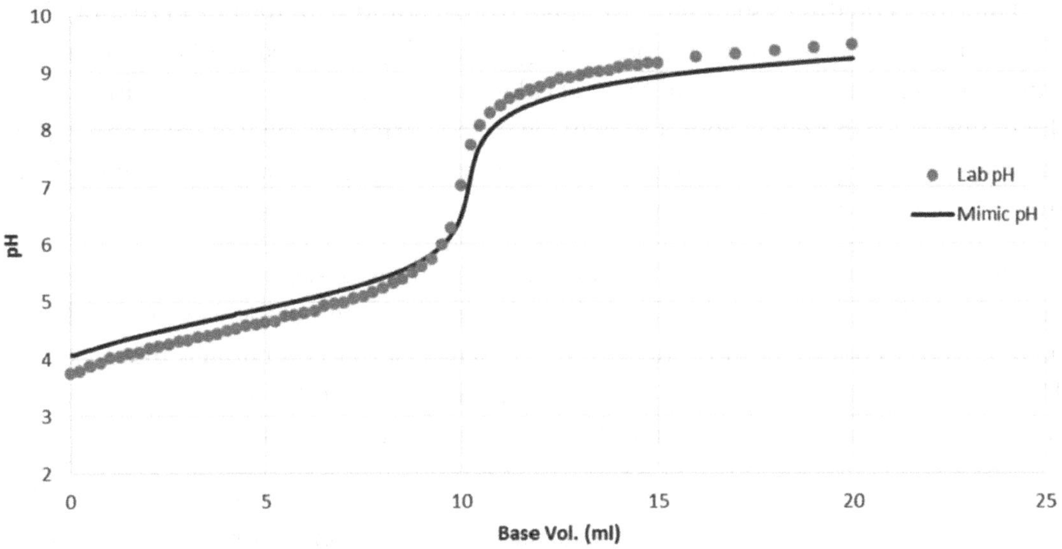

Figure 9-6. Comparison of pH values between simulated and lab titration. [7]

by a weak acid and a conjugate salt. It is conceptually useful to consider Le Chatelier's principle under the Brønsted–Lowry acid base theory. Le Chatelier's principle states that a change on one side of an equilibrium expression (e.g., a change in concentration) results in a correction mitigating that change. Consider the dissociation of acetic acid as shown in Equation 9-6g:

$$H_2O + HCH_3CO_2 \Leftrightarrow H_3O^+ + CH_3CO_2^- \tag{9-6g}$$

By adding a base to the system (e.g., NaOH), more hydroxide ions enter the system. These ions can then neutralize some of the hydronium ions in the solution. The net effect of this neutralization is that there are fewer hydronium ions in the system (on the right side of Equation 9-6g). Le Chatelier's principle then mitigates this effect by causing the reaction to proceed further from left to right. In Figure 9-6, around the 10 mL mark, there is no longer enough undissociated acetic acid to mitigate the base addition. Thus, the buffer is broken and a rapid change in pH results. This precipitous change in pH poses unique challenges in pH control. When the simulation includes these components, it enables realistic testing for control performance and builds the operator's understanding of the system's response.

9.7 Equipment Modeling

There are several pieces of important equipment within a bioprocess that can be simulated. Centrifuges and chromatographs are important pieces of equipment to consider whenever modeling.

Filtering centrifuges separate a wet cake from a liquid with the liquid (liquor) flowing through the porous wet cake out through the holes in a basket or bowl during the wet spin. Some of the liquid is in the wet cakes, and some of the solids (as fines) are in the liquor. A wash flow is added to help clean the surface of the wet cake and reduce the fines in the liquid. The resistance to the flow of liquor through the wet cake is a function of porosity, densities, particle size, viscosity, radii, and speed.

In sedimenting centrifuges, there are no holes in the basket or bowl and no liquor flows through the wet cake. The liquor layer builds up on top of the wet cake and must be removed. In liquid–liquid centrifuges, the heavy liquid is separated like wet cake. The light liquid forms the outer layer. Often the light liquid is a solvent, and a dissolved component in the heavy liquid moves to the light liquid solvent.

Centrifuges can be operated in batch or continuous mode. A nuclear meter or other measurement device to detect the layers is useful for batch operation to determine optimum loading and effectiveness of centrifuge operation, and is essential for continuous operation to control the feed and wash flows and the wet cake and liquor discharge.

An advanced modeling object can model the batch and continuous operation of these centrifuges using first-principle calculations of resistances and driving forces for separation.

The objective is the flexibility to model combinations of liquid and solid phases with and without fines and dissolved components.

Chromatographs are used to separate a solute from a liquid by the solute being captured in the stationary solid phase as the mobile phase of the feed flow moves through the chromatograph. They are used to purify products and to function as an analyzer to measure product concentration. An elution flow is used to remove the solute from the solid stationary phase. The separation of distributions of the solute concentration in the elution flow determines the ability to purify and measure the product concentration. The idea is to maximize peak height and distance between any measurable distributions of the components. While a Gaussian distribution is shown in the literature, there can be an area that is either larger before or after the peak. The model should show the peak height and the distribution shape and area accurately, as well as the time delays associated with the operation.

An advanced modeling object can model the many types of chromatographs using five or more volumes in series for the stationary and mobile phase. The material balance for each volume includes the adsorption rate, eddy diffusion rate, longitudinal diffusion rate, and desorption rate as a function of first-principle driving forces based on equilibrium concentrations and velocity based on void fraction.

9.8 Sparge Modeling

The bioreactor can be envisioned as a grouping of three compartments for mass: the liquid holdup, the mass held up in the vapor space, and the mass of the bubbles as they travel up through the liquid broth into the headspace. Due to the relatively small time constant of vapor, the simulation of the vapor inside the object falls victim to the phenomenon known as *numerical stiffness*. When simulating a bioreactor where the sparge is considered, it is imperative to be vigilant for instability. The tell-tale sign for this instability is the dramatic oscillation of variables (particularly pressure and variables linked with pressure).

Sparge modeling can be conceptualized as the interplay of equilibrium and mass transfer expressions.

The vapor bubbles through the liquid and transfers into the liquid broth or media. Additionally, an impeller within the vessel provides agitation that facilitates more oxygen transfer into liquid. There are two primary types of equations involved in this transfer action: the *equilibrium expressions* and the *mass transfer expressions*. The *equilibrium expressions* dictate the concentration of the dissolved gas (in this case, oxygen) at the interface of the liquid broth or media and the gas. The *mass transfer equations* use the equilibrium concentrations as an input to determine the amount of gas that moves from the gas phase into the liquid phase [5]. The equilibrium concentration can be approximated by the application of Henry's law as shown in Equation 9-8a.

$$X_{eO_2} = mw_{O_2} \bullet \left(\frac{P_{O_2}}{H_{O_2}} \right) \tag{9-8a}$$

where

X_{eO_2} = equilibrium concentration of oxygen (kg/m^3)
mw_{O_2} = molecular weight of oxygen (kg/kmol)
P_{O_2} = partial pressure of oxygen (kPa)
H_{O_2} = Henry's constant (kPa per kmol/m^3)

Transfer occurs according the rate expression per Equation 9-8b:

$$z_{bO_2} = kLa_{bO_2} \bullet \left(X_{eO_2} - X_{dO_2} \right) \tag{9-8b}$$

where

X_{dO_2} = DO concentration (kg/m^3)

kLa_{bO_2} = overall mass transfer coefficient of oxygen from bubbles (sparge gas) (1/h)
z_{bO_2} = mass transfer rate of bubble oxygen (kg/m³ per h)

This transfer rate dictates how much oxygen moves from the gas into the liquid. For the bioreactor to be viable, the oxygen transfer into the system must be sufficient to support the growth of new cells as well as the production of product. The key to tuning the model parameters for the sparge systems is gathering an accurate profile of the key variables.

The main variables of consequence in a sparged system are the DO concentration, cell turbidity, reagent feed rates, and (if available) the concentration of product in the broth. Of these variables, estimation requires the cell turbidity and DO.

After collecting these values, it is possible to estimate the change in oxygen using a finite differencing formula as detailed in Equation 9-8c.

$$\frac{dX_{dO_2}}{dt} = \frac{X_{dO_2,i+1} - X_{dO_2,i}}{t_{i+1} - t_i} \tag{9-8c}$$

where

$X_{dO_2,i+1}$ = DO concentration at moment $i + 1$ (kg/m³)
$X_{dO_2,i}$ = DO concentration at moment $i + 1$ (kg/m³)
t_{i+1} = time at moment $i + 1$ (h)
t_i = time at moment i (h)
$\frac{dX_{dO_2}}{dt}$ = change in DO concentration with respect to time $\left(\frac{kg}{m^3 h}\right)$

The change in oxygen can be attributed to the oxygen that enters from transfer and the oxygen that is consumed by the organism growing within the reactor. Using the collected profiles in tabular form, finite differencing formulas on the cell turbidity, product, and bioproduct concentrations yield a list of derivatives. From the yield factors, the rate of oxygen consumption by the cells can be approximated by Equation 9-8d:

$$r_{O_2} = \frac{r_v}{Y_{O_2,v}} + \frac{r_p}{Y_{O_2,p}} + \frac{r_a}{Y_{O_2,a}} + \frac{r_b}{Y_{O_2,b}} + m_{O_2,v} \bullet X_v \tag{9-8d}$$

where

r_{O_2} = oxygen consumption formation rate $\left(\frac{kg\,O_2}{m^3 h}\right)$

$$r_v = \text{viable cell mass growth rate}\left(\frac{kg\,cell}{m^3 h}\right)$$

$$Y_{O_2 v} = \text{viable cell yield on oxygen (0.5 g cell/g oxygen)}$$

$$r_p = \text{product formation rate}\left(\frac{kg\,product}{m^3 h}\right)$$

$$Y_{O_2,p} = \text{product yield on oxygen (0.2 g product/g oxygen)}$$

$$r_a = \text{acid by-product formation rate}\left(\frac{kg\,acid}{m^3 h}\right)$$

$$Y_{O_2,a} = \text{acid by-product yield on oxygen (0.2 g acid/g oxygen)}$$

$$r_b = \text{basic by-product formation rate}\left(\frac{kg\,base}{m^3 h}\right)$$

$$Y_{O_2,b} = \text{basic by-product yield on oxygen (0.2 g base/g oxygen)}$$

$$m_{O_2,v} = \text{maintenance factor (0.0001 g } O_2\text{/g cell per h)}$$

$$X_v = \text{viable cell mass concentration}\left(\frac{kg\,cell}{m^3}\right)$$

The change in the oxygen can then be treated as the sum of this rate and that which is transferred into the media from oxygen as well as the rate consumed in Equation 9-8e:

$$\frac{dX_{dO_2}}{dt} = z_{bO_2} + r_{O_2} \tag{9-8e}$$

The kLa can then be estimated. Computing the kLa around the middle of the profile and determining an average can be used in combination with a kLa correlation to tune the parameters for the system under study. Using a standard optimization technique (e.g., bound optimization by quadratic approximation—BOBYQA or even functions built into most spreadsheet software) will help tune the expression.

9.9 Kinetics Modeling

The task of simulating enzyme-catalyzed chemical reactions considering the total mechanism and kinetics at real time is challenging [11]. For example, the famous glycolytic pathway involves 10 main steps, each of which is performed by a different enzyme and several intermediate chemical species. The full discovery of this metabolic pathway converting glucose to pyruvate took almost 100 years of research [12]. Fortunately, breakthroughs offer friendly generic models [1-3].

Extremely important is the modeling of cell growth rate kinetics especially those associated with the profound effect of pH and temperature. Convenient cardinal equations offer the ability to model these kinetics with easily set parameters.

The convenient cardinal equations provide a simple means to approximate the profile of industrial bioreactors in real time.

The form of the bioreaction kinetic expressions is selected based on the objectives of a real-time dynamic process simulator—which are typically control system testing and improvement, operator training, and process dynamic studies. As discussed in Chapter 6, an extended Michaelis-Menten kinetic expression including terms for rate inhibition and limitation is found to fit well with industrial bioreactor profiles [1, 2]. Simulation models based on other cell growth rate expressions such as Tessier (Equation 9-9a) and Moser (Equation 9-9b) are also available [13]. For a given substrate S (g/L),

Tessier expression:
$$\mu_v = \mu_{v,max} \bullet \left[e^{\left(\frac{-S}{K_{1,v,S}}\right)} - e^{\left(\frac{S}{K_{2,v,S}}\right)} \right] \tag{9-9a}$$

Moser expression:
$$\mu_v = \mu_{v,max} \bullet \left[1 + \frac{K_{2,v,S}}{S^{\lambda}} \right]^{-1} \tag{9-9b}$$

where

μ_v = specific cell growth rate $\left(\frac{g\,cell}{g\,cell\,h}\right)$

$\mu_{v,max}$ = maximum specific cell growth rate $\left(\frac{g\,cell}{g\,cell\,h}\right)$

$K_{1,v,S}$ = substrate inhibition parameter for cell growth rate (g/L)

$K_{2,v,S}$ = substrate limitation parameter for cell growth rate (g/L)

λ = Moser inhibition exponent

Parameters of a high-fidelity bioreactor model are configured based on the process description, the equipment process data sheet, the plant plot plan, lab data, and time-evolution of transmitter readings. A profile fitting tool can be used to deduce maximum rate, inhibition, and limitation effect parameters for cell growth, product, and by-product formation from profiles of analytical measurements as exemplified by Figure 9-9a for optical density (OD) from a turbidity measurement.

The available input data on gas flows such as the O_2 inlet flow rate, off-gas flow rate, and composition is used in tuning the sparge, O_2 uptake, and CO_2 production model parameters. Today, highly precise real-time online bioreactor off-gas analyzers

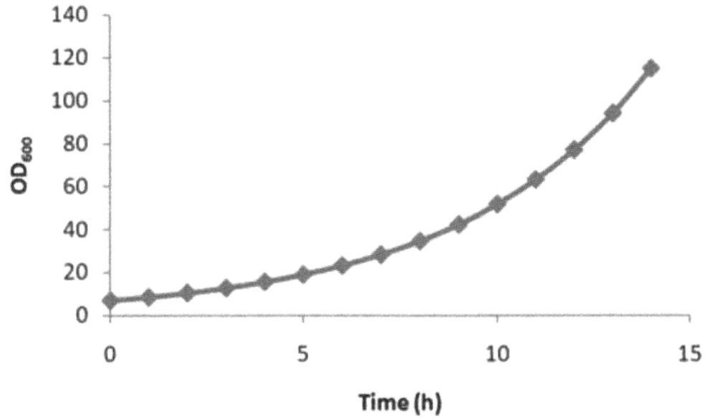

Figure 9-9a. Example bioreactor OD profile.

are available. Gas analyzers based on both gas chromatography and mass spectrometry are considered "top notch" in substance identification. Real-time online gas analyzers can operate at 1 to 20 s per point, accurately tracking the process dynamics.

The effects of temperature and pH on bioreaction are simply captured by the convenient cardinal equations as described in Chapter 6. The minimum and maximum temperature and pH set points for the bioreactor are readily available in the equipment data sheet, process description, and operating procedure [3].

The following figures (9-9b to 9-9g) depict model configuration using the Mimic advanced bioreactor object.

The component set selection in the bioreactor model is accompanied by the component role definition. Stream components (e.g., oxygen, cell, and glucose) are assigned available roles such as viable biomass, lysed biomass, product, by-product, utilization, growth gas, and waste gas (Figure 9-9b).

Finally, the inputs and outputs (streams) to the bioreactor must be configured to interact with the rest of the simulation (Figure 9-9h).

Figures 9-9b through 9-9g show snapshots of Mimic bioreactor object kinetics configuration based on the expressions and parameter schemes discussed in Chapter 6.

The objective of tuning and testing of the bioreactor model kinetics is to match the plant reactor cell growth, production, and substrate consumption profiles within acceptable tolerances.

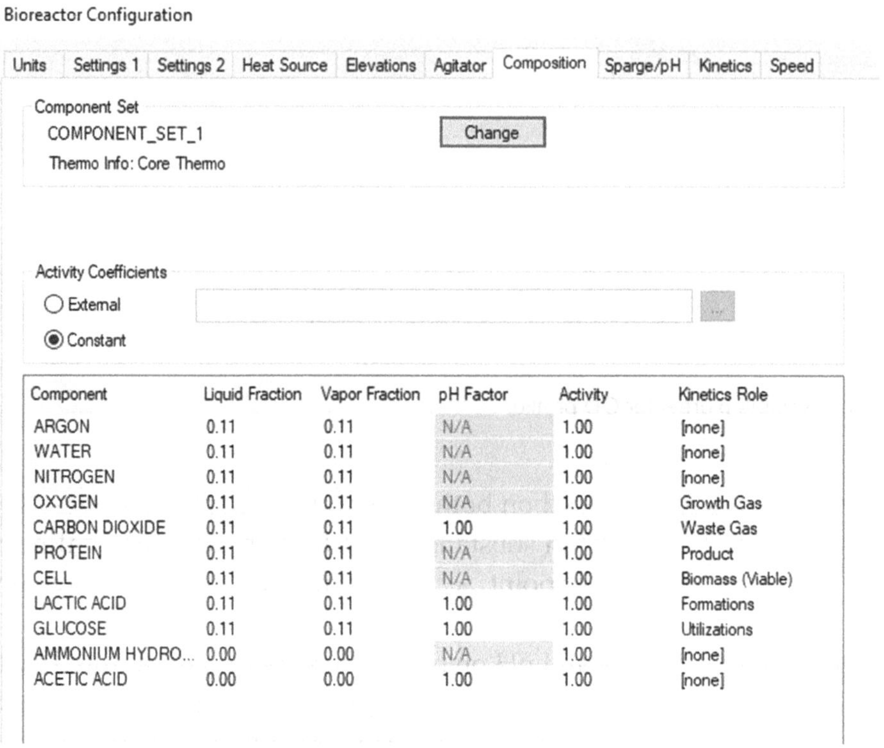

Figure 9-9b. Bioreactor object configuration—composition.

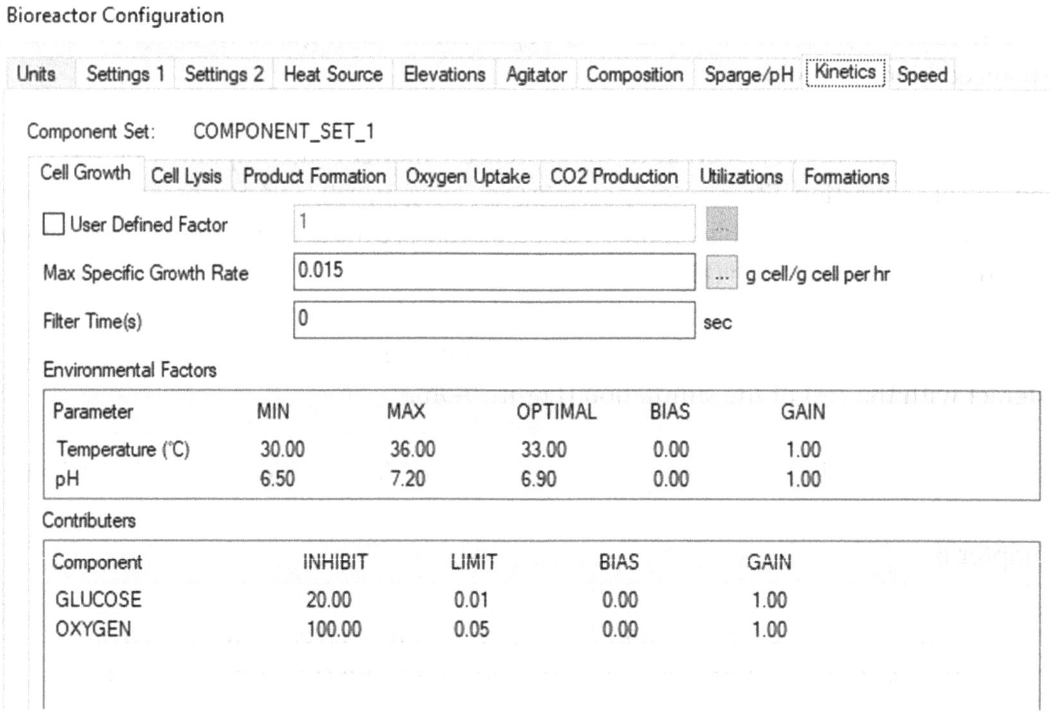

Figure 9-9c. Bioreactor object configuration—kinetics: cell growth.

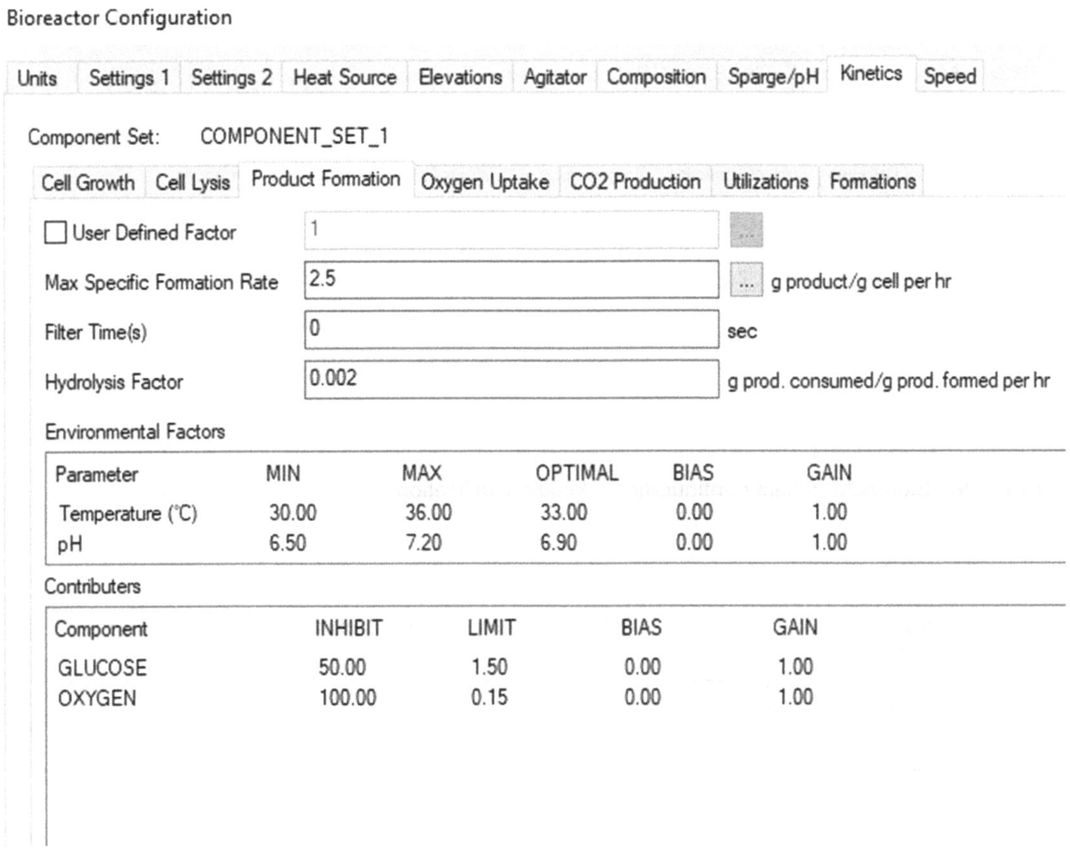

Figure 9-9d. Bioreactor object configuration—kinetics: product formation.

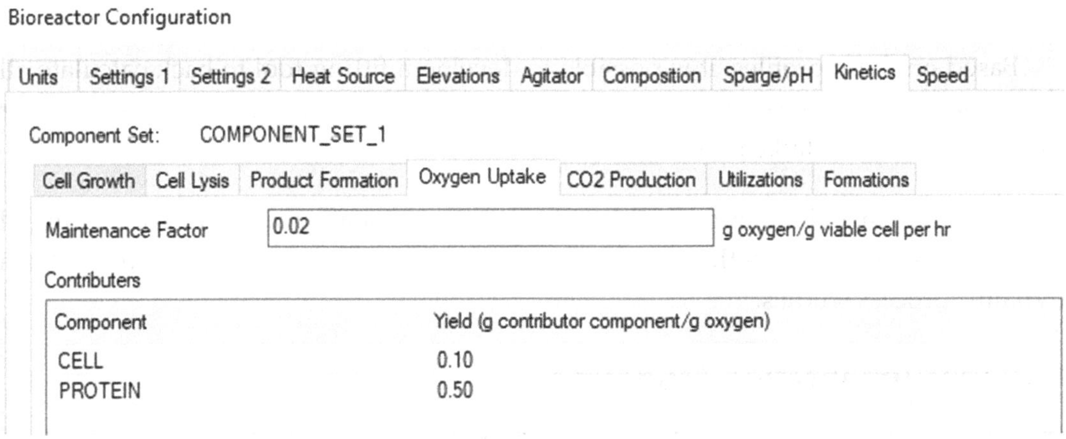

Figure 9-9e. Bioreactor object configuration—kinetics: oxygen uptake.

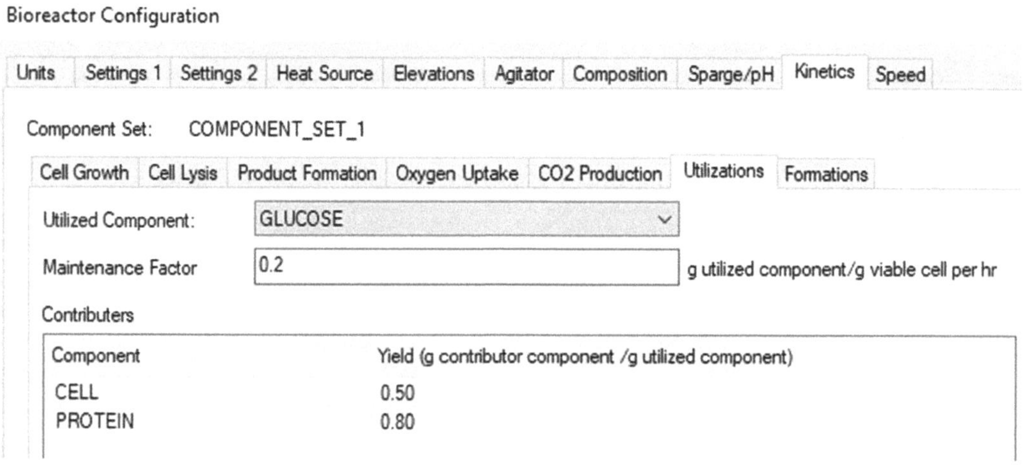

Figure 9-9f. Bioreactor object configuration—kinetics: utilization.

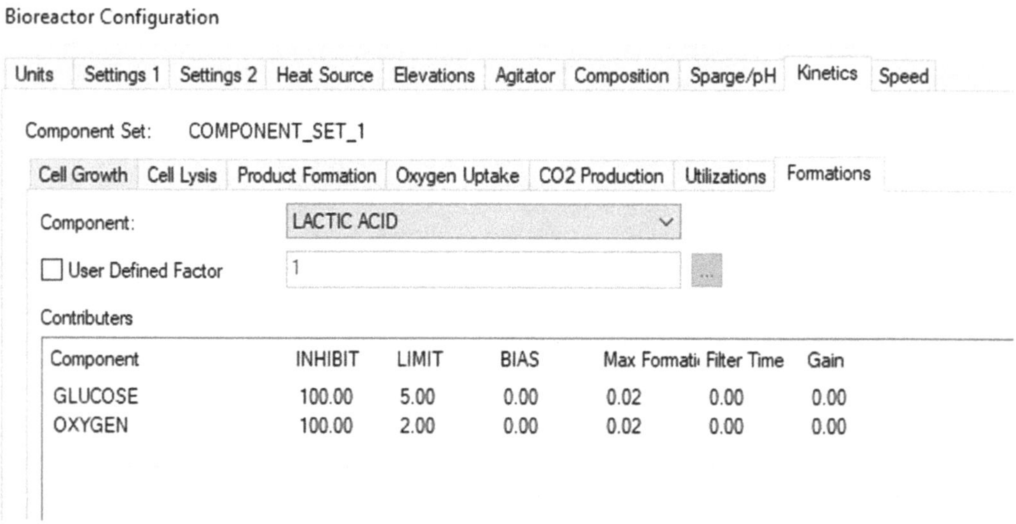

Figure 9-9g. Bioreactor object configuration—kinetics: formation.

Based on these profiles, it is possible to develop a fitting tool to back calculate all major model parameters, although reasonable assumptions might be necessary. Refer to Section 9.13 for details on fitting of profiles.

The target model accuracy is dictated by the functional specification of the digital twin solution, that is, whether to test the control system, train operators, or perform dynamic process studies.

9.10 Instrumentation Modeling

The "dynamic" aspects of modeling are particularly important for bioprocesses because a steady state is rare, as most unit operations are batch. Even perfusion processes start out

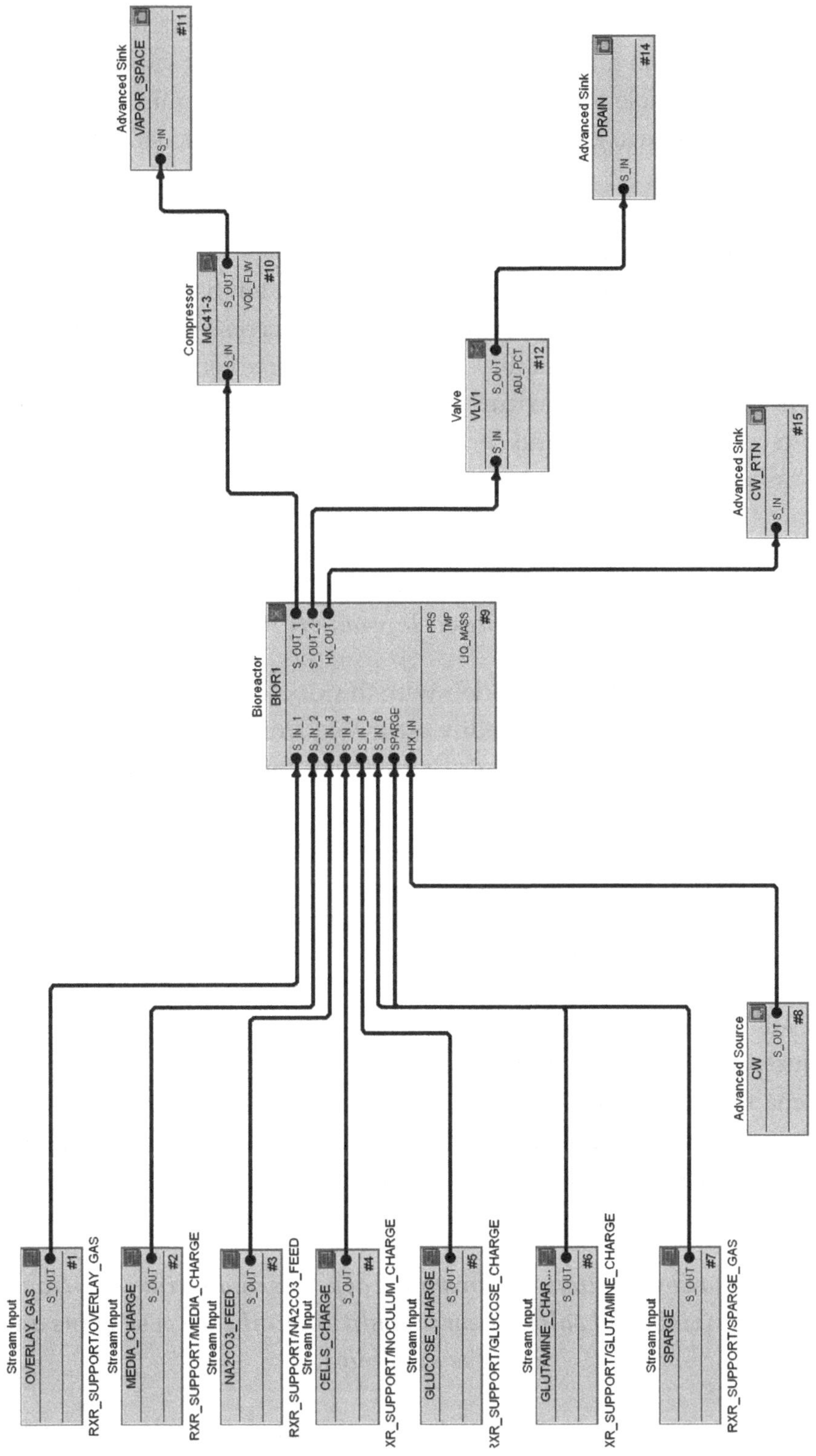

Figure 9-9h. Bioreactor object configuration—streams.

as a batch and remain continuous by recycle and purification only until contaminants build up, requiring a restart as a batch. By computing the slope of the cell growth rate and the product formation rate, you can achieve a pseudo steady state and an insightful and useful process variable for batch profile control, but this requires a good dynamic model and a computation (e.g., future value block) that has a good signal-to-noise ratio.

The dynamic response of cells is critical. Cell health, growth, and productivity are constantly changing and are a function of dynamic response and tightness of control of key process variables. A process control error of a few tenths of a pH or temperature can have a significant detrimental effect seen as an increase in cell death (e.g., lysis) and a decrease in growth rate, viability, and product formation. A simulator of insufficient dynamic fidelity results in an unrealistic system response, which provides a poor basis for both operator training and controller tuning. However, a high-fidelity model provides a basis for operator training, offline optimization and experimentation, and gauging controller performance.

A good dynamic model is needed for operator training and control that promotes cell productivity.

Dynamic high-fidelity process models with the kinetics and mass transfer detailed in Chapter 6 and with parameters readily estimated from batch profiles per methods given in this chapter provide the process gains, time constants, and dead time. What most do not realize is that industrial proportional-integral-derivative (PID) and model-predictive control (MPC) algorithms work in percent rather than engineering units, most of the total dead time comes from instrumentation, and most disturbances are on an input to the process instead of on the process output. This missing understanding has profound implications.

What really determines the gain and dynamics seen by the controller is much more than just the process gain or process dynamics. The controller sees an open-loop gain as the change in percent process variable (controller input) divided by the percent change in the manipulated variable (controller output), and that this gain and the total loop dead time, rather than just process dead time, determine tuning and control loop performance for process disturbances as detailed in Section 3-6 of Chapter 3.

Tuning and control loop performance depends on instrumentation dynamics, sizing, and calibration in addition to the process gains and process dynamics.

The open-loop gain is a product of the manipulated variable gain (e.g., control valve or variable speed pump flow span divided by 100% for a linear installed flow characteristic), a hidden factor that is the inverse of the feed flow, a process gain that is the change in the process variable for a change in ratio of manipulated flow to feed flow, and a measurement gain that is 100% divided by the measurement span as detailed in Chapter 3 Equations 3-6a through 3-6j. Assuming the piping pressure drops and inherent valve or pump flow characteristics are properly modeled, a digital twin automatically provides the proper open-loop gain because it uses the actual control system with its input and output scales.

A digital twin gives the controllers scales that—combined with valve or pump inherent flow characteristic with a dynamic piping pressure drop—provide a proper model of the open-loop gain needed for tuning and control loop performance.

The total dead time is the sum of the dead times that are pure delays and small lags that are effectively dead time in manipulated variable response (e.g., valve position or pump speed), manipulated flow transportation (e.g., reagent injection delay), process transportation to sensor (e.g., sample delay), sensor response, transmitter response (e.g., damping and update rate), and controller response (e.g., filter time, scan time, and execution rate). Transportation delays of manipulated flows by valves and pumps and process flows to sensors are often missing in the process model. There are also dead times that are the resolution, sensitivity, or deadband divided by the rate of change of the corresponding signal.

Models of instrumentation response and transportation delays of associated manipulated and process flows provide the total dead time needed for tuning and control loop performance.

A control loop cannot do anything to correct for an unmeasured load disturbance until it starts to see the disturbance and correction arrive at the same point in the loop as the disturbance. The total loop dead time from the point of disturbance entry through the process, measurement, controller, valve or variable frequency drive (VFD), and piping to the point of entry determines what a controller can do. In a first-order plus dead time approximation, the time constants in series as you traverse the loop that create an equivalent dead time are added to the pure dead times from transportation delays; digital device update and execution rates; analyzer cycle time; and instrumentation sensitivity, resolution, and deadband.

The best controller gain for load disturbance rejection is inversely proportional to the open-loop gain and total loop dead time. The minimum peak and integrated error are proportional to the dead time and dead time squared, respectively. Blocks that model each of the major sources of dead time should be included in the model to achieve the dynamic fidelity needed.

The minimum peak error and integrated error for a load disturbance are proportional to the total dead time and dead time squared, respectively.

Instrumentation characteristics (i.e., response time, sensitivity, resolution, and deadband) depend on the valve's operating point. These characteristics are generally worse for low flows. In low-flow situations during operation, valves experience higher stiction near the seat as well as low velocities at the sensors (especially less expensive actuators, positioners, and valves). Many positioners designed in the 1970 to 1990 time frame have a response time that increases by two orders of magnitude (e.g., from 0.5 to 50 s) for step sizes less than 0.25%. Even more recently designed positioners have this problem, which remains unrecognized due to the traditional step test of 5% or more. The less expensive positioners (e.g., spool) have a particularly slow response time. Thus, a variable filter block and dead time block should be used that enables the time constant and dead time to be bumplessly updated online. A control valve designed for throttling (e.g., diaphragm actuator, digital valve positioner, and sliding stem globe body) has a sensitivity resolution approaching 0.1% and a backlash deadband of 0.2% above 20% open. A control valve designed for tight shutoff and on-off control with a piston actuator and spool positioner has a resolution that is 1% to 10% and a backlash deadband that is 2% to 20%. The lower price tag, lower leakage, and greater capacity of the on-off valve posing as a throttle valve and the lack of requirements for sensitivity, resolution, or response time on valve specifications makes this a common problem [8].

The slew rate limit of a throttling valve can be kept at a rate of 50% per second or faster (100% stroke time less than 2 s) by the use of a volume booster on the positioner output. The omission of the volume booster for large valves can slow down the slew rate by two orders of magnitude. These are important factors to include when modeling valves.

The time constant of a new pH electrode in a buffer solution for a decrease in pH is more than twice what it is for an increase in pH. The time constant is two to four times larger for a strong acid and strong base solution. The time constant increases by an order of magnitude for a decrease in pH for 0.5 fps or less velocity at the electrode. Almost imperceptible coatings cause a dramatic increase in the electrode time constant. A slime coating of 1 mm thickness can increase the electrode 98% response time from 10 s to 7 min. Premature aging of the glass due to prolonged exposure to process temperatures greater than 50°C can increase the 98% response time to 160 min. Middle signal selection of 3 pH measurements will inherently ignore the slowest electrode, including a single failure of any type, as well as reduce the effect of noise and provide much more intelligent diagnostics and a calibration schedule. An 86% response time is a better metric because 98% response time takes more than twice as long and is much more susceptible to errors in response time from noise and disturbances.

A temperature measurement time constant is mostly governed by the thermowell design and the fit of the sensor, but it is also adversely affected by a low fluid velocity. Air acts as a great insulator. If the sensor tip is not touching the bottom of the thermowell, the increase in time constant can be orders of magnitude. An annular clearance of 0.055 in (very loose fit) instead of 0.005 in (extremely tight fit) can increase the largest time constant for even a high velocity liquid stream from 4 to 228 s. The time constant for 1/4-inch sheathed bare elements is much less than the largest thermowell time constant.

The time constants for electrodes and thermowells can vary from a few seconds to several minutes depending on operating conditions and the sensors' design and health, and how the electrodes and thermowells are installed.

A block used to model a discrete device or analyzer should have sensitivity and/or resolution input parameters, a cycle time or update time, and an analysis time or latency. The resulting dead time is the sensitivity and/or resolution divided by the process variable rate of change plus the dead time from periodic updates that is the sum of ½ the cycle or update time and the total analysis time or latency. The analysis time is typically the cycle time for analyzers, and the latency is typically near zero for digital devices not overloaded so that the dead time is 1.5 times the cycle time and 0.5 times the update time, respectively. The resolution of older thermocouple and resistance temperature detector (RTD) input cards (e.g., 1980s) caused a noticeable problem with dead time and a poor signal-to-noise ratio. The sensitivity of analyzers is often a concern.

The dead time from periodic updates, from analyzer sensitivity, and from many less expensive actuator and positioner designs is a significant problem best recognized and corrected by the use of models with instrumentation dynamics.

A backlash stiction block as shown in Figure 9-10a enables the nonideality of valves to be encapsulated simply. The signal represents the input from the control system. It is a value ranging from 0% to 100%. The backlash represents the actuator's deadband. The resolution represents the minimum change in requested position the actuator can manifest. A prestroke delay enables modeling of the pure delay between when a signal is requested from the control system and when it reaches the actuator. Additionally,

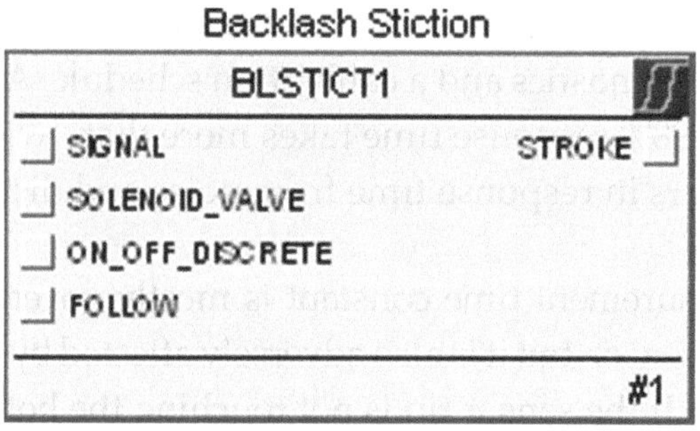

Figure 9-10a. Backlash stiction modeling block.

two stroke lags enable modeling of the change as a second-order process. If additional restrictions on the rate of change are required, a slewing rate model directly limits the rate of change that can pass into the backlash stiction object. To summarize, the valve signal passes through a dead time block, two lag blocks, a slew rate block, and finally the backlash stiction block before becoming the stroke. The stroke represents the actual opening of the valve after accounting for all the nonidealities explained. If there is a positioner whose sensitivity drastically deteriorates for small signal changes, particularly in the opposite direction, the dead time parameter can be increased accordingly. For example, consider a positioner that exhibited a dead time of only 0.5 s for a 2% or larger signal change, the dead time increased to 50 s for a 0.25% signal change in the same direction and increased to 100 s for a 0.25% signal change in the opposite direction [8].

A slew rate can be applied via the block shown in Figure 9-10b to the signal resulting from the stroke or used to intercept the signal to the block. A slew rate provides a maximum rate of change per model execution in the upward (increasing slew rate) and downward (decreasing slew rate) directions. The combination of the slew rate and backlash stiction models allows for full modeling of valve nonidealities. Determining these parameters may require step tests if the measurements are unavailable from the manufacturer [8].

Valves play an important role in capturing the dynamics critical for modeling bioprocesses. A control valve exerts a profound impact on the control strategy in all processes, especially bioprocesses where a control valve might be responsible for modulating sensitive variables like pH.

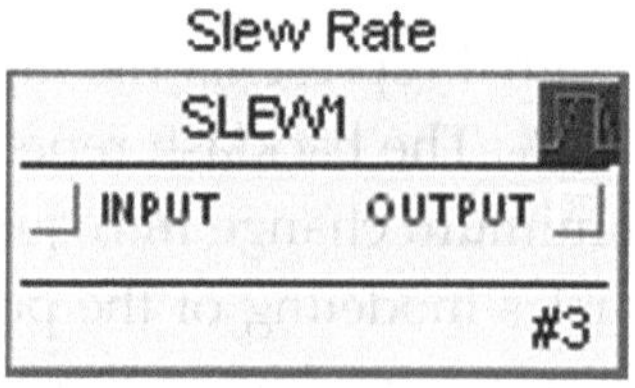

Figure 9-10b. Slew rate modeling block.

In a simulation, detailed information is available instantaneously. This is a luxury that real processes do not enjoy. The analyzer block shown in Figure 9-10c allows for simulating an off-line analyzer. The simulated value is transmitted through the measurement input. The sample time average represents the expected time between samples taken by an analyzer or a technician. The analyze time average represents the expected time from a sample being taken to the result being delivered to the control system. Additionally, a deviation for both the sample time and the analyze time can be added to incorporate the nonuniformity inherent in a real system. Analysis noise provides a random fluctuation in the measurement. Resolution dictates the minimum amount of change in the measurement that is required to take a sample. Once the process is complete, the sample with noise injected is returned as the new value. The new flag becomes true for one execution to signal to other parts of the model whenever a fresh sample's analysis is complete.

The use of an enhanced PID block (e.g., PIDPlus) as shown in Figure 9-10d with external-reset feedback (e.g., dynamic reset limit) that suspends integral action between periodic updates and uses time between updates for derivative calculation can enable much

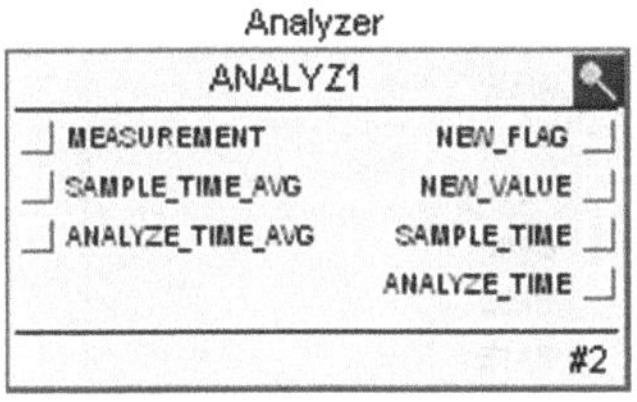

Figure 9-10c. Analyzer modeling block.

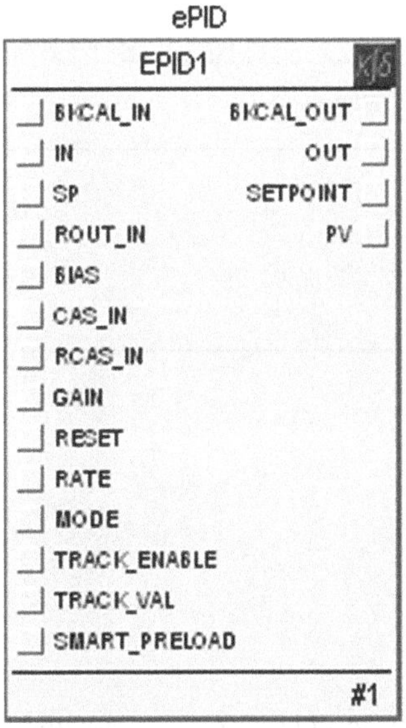

Figure 9-10d. Example PID block.

more aggressive tuning independent of changes in time between updates when this time is larger than the 63% response time of the process [9]. This capability is achieved by simply enabling the external-reset feedback mode and connecting the actual process variable result of periodic updates from a digital device or analyzer as the external-reset feedback signal. The controller should automatically become more stable, and in many cases the controller gain increased and reset time decreased. If the time between period updates is more than the 63% response time, increases in the time between updates do not require retuning, which means that incredibly long analyzer cycle times possibly from off-line analysis in labs do not require retuning or cause oscillations.

The PID block's in and out can be connected to a Rough N Ready tuner as exemplified in Figure 9-10e. A PID can be placed in remote output mode to enable the Rough N Ready tuner to perform a series of step tests. This allows for rapid identification of

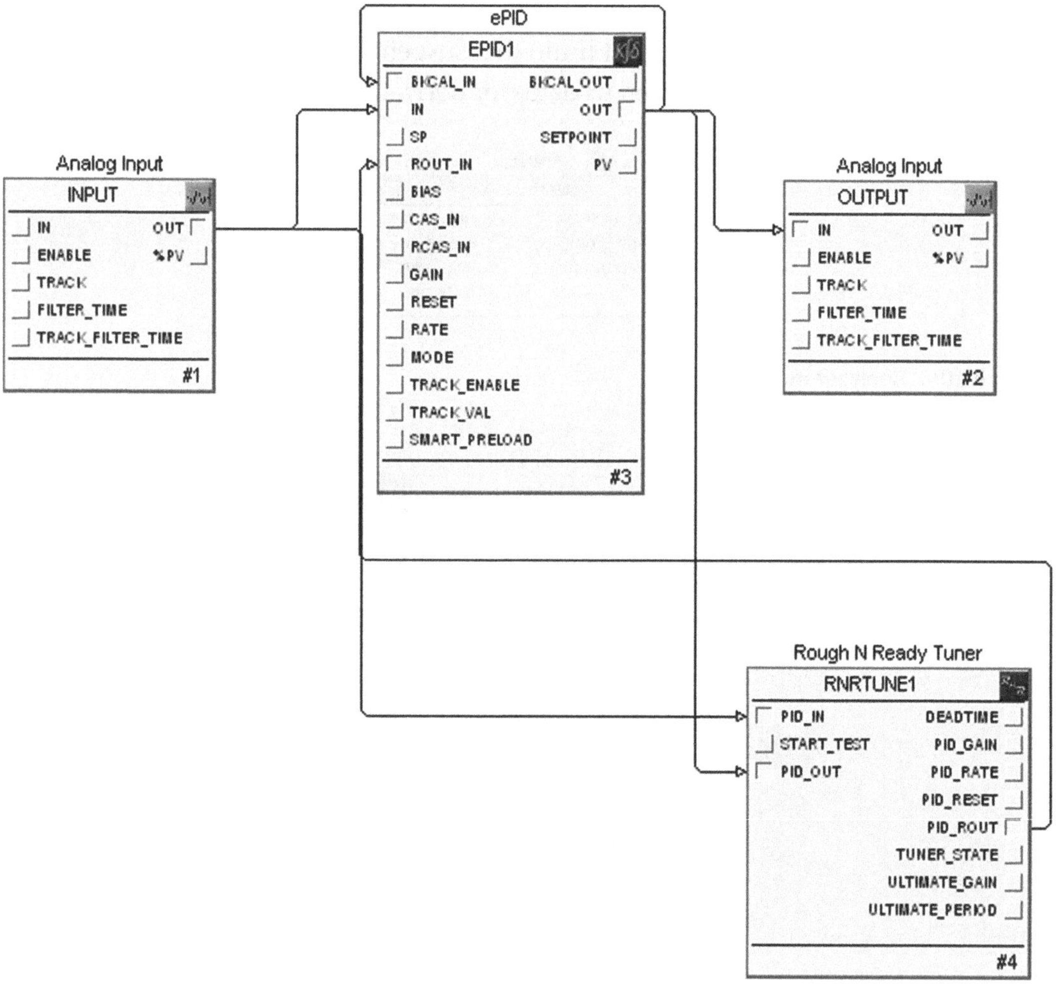

Figure 9-10e. Example of a Rough N Ready tuner in a simulated loop.

initial tuning settings. It is important to keep the scales and control actions synchronized in both blocks. The Rough N Ready tuner only works whenever the system has settled. A set of initial tuning settings for PID control can assist when prototyping configurations in the digital twin. Some initial settings are shown as the first number outside the range of settings in parentheses for various applications are listed in Table 9-10. The initial settings and ranges of settings shown are not meant to be indiscriminately used in the actual system deployed. They can be used as a check for unusual settings to trigger investigation as to whether a tuning mistake has been made or there are some unusual application conditions. There are exceptions to every rule, especially when it comes to the dynamics of industrial processes, which are usually severely nonlinear and a function of raw materials, production rate, cycle time, by-products, and operating/equipment conditions.

In bioreactor control, biomass measurement is one of the most critical process variables. Various online in situ measurement techniques are available, such as turbidity, dielectric spectroscopy, conductivity spectroscopy, and fluorescence [14, 15].

In industrial bioprocess, the most common and widespread method to measure cell growth and biomass concentration is the turbidity technique using OD measurement.

Table 9-10. Typical PID execution rates and tuning settings for typical loops.

Application Type	PID Execution Rate (s)	PID Gain	Reset Time (s)	Rate Time (s)
Liquid Flow	0.5 (0.2–1)	0.3 (0.2–0.8)	6 (1–12)	0 (0–2)
Liquid Pressure	0.1 (0.02–0.5)	0.5 (0.2–1.0)	3 (1–6)	0 (0–2)
Tight Liquid Level*	5 (1.0–30)	10 (5–50)	600 (120–6000)	0 (0–60)
Gas Pressure (psig)*	0.2 (0.1–0.5)	5.0 (0.5–20)	300 (60–600)	3 (0–30)
Reactor pH*	1 (1.0–5)	1 (0.01–50)	120 (60–600)	30 (6–30)
Neutralizer pH	2 (1.0–5)	0.1 (0.001–5)	300 (60–600)	60 (6–120)
Inline pH	1 (0.2–2)	0.2 (0.01–0.4)	30 (15–60)	0 (0–3)
Liquid Reactor Temperature	5 (2.0–15)	5.0 (1.0–50)	60 (30–300)	12 (12–60)
Gas Reactor Temperature	1 (0.5 –2)	1.0 (0.5–5)	15 (5–30)	3 (1–6)
Inline Liquid Temperature	2 (1.0–5)	0.5 (0.2–2)	30 (15–60)	6 (3–15)
Distillation Column Temperature	10 (2.0–30)	1.0 (0.2–10)	3000 (600–12,000)	300 (60–1200)

Turbidity of a medium is an expression of its optical property to scatter and absorb light. Turbid fluid will be less transparent because of the presence of suspended particles; thus, turbidity is the opposite of clarity. Turbidity relates to measurement of the degree of attenuation of a beam of light of known initial intensity. The mass concentration of suspended particles in a media sample, such as cells in broth culture, is related to its turbidity [16].

In OD measurement, the turbidity of a material sample is related to light transmittance at a given wavelength. The transmittance will depend on light scattering and absorbance. For example, OD_{600} is an OD reading using a light of wavelength 600 nm.

The sensor light beam wavelength is selected considering the two competing measurement requirements, sensitivity and robustness. The probe geometry and angle between the light source and detector affects the sensitivity and specificity of the measured signal [18].

For measuring biomass, OD measurement wavelength is selected so that attenuation by scattering is predominant. This is because light scattering is relatively constant for a biomass media due to consistent cell composition and associated refractive index.

The Beer-Lambert law, which relates attenuation or transmittance of light through a material sample to its properties, is applied in turbidimeters. The law relates light absorbance to the concentration of the attenuating species and the light path length in the sample, which leads to the following [17],

$$\mathrm{OD} = -log\left[\frac{I}{I_o}\right] = \varepsilon \bullet C \bullet L \qquad (9\text{-}10a)$$

where

I = intensity of the light transmitted through the sample
I_o = incident intensity of the light
ε = attenuation coefficient (L/(mol cm))
C = concentration of attenuating species in the sample (mol/L)
L = light beam path length (cm)

The linear relationship depicted in Equation 9-10a between the OD and the concentration of suspended particles is complicated by several factors, such as:

- Scattering and reflection of light by particles
- Interaction between particles, such as aggregation or chemical reaction, usually at high concentration
- Changes in the refractive index of the material sample
- Fluorescence of the sample
- Nonmonochromatic light beam; ε varies over the bandwidth

Typical industrial OD meters are designed and operated to follow the linear OD to the biomass (such as cell, protein) concentration, with constant attenuation coefficient and path length. This general relation is of the form shown in Equation 9-10b:

$$\mathrm{OD} \bullet f_D = f_C \bullet X_v \tag{9-10b}$$

where

f_D = dilution factor
f_C = correlation coefficient (L/g)
Xv = biomass concentration (g/L)

For a given material sample and spectrophotometer, in cases such as the physiological growth conditions ($\{z_i\}$) of the biomass changes substantially, the OD measurement is nonlinear. This will require defining the correlation coefficient (f_C) as a function as shown in Equation 9-10c, such that

$$\mathrm{OD} \bullet f_D = f_C\left(\{z_i\}\right) \bullet X_v \tag{9-10c}$$

In the case of a limitation to the sample dilution, a short light beam path length can provide accurate online OD measurement. It will ensure that sufficient light through the sample cell reaches the detector.

The advantages and disadvantages of turbidity measurement are as follows.

Advantages:

- Sensitive, specific, and noninvasive measurement
- Applicable for a wide range of microorganisms and cell types
- Continuous bioprocess monitoring in situ

Disadvantages:

- A lack of standardization in units, measurement devices, and calibration. The nonlinear dependency of light scattering and absorption on cell density makes it difficult to set correct regression and calibration procedures.
- Interferences caused by bubbles, foam, colored substances, solids, and cell aggregation.
- Error due to biomass and protein buildup on optical components.
- The inability to distinguish between viable and dead cells or noncellular particles in the sample media.

9.11 Speedup

Bioprocess batch operations can have incredibly long batch cycle times (e.g., 10 days). The speedup factors shown in Figure 9-11 can be used to complete a batch in hours rather than days. The speedup factor for media speedup can be as fast as 20 times real time, and the speedup factor for kinetics can be 10 or more. The total speedup for media response is the product of the two factors, which means a speedup of 200 or more times real time is possible. The flow ranges of valves, variable speed drives (VSDs), and flow transmitters should be increased by the kinetic speedup factor. If both the capacity and flow span are adjusted correctly, the flow controller tuning should not change much. However, the primary process time constant and integrating process gain will be faster for composition, DO, pH, and temperature control. The dead time of instrumentation should be decreased by the total speedup factor to keep about the same ratio of the total dead time to the actual primary process time constant or the equivalent primary process time constant that is the inverse of an integrating

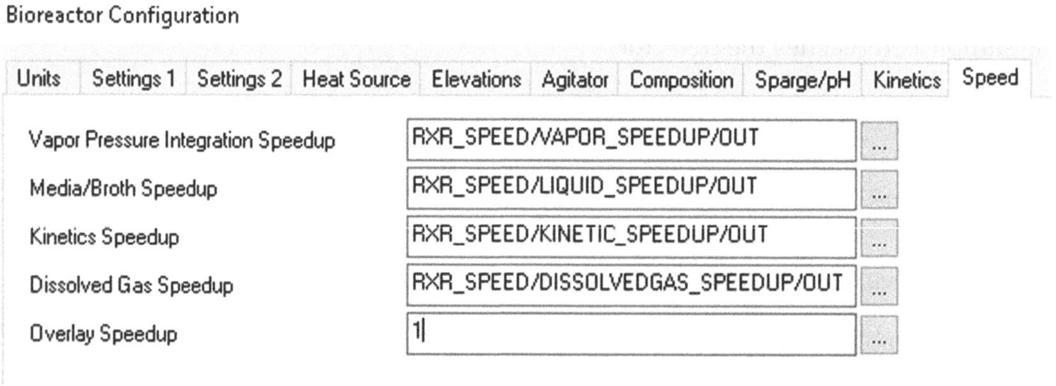

Figure 9-11. Configuration of individual speedup parameters.

process gain. For composition, DO, pH, and temperature control, the PID gain and reset time must be decreased based on auto tuner results. Many batch processes have too small a reset time even without speedup, as seen by an overshoot and delayed correction of PID output until after the process variable crosses the set point. Also, some sources of dead time cannot be reduced enough to match speedup. Consequently, the PID gain will need to be decreased and the decrease in reset time moderated. A reset time smaller than four times the maximum dead time or an oscillation period greater than six dead times is indicative of a reset time that is too small. A faltering (temporary settling of PID output just before reaching the set point) on the approach to the set point or oscillations with periods about four times the dead time are indicative of the need to decrease the PID gain. An auto tuner should be used. The Rough N Ready auto tuner can be used to quickly identify the controller settings at different batch times.

The batch cycle time can be speeded by the product of kinetic and model-media speedup factors, but flow capacities of valves and VSDs and flow measurement spans must be increased in proportion to the kinetic speedup factor. The instrumentation dead times should be decreased by the total speedup factor. Flow controller tuning should not need to be adjusted if capacities and spans are adjusted properly, but for composition, DO, pH, and temperature control, the PID gain should be decreased and possibly the reset time should also be decreased based on the auto tuner results.

9.12 Performance Monitoring

Performance metrics can be automatically computed by a block shown in Figure 9-12 for process input (load) disturbances and for set point changes so that the performance of control strategies, instrumentation, and tuning can be assessed and improved.

If load metrics are enabled, this block continually computes the peak error and integrated absolute error. The load metrics are reset to zero if this metric disabled. It is up to the user to periodically disable these metrics or apply forgetting logic. If the set-point metrics are enabled and a set-point change larger than the threshold setting occurs, the rise time (time for the process variable to first reach within settling error of the set point) is computed with subsequent metrics. The maximum error in the same direction of the set-point change after the first crossing of the set point is the overshoot. The maximum error in the opposite direction after the second crossing of the set point is the undershoot. The time after overshoot for the filtered error to stay within the settling error is the settling time. Note that error filter is used to prevent reversals in the sign of direction of the error change. The set point metrics are reset to zero if

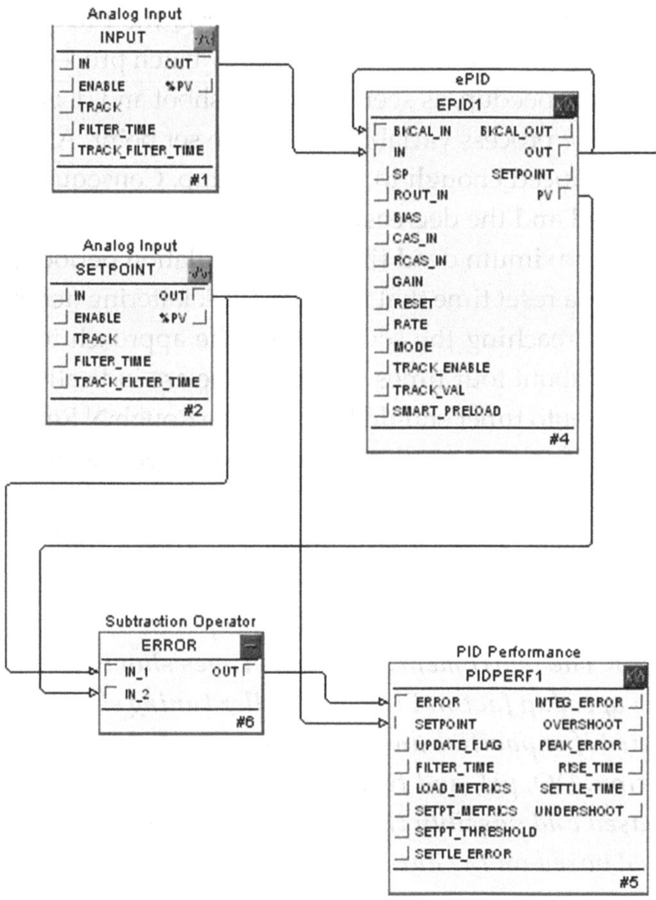

Figure 9-12. Example of a PID performance metrics setup.

there is another set-point change or the set-point metrics are disabled. Thus, there is no need for a period reset or forgetting factor for set-point metrics. The peak error is not useful for set-point metrics because the peak error will simply be the set-point change plus the overshoot if it is larger than the previous value of peak error.

9.13 Generation and Fitting of Profiles

The future value block shown in Figure 9-13 provides a continuously updated rate of change with a good signal-to-noise ratio for each model execution. This block allows computation of the slope of cell and product concentration profiles (i.e., the cell growth rate and product formation rate, respectively) and the slope of profiles for any type of batch operation. The slope can be used as an MPC or PID controlled variable changing a batch process effectively into a continuous process. The Future Value block uses the rate of change to predict a value that is one dead time or more into the future to enable preemptive evaluation and adjustments to improve batch operation typically at the

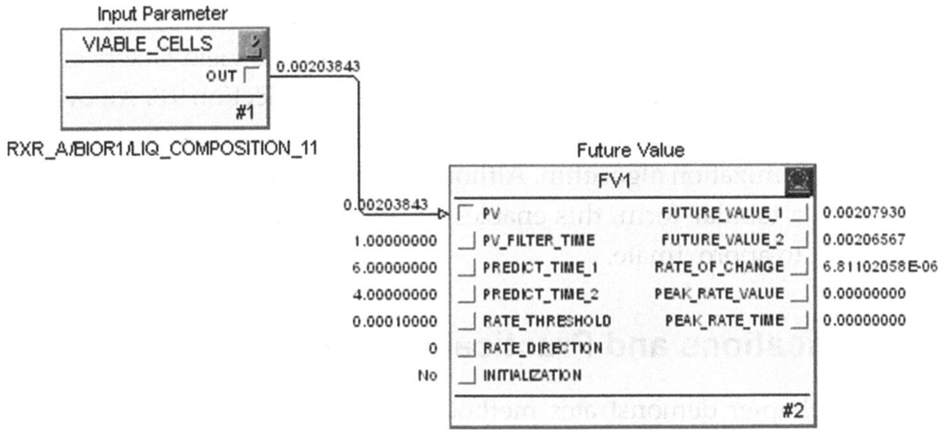

Figure 9-13. Example of future value to estimate cell compositions.

midpoint and end point of the batch cycle. The results of a Future Value block can be used to improve the set-point response and operator interface of most loops by showing where a key process variable will be 1 to 1.5 dead times into the future.

A Future Value block can provide a continuous update of batch profile slopes that can be used for model fitting and as a controlled variable, enabling a batch process to have a continuous response.

A Future Value block can provide predictions one or more dead times into the future of key process variables for greater operational and process understanding and a better set-point response.

The future value block should be tuned for the dead time of the system for the predict time. With a profile in place, the predicted value can be compared to an expected value to determine when a batch might be failing or simply acted on directly. Observing a full run of values as well as change rates enables fitting of a profile.

To fit a profile, the agitation, cell concentrations, liquid volumes, glucose addition rates, pH, DO, temperatures, and yield factors must be known.

Additionally, the end points, inoculum introduction time, and batch end point must be known. A spreadsheet can be constructed by placing cell concentration and time in two columns. By using a finite difference between subsequent points, a cell growth rate can be approximated. The convenient cardinal equations may be computed alongside the historical data. Yield factors coupled with these finite differences (as well as the oxygen and substrate addition integrated over time) provide a means to compute the other variables. A set of constants for tuning (such as K_1 and K_2 for each of the rate expressions) are set at some initial but reasonable values. The computation is carried out for each point in time. For an example of how the yield factors and oxygen rates are interrelated (and their derivatives computed), see Section 9.8. An overall error function is used for an objective. The tuning constants are used as the parameters for a multivariate optimization algorithm. Although it takes time to acquire such data and assemble it into a tabular form, this enables tuning of parameters that would otherwise be difficult to approximate.

9.14 Simplifications and Practical Solutions

Although this chapter demonstrates methods to develop rigorous dynamic models for various systems, it is also important to understand potential simplifications and their applicability in different use cases as simulations can make otherwise difficult-to-simulate processes relatively straightforward.

The first step in the process of simplifying models is to consider the model objectives.

As an example, for training operators, one may ask whether the model is intended to train on process understanding or interacting with the control system interfaces. In the case of the latter, the model may be able to be simplified to minimally satisfy the controls of interest, but for operators or control engineers to build a full mental picture of a process, such simplifications may be misguided. As for automation and control, one may ask whether the model is intended to be used for controller tuning or for batch logic development and management of change. The latter is much more forgiving of model simplifications. These are some examples of surface-level questions to ask in determining where simplifications can be made. Readers are encouraged to dive deeper into these objectives to fully define model requirements on a unit-by-unit basis.

Interest in simplification is often driven by a desire either to avoid modeling complexities or to address unavailability of data for more intricate unit operations.

As outlined in this chapter, to develop a comprehensive bioreactor model, one should consider titration curves, sparging, kinetics, and instrumentation, among other details; however, this could be simplified if the consequences of doing so are understood and acceptable. One such simplification may be the reduction of the complex kinetics to a conversion rate, which can be correlated to a life-death curve generated using Monod kinetics. This simplification must be applied with caution as it removes the cell balance from the equation and ignores cellular behaviors away from expected operating conditions, but it is able to provide reasonable batch profiles for the reaction products and by-products, which would enable suitable models of the pH profile over the course of a batch. This model simplification is particularly useful when looking to simplify the bioreactor itself without inadvertently simplifying associated models for other unit operations.

Taking the simplification of the bioreactor model a step further, the kinetics could be ignored entirely, and the reaction could be simplified to fitted batch profiles for cell growth, media consumption, and production. This approach does not consider a component-level material balance for the reactor, which means that the behavior will not predict process performance. The advantage, however, is that this approach is easy to implement with available production data and without detailed process knowledge, while also retaining an overall mass balance. A similar idea can be applied to the simulation of chromatography columns based solely on bed volumes.

This leads to the idea of a holistic simplification that can be applicable across upstream and downstream processing units when modelers are not interested in predicting process performance, controller tuning, or control system capability. Because these processes are predominantly single-phase, major process units ranging from tanks to bioreactors to chromatography columns could be reduced to liquid holdups connected with supporting models for pumps, valves, transfer panels, and heat exchangers with empirical relations for pH, conductivity, turbidity, or other instrumentation. This provides an overall material and energy balance and leverages a pressure-flow solver with limited process information, which often can meet modeling requirements from a simplified batch perspective at the expense of predictive capability. This can extend to other units such as centrifuges and filters, but special handling is necessary for determining outlet flows.

Here are some other practical solutions to consider when developing dynamic models:

- Effects of instrumentation and transfer delays may be neglected in cases where the dead time and time constant are small compared to the process dead time or the model is not intended to be used for control studies, instrumentation type and placement, or process improvement.

- Holdups may be considered perfectly mixed if the residence time or turnover (mixing) time is small or the model is not intended to be used for control studies or process improvement. Alternatively, the effect of mixing may be simplified and modeled at the instrumentation level.

- Inline mixers and transportation delays may be neglected in the case that the residence time is small or the model is not intended to be used for control studies or process improvement.
- Solids may be simulated as a liquid assuming their impact on the process density and viscosity is small or negligible (i.e., they do not account for a large amount of the stream's mass).

Remember that there is no silver bullet. While this discussion highlights some simplifications that can be applied when developing practical simulation solutions, the assumptions associated with these simplifications must always be considered and weighed against the objectives of the models.

9.15 Best Practices

1. Use a digital twin with the actual set points, control capability, a historian, and an operator interface.
2. Explore how turbidity, mass spectrometers, DO, dissolved carbon dioxide, dielectric spectroscopy, BioProfile FLEX analyzer, near-infrared spectroscopy (NIR), and fast protein liquid chromatographs can be used as noted in Chapter 7 to improve modeling and control.
3. Use a digital twin to test out new strategies and process metrics when a good model is available.
4. Use historized data to tune a model by matching model operating points, manipulated flows, and batch slopes to plant. Model the effect of temperature and pH on growth and product formation rates using set points to help initialize convenient cardinal equation parameters. Model other dynamics using Michaelis-Menten equations by using low and high alarm settings for limitation and inhibition parameters. Fine-tune the model by subsequently adjusting parameters including yields and by-product formation.
5. Account for the nonidealities of valves and sensors, including slow response time, deadband, and resolution limits when building models and tuning loops.
6. Take advantage of the capabilities of the future value algorithm to compute rates of change (e.g., batch profile slopes) and to predict key process parameters

(e.g., batch end points and set-point responses) 1 to 1.5 dead times into the future.

7. Include the buffering effect of salts, weak acid, weak bases, and dissolved carbon dioxide, as well as electrode lags, and reagent injection and mixing delays whenever modeling and controlling pH.

8. Use speedups to quickly test new strategies and model parameters realizing that model dead time must be reduced proportionally for existing tuning settings to be reasonable.

9. Model wireless update rates and analyzer sample transportation delays and cycle times, and use an enhanced PID block (e.g., PIDPlus) with external-reset (e.g., dynamic reset limit) to accommodate wireless updates and analyzer cycle times. Simply turning on external-reset can stop oscillations from deadband, stiction, and slow secondary valves, VFDs, and secondary loops by connecting the manipulated stroke, speed, or flow as external-reset feedback.

10. Use Rough N Ready tuning settings as a first step to check PID tuning and response capability.

11. Set up the control system and metrics to understand possible opportunities and loop performance.

12. Build all stakeholders' understanding by using the digital twin as a laboratory for exploration, discovery, education, innovation, and excellence using "before" and "after" online metrics.

13. Run a real-time modeling system (RTMS), such as that described in Appendix N, that deals with all the data, complexity, and variability affecting the production capability in the plant. Such a system enables a consequential identification of the location and impact of bottlenecks.

14. Use the digital twin and first principle models with automation system dynamics to do a nonintrusive design of experiments (DOE) to find the causes of the bottlenecks identified by RTMS. Use these identified bottlenecks to design and test improvements inserted into the RTMS to quantify the impact on overall plant production.

15. Train operators and engineers on the use, results, and synergy of the RTMS and digital twin.

References

1. J. E. Haag et al., "Dynamic Modeling of Complex Biological Systems: A Link between Metabolic and Macroscopic Description," *Mathematical Biosciences* 193 (2005): 25–49.

2. J. E. Haag et al., "A General Model for Reaction Kinetics in Biological Systems," *Bioprocess Biosystems Engineering* 27, 2005: 303–309.

3. L. Rosso et al., "Convenient Model to Describe the Combined Effects of Temperature and pH on Microbial Growth," *Applied and Environmental Microbiology* (February 1995): 610–616.

4. Gregory K. McMillan and Robert Cameron, *Advanced pH Measurement and Control*, 3rd ed. (Research Triangle Park, NC: ISA [International Society of Automation]), 2005.

5. M. Moo-Young, "Section 2: Transport Phenomena in Bioprocesses," in *Comprehensive Biotechnology* (Oxford, England: Pergamon Press Ltd., 1985): 30–33.

6. Gregory K. McMillan and M. E. Berutti, "Virtual Plant Virtuosity," *Control* (August 2017), https://www.controlglobal.com/articles/2017/virtual-plant-virtuosity/?stage=Live.

7. Christopher Stuart, Rehman Fazeem, and Gregory K. McMillian, "Improve pH Control," *Chemical Processing* (March 2016), https://www.chemicalprocessing.com/articles/2016/improve–ph–control/.

8. Gregory K. McMillan and P. Wu, "How to Specify Valves and Positioners That Don't Compromise Control," *Control* (March 2016), http://www.controlglobal.com/articles/2016/how-to-specify-valves-and-positioners-that-dont-compromise-control/.

9. Gregory K. McMillan, "Wireless: Overcoming Challenges in PID Control & Analyzer Applications," *InTech* 57, no. 6 (July/August 2010), https://www.isa.org/link/wireless/.

10. Gregory K. McMillan, Rehman Fazeem, and Christopher Stuart, "Bioreactor Control Breakthroughs: Biopharm industry turns to advanced measurements, simulation and modeling," *Control* (February 2020).

11. F. Mazzocchi, "Complexity on Biology. Exceeding the Limits of Reductionism and Determinism Using Complexity Theory," *EMBO Reports* 9, no. 1, (January 2008): 10–14.

12. J. A. Barnett, "A History on Research on Yeasts 5: The Fermentation Pathway," *Yeast* 20, no. 6 (April 2003): 509–543.

13. M. S. M. Annuar et al. "A Kinetic Model for Growth and Biosynthesis of Medium–Chain–Length Poly–(3–Hydroxyalkonates) in Pseudomonas Putida," *Brazilian Journal of Chemical Engineering* 25, no. 2 (June 2008): 217–228.

14. K. Kiviharju et al. "Biomass Measurement Online: The Performance of in Situ Measurements and Software Sensors," *Journal of Industrial Microbiology and Biotechnology* 35 (July 2008): 657–665.

15. J. M. Sandnes et al. "Real-Time Monitoring and Automatic Density Control of Large-Scale Microalgal Cultures Using Near Infrared (NIR) Optical Density Sensors," *Journal of Biotechnology* 122 (2006): 209–215.

16. D. M. Lawler, "Turbidity, Turbidimetry and Nephelometry," *Encyclopedia of Analytical Science*, 3rd ed. (Amsterdam: Elsevier, 2016), 152–163.

17. J. A. Myers et al. "Improving Accuracy of Cell and Chromophore Concentration Measurements Using Optical Density," *BMC Biophysics* 6, no. 4 (2013): 1–15.

18. C. Ude et al. "Application of an Online-Biomass Sensor in an Optical Multisensory Platform Prototype for Growth Monitoring of Biotechnical Relevant Microorganism and Cell Lines in Single-Use Shake Flasks," *Sensors* 14 (2014): 17390–17405.

19. Gregory K. McMillan and C. A. Stuart, "Simulation Breeds Innovation," *Control* (November 2019), https://www.controlglobal.com/articles/2019/simulation-breeds-innovation/.

20. Gregory K. McMillan, Chris Stuart, and Thomas Hildebrand, "Compressor Surge Control: Deeper Understanding, Simulation Can Eliminate Instabilities," *Control* (October 2018), https://www.controlglobal.com/articles/2018/compressor-surge-control-deeper-understanding-simulation-can-eliminate-instabilities/?stage=Live.

Appendix A: Automation System Performance Top 10 Concepts

Automation system performance depends not only on tuning but also on the performance of each component in each control loop. System performance can be maximized by attention to design and implementation details that affect the ability of the controller, final control element(s), equipment, piping, sensor, and transmitter to do their job. Here is a summary of the Top 10 Concepts to guide the automation engineer in this endeavor. We finish with a perspective on the relative importance of accuracy and precision.

A.1 Delay

- **"Without dead time I would be out of a job."**
- **Fundamentals**
    - A more descriptive name would be *total loop dead time*. The total loop dead time is the amount of time required for the start of a change that has occurred

in the loop to travel around the complete control loop and end up at the point of origin. For example, an unmeasured disturbance cannot be corrected until the change is seen in the controlled variable and the correction arrives in the process at the same point as the disturbance.

- Process dead time offers a continuous train of values, whereas digital devices and analyzers offer noncontinuous data values at discrete intervals. The resulting delays add a phase shift and increase the ultimate period (decrease the natural frequency) like process dead time.

- **Goals**
    - Minimize delay (the loop cannot do anything until it sees and enacts change).
- **Sources**
    - Pure delay from process dead times and discontinuous updates
        - Piping, duct, plug flow reactor, conveyor, extruder, spin-line, and sheet transportation delays (pure process delays set by mechanical design; remaining pure delays set by automation system design)
        - Digital device scan, update, reporting, and execution times
        - Analyzer sample processing, analysis cycle time, and multiplex time
        - Sensor threshold sensitivity
        - Signal resolution
        - Valve stiction
        - Valve backlash
        - Variable speed drive deadband
        - Wireless trigger level
        - Wireless update rate
    - Equivalent delay from lags
        - Mixing, column trays, dip tube size and location, heat transfer surfaces, and volumes in series (process lags set by mechanical design; remaining lags set by automation system design)
        - Thermowell lag
        - Electrodes

- Transmitter damping
- Signal filters

A.2 Speed

- **"Speed kills"—high-speed processes and disturbances with low-speed control systems can kill performance.**
- **Fundamentals**
  - The rate of change in four dead time intervals is most important. By the end of four dead times, the control loop should have completed most of its correction. Thus, characterizing self-regulating processes with a large time-constant to dead-time ratio as near-integrating processes is consistent with performance objectives.
- **Goals**
  - Make control systems faster and make processes and disturbances slower.
- **Sources**
  - Control system
    - Proportional-integral-derivative (PID) tuning settings (gain, reset, and rate)
    - Slewing rate of control valves and velocity limits of variable speed drives
  - Disturbances
    - Steps: batch operations, on-off control, manual actions, safety instrumented systems (SISs), start-ups, and shutdowns
    - Oscillations: limit cycles, interactions, and excessively fast PID tuning
    - Ramps: reset action in PID controller
  - Process
    - Degree of mixing in volumes due to agitation, boiling, mass transfer, diffusion, and migration

A.3 Gains

- **"All is lost if nothing is gained."**
- **Fundamentals**

- Gain is the change in output for a change in input to any part of the control system. Thus there is a gain for the PID controller, valve, disturbance, process, and measurement. The open-loop gain that is the product of the valve, process, and measurement gain is needed for controller tuning. Knowing the disturbance gain (e.g., the change in manipulated flow per change in disturbance) is important for sizing valves and for feedforward control).

- **Goals**
  - Maximize control system and process gains.
    - Maximize valve sensitivity.
    - Maximize sensor sensitivity.
    - Maximize transmitter gain (minimize calibration span).
    - Maximize controller gain.
    - Maximize process sensitivity (e.g., find best distillation column tray).
  - Minimize disturbance gain (size and impact).
    - Minimize size of upstream disturbances by automation (e.g., elimination of oscillations and transfer or absorption of variability by control).
      1. Decrease limit cycles.
      2. Decrease interactions.
      3. Replace operator actions and on-off control with throttling control.
      4. Increase absorption of flow variability by surge tank level tuning.
      5. Use redundancy to minimize failures (e.g., middle signal selection).
    - Minimize size of process disturbances by enabling process self-regulation (e.g., cancel or moderate disturbances).
      1. Get internal reflux control by tight distillate level control.
      2. Use weak acid and base reagents to buffer pH systems.
- **Sources**
  - Abnormal operation and failures (e.g., pump, measurement, and valve failure)
  - PID controller gain
  - Inferential measurement location (e.g., sensitivity of temperature change as an inference of composition change in the distillation column)
  - Sensor type (e.g., threshold sensitivity of radar versus differential pressure (d/p) level measurement and resistance temperature detector (RTD) versus thermocouple)

- Control valve or variable speed drive installed characteristic (gain is slope that includes the effect of inherent characteristic and system pressures)
- Measurement calibration (gain is 100%/span)
- Process design
- Backlash and deadband (limit cycle amplitude)
- Stiction and resolution limits (limit cycle amplitude)
- Interaction (fighting loops oscillation amplitudes)
- Operator actions
- Safety instrumented systems (SISs)
- Sequences
- Variations in stream composition, density, temperature, and physical properties particularly for recycle, heat integration, and stressed utility systems.

A.4 Resonance

- **"Don't make things worse than they already are."**
- **Fundamentals**
  - An oscillation period that is close to the ultimate period can be amplified by the PID controller as shown in Figure A-4.
- **Goals**
  - Make the disturbance oscillation period slower or control loop faster.
  - For interacting loops with similar priorities, make the fast loop faster.
  - For interacting loops with different priorities, make the least important loop slower.
- **Sources**
  - Control loops in series with similar dead times (e.g., multiple stage pH control)
  - Control loops in series with similar tuning, valve stiction, and backlash
  - Control loops that affect each other with similar dynamics

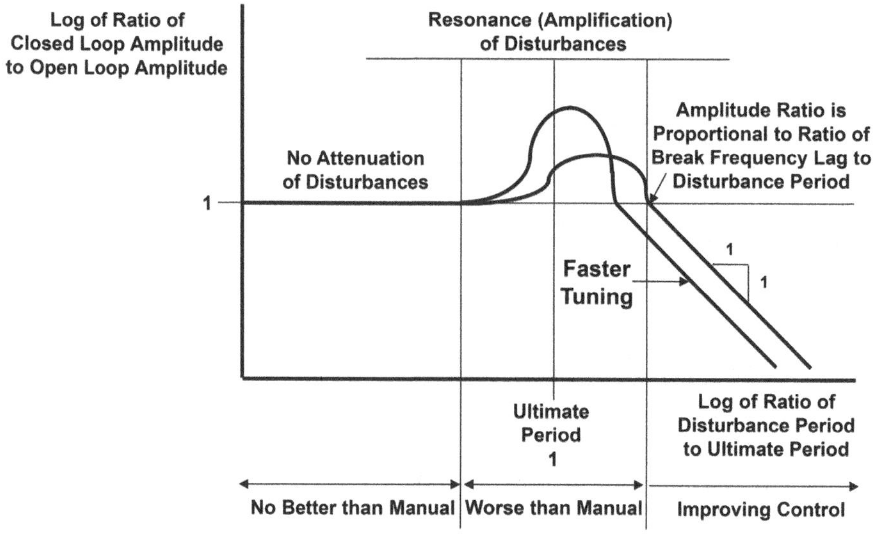

Figure A-4. An oscillation with a period near the ultimate period (frequency near the natural frequency) is amplified by the PID controller due to *resonance*.

- Control loops in series with improperly set priorities and level of importance
- Day to night ambient changes to slow loops (e.g., column temperature control)

A.5 Process Attenuation

- **"If you had a blend tank big enough, you would not need control."**
- **Fundamentals**
  - Attenuation increases as the volume of the blend tank increases and the ultimate period of the control loop decreases (filter time constant is back mixed volume residence time).
- **Goal**
  - Maximize attenuation by increasing volume and mixing and making loops faster.
- **Sources**
  - Mixed volume size and degree of back mixing (lower feed rate increases attenuation by increasing residence time for continuous processes and makes rate of change more gradual for batch processes)

$$A_f = \frac{A_o}{\sqrt{1+(\tau_f \bullet \omega)^2}} \tag{A-5a}$$

The attenuated (filtered) amplitude (A_f) for original amplitude (A_o) and attenuation (filter) time constant (τ_f) that is much greater than the oscillation period (T_o) using the relationship between oscillation frequency and period ($\omega = \frac{2 \bullet \pi}{T_o}$):

$$A_f = \frac{A_o \bullet T_o}{2 \bullet \pi \bullet \tau_f} \quad \text{(A-5b)}$$

Equation A-5b is also useful for estimating the filtering effect of the thermowell lag, electrode lag, transmitter damping, and signal filtering on process oscillation. The equation can also be used to estimate original process amplitude given measurement amplitude when the measurement lag is causing an attenuated version of the real process variability.

A.6 Threshold Sensitivity and Resolution

- **"You cannot control what you cannot see."**
- **Fundamentals**
  - Minimum possible change measured or manipulated is set by sensitivity and resolution: Once past the threshold sensitivity limit, full change is seen or used as depicted in Figure A-6a, but the resolution limit will quantize the

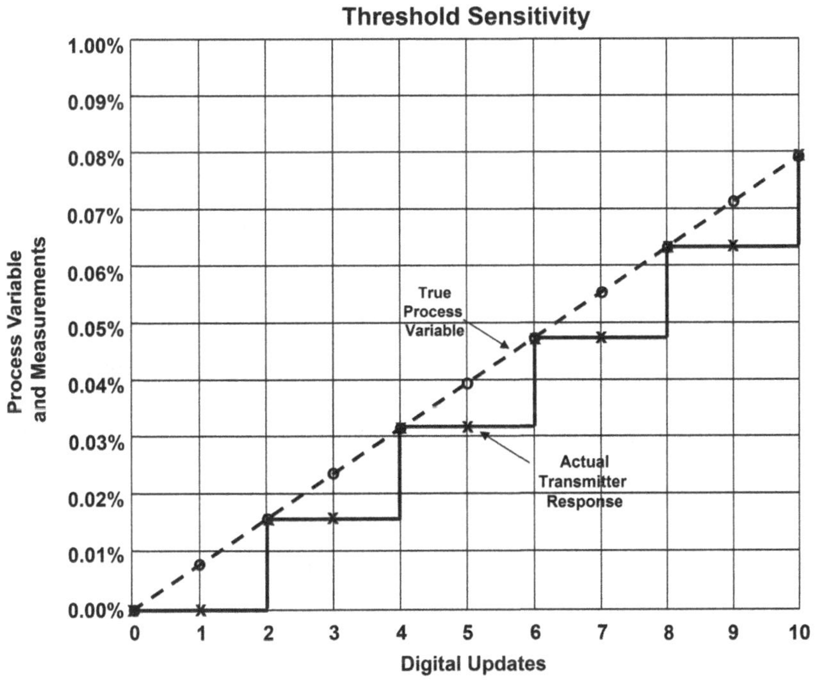

Figure A-6a. When the change in the process variable exceeds the *threshold sensitivity limit*, the full change in the process variable is seen.

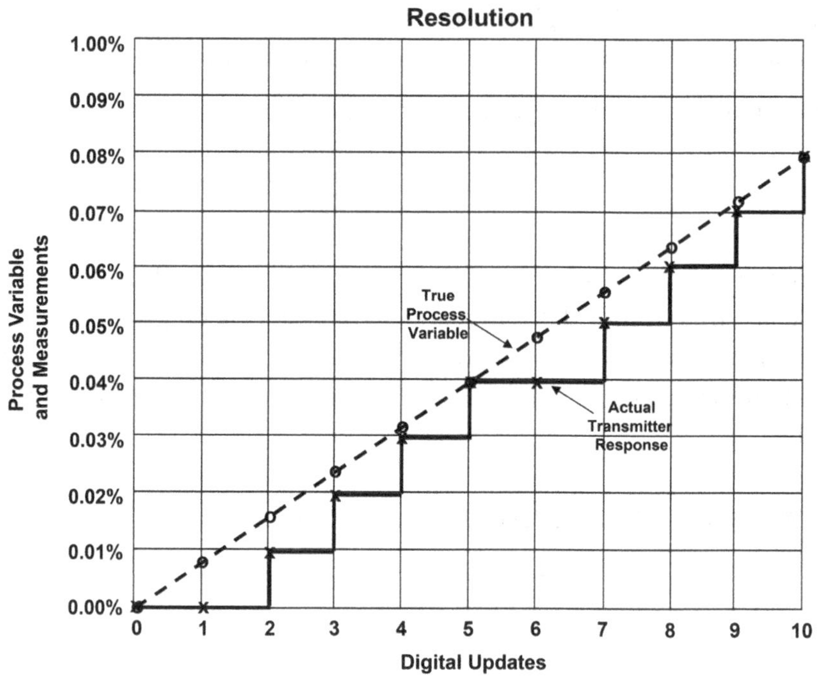

Figure A-6b. When the change in the process variable exceeds the *resolution limit*, the change in the process variable is seen as a step the size of the resolution limit.

change (stair step where the step size is the resolution limit, as depicted in Figure A-6b). Both will cause a limit cycle if there is an integrator in the process or control system.

- **Goal**
  - Improve threshold sensitivity and resolution.
- **Sources**
  - In measurements, minimum change detected and communicated (e.g., sensor threshold and wireless update trigger level) and quantized change of analog to digital (A/D) and digital to analog (D/A) converters
  - Minimum change manipulated (e.g., valve stick-slip and speed resolution)

A.7 Backlash Deadband and Hysteresis

- **"No problem, man—if you don't ever change direction."**
- **Fundamentals**
  - Hysteresis is the bow in a response curve between full-scale traverses in both directions, as depicted in Figure A-7a. It is normally much smaller and less disruptive than backlash deadband.

- o Backlash deadband is the minimum change measured or manipulated once the direction is changed. Once past backlash deadband limit, you get full change, as depicted in Figure A-7b.
- o Backlash deadband will cause a limit cycle if there are two or more integrators in the process or control system.

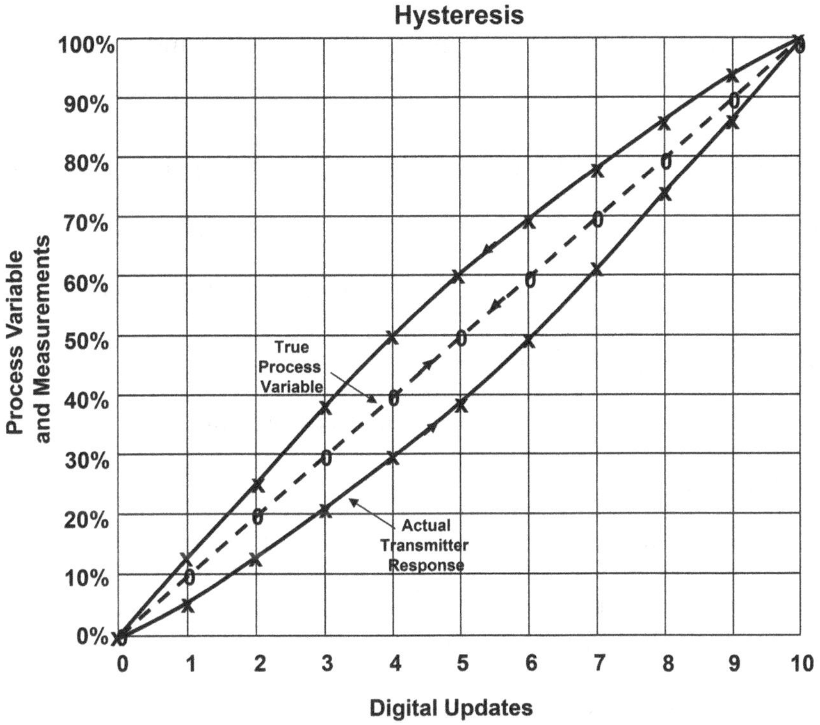

Figure A-7a. When the controller output changes direction, the path is bowed by the *hysteresis* but the change is immediately and fully seen.

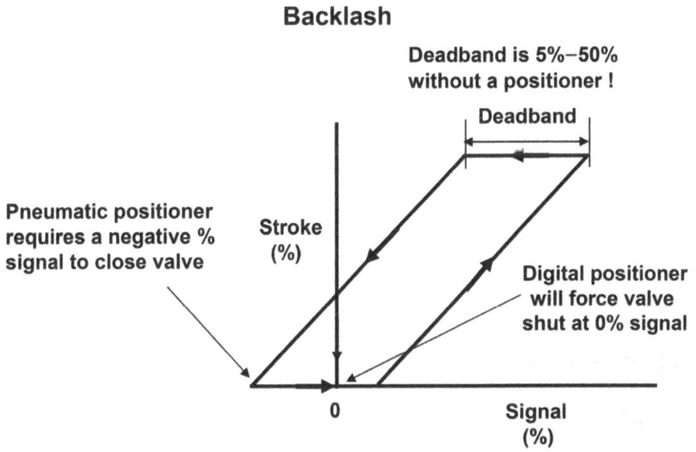

Figure A-7b. When the controller output changes direction and the change exceeds the *deadband*, the full change in the valve stroke or drive speed is seen.

- Deadband can be a parameter in a PID controller or a variable frequency drive (VFD).

- **Goals**
  - Minimize hysteresis in measurements (eliminate mechanical components).
  - Minimize backlash deadband in final control elements.
  - Minimize deadband settings in the PID controller and VFD.
- **Sources**
  - Pneumatic instrument flappers, links, and levers (hopefully these are long gone)
  - Rotary valve and damper links, connections, and shaft windup
  - VFD setup parameter to eliminate hunting and chasing noise

A.8 Repeatability and Noise

- **"The best thing you can do is not react to deception."**
- **Fundamentals**
  - Noise is extraneous fluctuations in measured or manipulated variables.
  - Repeatability is the difference in readings for the same true value in the same direction, as depicted in Figure A-8.
  - Repeatability is often confused with noise.
- **Goals**
  - Minimize the size and frequency of noise.
  - Do not transfer noise to the process.
- **Sources**
  - Noise
    - Bubbles
    - Concentration and temperature nonuniformity from imperfect mixing
    - Electromagnetic interference (EMI)
    - Ground loops
    - Interferences (e.g., sodium ions on pH electrodes)

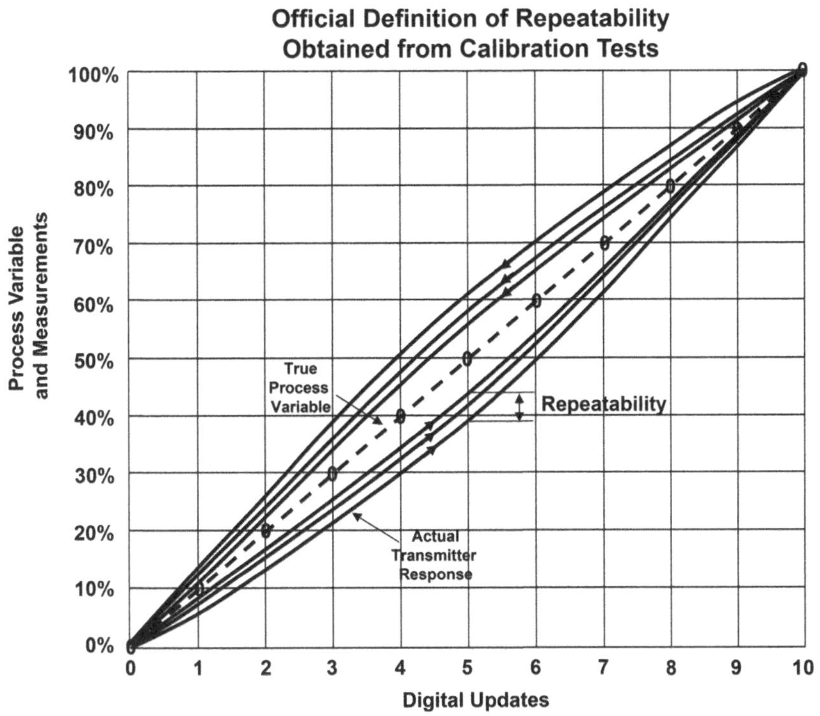

Figure A-8. The maximum separation of the traverses of the scale range in the same direction is the *repeatability*.

  - Velocity profile nonuniformity
  - Velocity impact on pressure sensors
- Repeatability
  - Threshold sensitivity, resolution, and extraneous effects

A.9 Offset and Drift

- **"There are always an offset and drift; it is matter of their size and consequence."**
- **Fundamentals**
  - Offset is the deviation of the peak in the distribution of actual values from true value, as depicted in Figure A-9a.
  - Drift shows up as a slowly changing offset, as depicted in Figure A-9b.
- **Goals**
  - Minimize offset and nonlinearity (see Concept 10) by using smart transmitters, sensor matching, and smart tuned digital positioners with accurate internal closure member feedback.

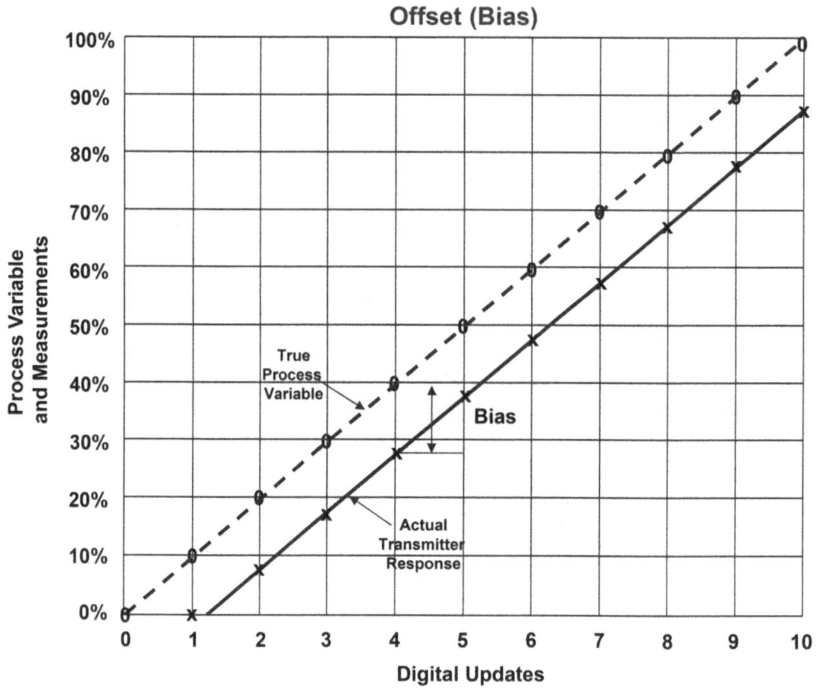

Figure A-9a. An *offset* (bias) is a constant difference between the process variable and the measurement in a complete traverse of the scale range.

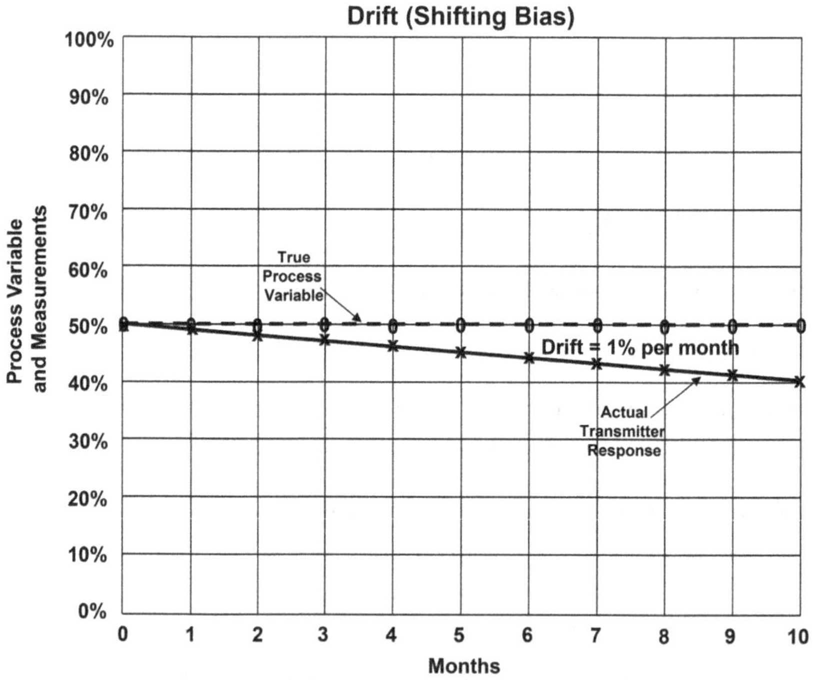

Figure A-9b. A shifting difference between the process variable and the measurement with time is *drift.*

- **Sources**
  - o Manufacturing tolerances, degradation, de-calibration, and installation effects (process and ambient conditions and installation methods and location)

A.10 Nonlinearity

- **"Not a problem if the process is constant, but then again if the process is constant, you do not need a control system."**
- **Fundamentals**
  - o While normally associated with a process gain that is not constant, in a broader view, a nonlinear system result occurs if a gain, time constant, or delay changes anywhere in the loop including measurement nonlinearity as depicted in Figure A-10. In addition, all process control systems are nonlinear to some degree.
- **Goals**
  - o Minimize nonlinearity by process and equipment design (e.g., reagents and heat transfer coefficients), smart transmitters and sensor matching, valve selection, signal characterization, and adaptive control.

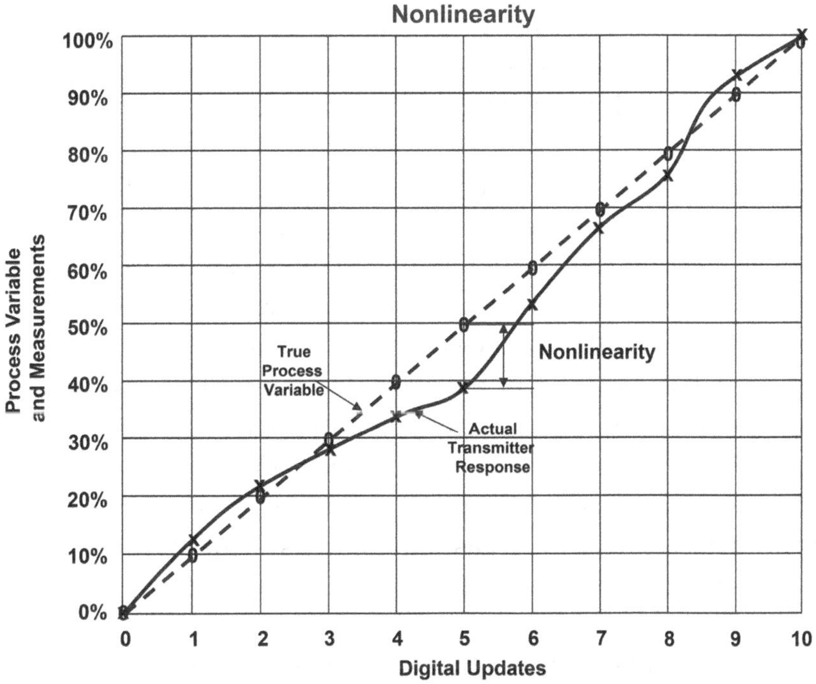

Figure A-10. A difference between the process variable slope and the measurement slope for a traverse of the scale range is *nonlinearity*.

- **Sources**
  - Control valve and variable speed drive installed characteristics (flat at high flows)
  - Process transportation delays (inversely proportional to flow)
  - Digital and analyzer delays (loop delay depends on when change arrives in discontinuous data value update interval)
  - Inferred measurement (conductivity or temperature versus composition plot is a curve)
  - Logarithmic relationship between sensor output and process concentration (glass pH electrode and zirconium oxide oxygen probe)
  - Process time constants (proportional to volume and density)
  - Shape of equipment volume (particularly cross-sectional area for level control)

A.11 Which Is More Important, Accuracy or Precision?

The accuracy of a measurement is determined as the difference between the true value and the measured value. The precision is how much the measured value changes for repeated measurements. In a plot of measurement frequency versus measured value for the same true process variable value (Figure A-11), a mean measurement value closer to the true process value is indicative of better accuracy. A tighter distribution of measurement values (small standard deviation) is indicative of better precision.

Inaccuracy can be corrected by changes in the set point, either manually or by upper-level loops, so that the effect is less on performance and more on process calculations and analysis. In contrast, a lack of precision adds an uncertainty for which there are few remedies for the resulting poor loop performance. For example, control valve stroke inaccuracy will introduce errors in system pressure drop and installed flow characteristic calculations, but the effect on loop performance will be minimal because the process control loop throttling the valve will adjust the signal to the valve as needed. Similarly, secondary flow measurement inaccuracy will create errors in material balance calculations, but the effect on process performance will be minimal because the primary controller in a cascade control system will eliminate the bias by adjusting the secondary controller set point.

Consequently, improving factors that affect precision, such as threshold sensitivity, repeatability, and resolution, and minimizing backlash, deadband, and noise, are

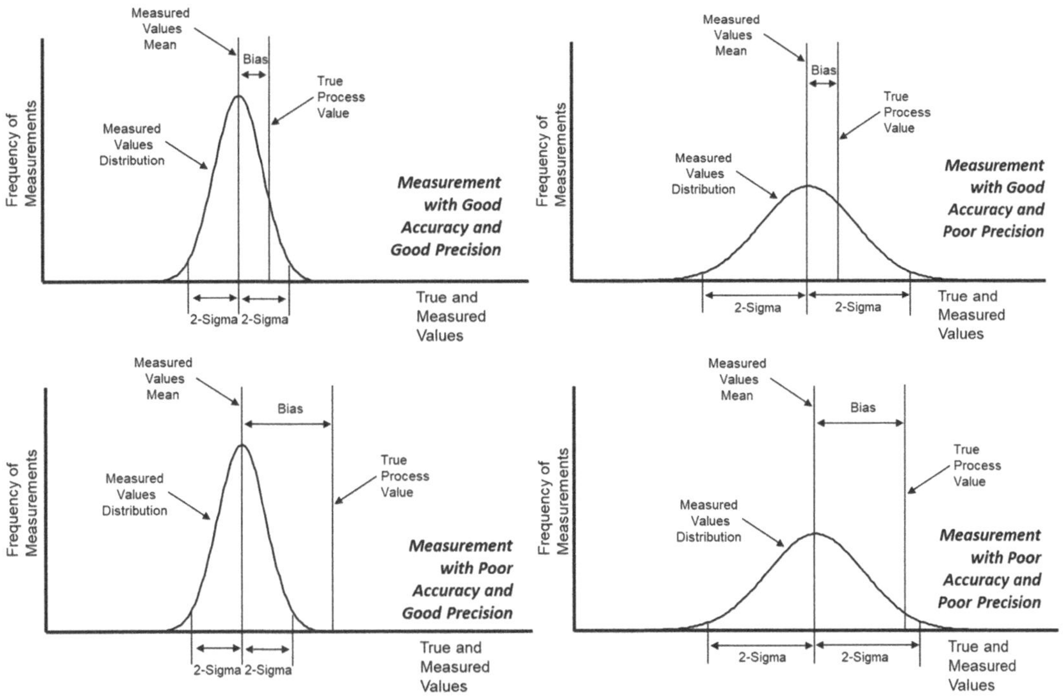

Figure A-11. *Accuracy* can be corrected by a set point, but *precision* is prevalent uncertainty.

more important than reducing factors that affect accuracy, such as offset, drift, and nonlinearity.

Besides the precision of measurements and final control elements, minimizing dead time and resonance, slowing down disturbances and making them smaller, optimizing process gains, and maximizing process attenuation are important for loop performance. What is left is controller tuning and possibly feedforward control.

Appendix B: Bioprocess Biology

B.1 Protein Production

In mammalian cells, once proteins are synthesized by ribosomes, they are passed through the rough endoplasmic reticulum (ER). After being modified by the rough ER, they are transported to the Golgi apparatus. The Golgi apparatus modifies the protein further and sorts and packages the protein in a vesicle. The vesicle buds off from the Golgi and moves to the plasma membrane. The vesicle then fuses with the plasma membrane and the protein is excreted into the extracellular space. At this point, the protein is correctly folded and modified, and completely functional. Hybridomas and other mammalian cells are ideal for bioreactor product formation because they can make, fold, package, and excrete the product protein in a functional form.

In bacterial cells, proteins are synthesized and folded in the cytosol. The bacterial cells lack the rough ER and Golgi apparatus that are present in mammalian cells. They often cannot excrete protein products. Because of this, recombinant bacterial cells commonly used in bioreactor product formation do not have the mechanism to excrete the large product or sometimes even fold it correctly. Often, bacterial cells must be broken open (lysed) once the product is formed in order to release the product into extracellular space and isolate the product out.

B.2 Temperature Effects

Temperature has a profound effect on the amount of energy that molecules have. The more energy a molecule has, the more it vibrates, bends, and moves. This movement affects the efficiency and shape of proteins in the cell, and leads to a change in functionality. Cell cultures have a large window of temperatures that they can survive in, but if the cell line is too far out of the optimal temperature range, the amount of product that the cell line will produce will be inhibited.

Primary and Secondary Metabolism

Cells metabolize food to make energy for cellular functions. Two types of metabolism are important in bioreactor cell lines, primary and secondary. Primary metabolism is energy use that is required to keep the cell alive and functioning. Any process that is essential for survival, growth, development, or reproduction can be considered primary metabolism. This type of metabolism is always carried out before anything else, as it ensures that when food, resources, or environmental conditions are not ideal, the cell can survive until they are. Only when conditions are ideal and the cell is at homeostasis can secondary metabolism begin. Secondary metabolisms are processes that are not vital to cell survival but can support and benefit the cell. For example, the product that recombinant *E. coli* produce in bioreactors is considered a secondary metabolite. Secondary metabolisms are carried out once the cell is "happy" and in optimal conditions.

Cold Temperatures

The colder the temperature of a cell culture, the slower the cell will carry out cellular processes. This can cause cells that are in a cold solution to grow very slowly, or become dormant. Cold temperatures are not usually lethal to most cell lines. However, freezing a cell culture can cause death due to fracture. The ice crystals that form when water freezes causes damage inside the cell and can kill the cell. A way around this is the addition of an agent such as DMSO (dimethyl sulfoxide). DMSO is an organic solvent that makes the ice crystals formed during water freezing much smaller, and can allow cell cultures to be frozen and thawed. Cells in cold temperatures may make product, but at a reduced rate. The slowing of cellular processes will mean the requirement of more energy for primary metabolism, and thus a slowing or stopping of secondary metabolites (product). If a cell becomes too cold, it may essentially stop cellular metabolic processes and go into a dormant stage until it is warm enough to begin growing again.

Hot Temperatures

Hot temperatures can be much more devastating to cell life. Unlike cold temperatures, extreme heat will cause bond breakage that leads to unfolding and denaturing of proteins. Protein denaturation leads to a loss of functionality that is usually irreversible. Cells have mechanisms that can help them survive extreme heats but they will severely reduce product formation. When the product is excreted by the cells as it is made, like in mammalian cells, it is subject to the high heat of the reactor, and temperatures above 41°C/314°K/105.6°F can cause loss of product in the solution.

At 41°C/314°K/105.6°F, proteins start to denature and lose functionality. Anything above this temperature is not ideal for most cells and can lead to cell death. Example: This is why fevers are so dangerous; a fever of 105.6°F will start to denature proteins in the brain and lead to severe damage and death.

Mammalian Cells

- Mammalian cells that are used in bioreactors usually fold and secrete product protein after it is made, leaving the protein to be affected by extracellular conditions. Temperatures above 41°C/314°K/105.6°F can easily denature extracellular product.

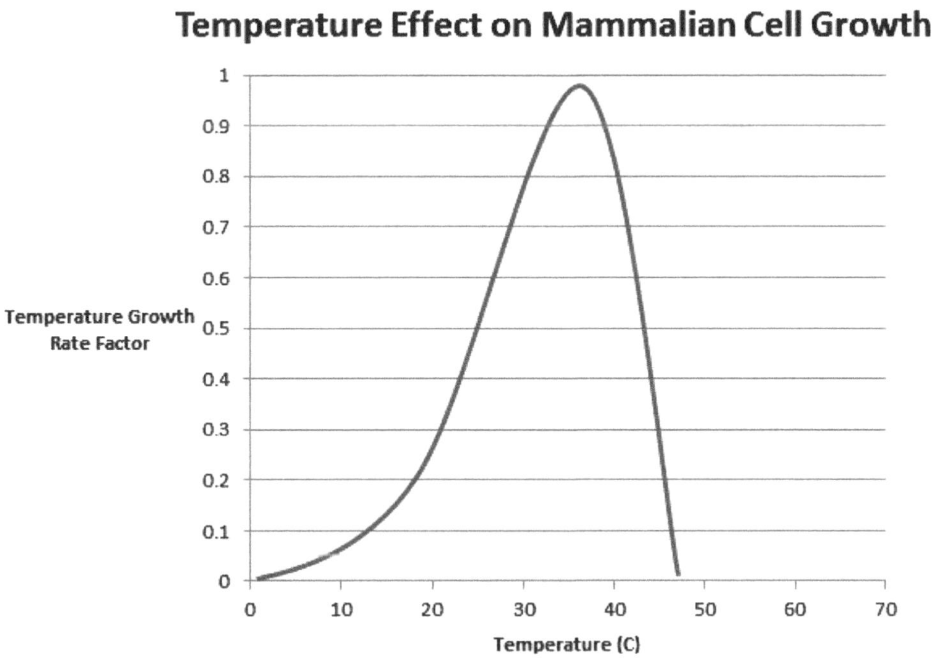

Figure B-2a. Effect of temperature on mammalian cell growth rate.

Ideal temperature for growth and product formation: 37°C

Acceptable product formation/growth range: 35°C to 40°C

Slow growth and little product below 30°C

No product and cell death above 41°C

Cells can freeze and die below 0°C

- Mammalian cells are more sensitive to heat than other kinds of cell lines are. They are used to working best at body temperature (37°C) and have a smaller range of acceptable temperatures, as exemplified in Figure B-2a. Because there is a cell membrane instead of a cell wall, they are more prone to physical damage from agitation.

Bacterial Cells

Bacterial cells can survive at temperatures above 41°C, but there is little to no growth and product.

- Product made by recombinant bacteria (*E. coli*) usually cannot be folded and secreted by the bacteria. This means that product has to be isolated after bacterial cell lysis. The product is better protected if the environmental heat does not exceed 41°C.
- Bacteria are much hardier and can survive well outside of their ideal temperature ranges. Product formation will be inhibited outside of ideal ranges, as exemplified in Figure B-2b.
- Bacteria are much better at living above 41°C/314°K/105.6°F than other cells are; however, the higher the temperature, the harder it is for the bacteria to grow, and they do start to die due to heat.

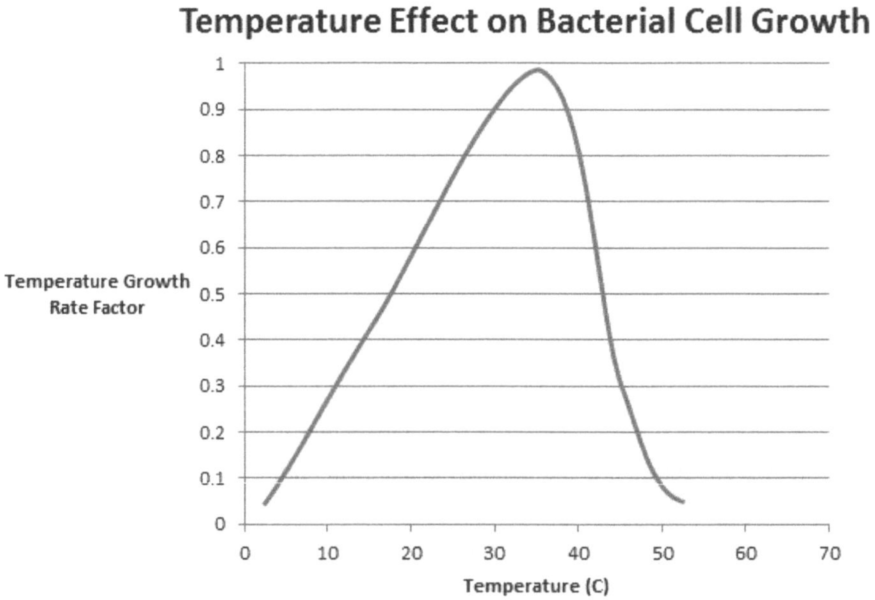

Figure B-2b. Effect of temperature on bacterial cell growth rate.

Ideal range: 25°C to 35°C

Freezes and dies: 0°C

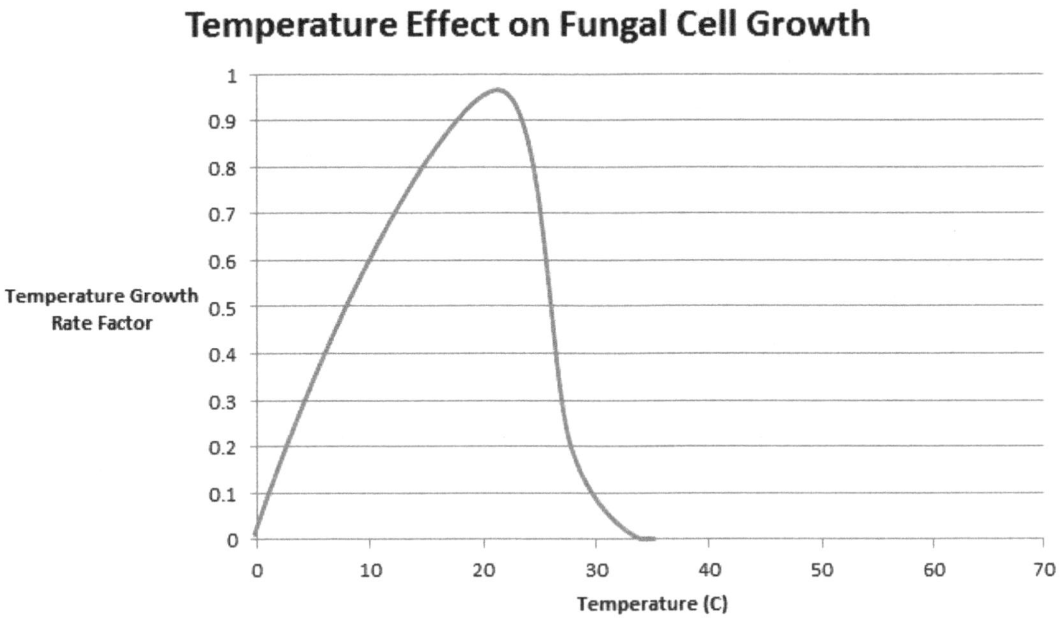

Figure B-2c. Effect of temperature on fungal cell growth rate.

Ideal range: 25°C to 35°C

Freezes and dies: 0°C

Fungal Cells

Fungal cells can survive at temperatures above 41°C, but there is little to no growth and product.

- Average fungal cultures prefer cooler temperatures as they thrive in the spring and fall temperatures of soils and dirt. They are able to survive in higher temperatures, but the best conditions for product formation occur at lower than body temperature, as exemplified in Figure B-2c. The ability to handle different temperatures is affected largely by fungal species, and a large amount of variation is common.

B.3 pH Effects

pH can drastically affect the ability of cells to survive, grow, and produce product. Putting a cell in an acidic condition will leave extra H+ ions floating around the cell membrane. With too many of these ions floating around, the proteins that are attached to the outside of the cell membrane or are part of the membrane itself will start to denature and break the hydrogen bonds that were keeping them folded. Once proteins in or on the outer cell membrane are denatured, pores start to form in the outer cell membrane due to the denatured proteins. These pores cause the cell to lose the ability to separate the outside extracellular environment and the inside cytosol, leading to cell lysis and death.

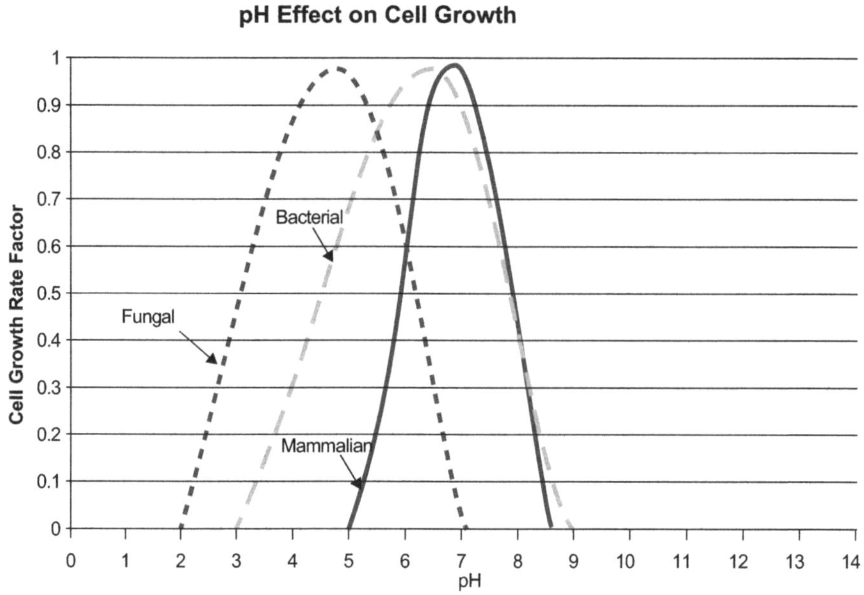

Figure B-3. Effect of pH on fungal, bacterial, and mammalian cell growth rates.

Too many OH– ions will cause protein denaturation as well. This time, the OH– ions will abstract hydrogens from folded proteins on the outer cell membrane, which will again denature the proteins. Pores form as a result of the dysfunctional and denatured proteins within/on the outer cell membrane, and the cell will lyse. Basic conditions are very harsh, and most species of microbes die if the pH gets above 8.5.

Cells have mechanisms that protect them from the effects of extreme acidic/basic conditions. The main way is by producing a cellular pH buffer that controls the pH around the cell to keep the pH as ideal as possible, as exemplified in Figure B-3. Keeping the extracellular pH at an ideal value is energetically costly to the cell. This prevents the cell from making product, and even if survival is possible at a harsh pH, product will be inhibited.

B.4 Glutamine Effects

Glutamine is an amino acid that is used for energy, protein synthesis, nitrogen donation, and carbon donation. Glutamine is commonly used by bioreactor cell cultures to support growth. The metabolism of glutamine can be either anaerobic or aerobic, and the presence of oxygen dictates the products of metabolism.

In the presence of O_2, glutamine can be completely metabolized into CO_2, H_2O, and NH4+ , through the Krebs cycle. However, even in the presence of O_2, ammonia is produced as a by-product, which is toxic to cell growth and product synthesis.

When no oxygen is present, glutamine can still be metabolized for energy. There are, however, different byproducts that can be made by the cell. Lactate, aspartate, and glutamate are all possibly formed after glutamine breakdown. Ammonia will still be released even in anaerobic breakdown. It is hard to say which one of these paths the glutamine will take, considering that glutamine is used for a variety of cell processes and is not limited to energy production. Even the by-products are useful in other aspects of the cell.

B.5 Osmolarity Effects

Osmolarity is a measurement of the amount of solute in a solution that contributes to the osmotic pressure of a chemical solution. It is defined as the number of osmoles of solute per liter of solution. In expression terms this is osmol/L or Osm/L. Osmolarity can be affected by all kinds of compounds. The most common are ionizable salts and non-ionizable species such as fructose/glucose. For example, 1 mole of NaCl dissolved in 1 L of water has an osmolarity of 2 osmol/L because the NaCl dissociates into 1 mole of Na+ and 1 mole of Cl– ions.

Salts in solution cause cells to shrink, grow, or stay the same size. The reason is that small charged particles that are considered salts are not able to pass freely across the cell's membrane. The inside of the membrane is nonpolar and stops the passing of charged ions through the cell membrane without the help of transporter proteins. Ion concentrations inside and immediately outside of the cell want to be at equilibrium. If there is a higher concentration of ions on one side of the membrane than on the other side, the ions will want to move across the membrane to even the concentrations. Because this is not possible, water (which can freely pass across the cell membrane) will start to move toward the area of high ion concentration. This happens because if the amount of water increases, the ion or salt concentration will decrease, bringing the salt concentration close to equilibrium. As you can guess, this can lead to problems in cell volume if there is too high or too low of an osmotic concentration.

When a cell is in an isotonic solution, the amount of solute (in this case salt or Na^+/CL^- concentration) is at the same concentration inside and outside the cell membrane. This causes a net movement of water across the membrane, and the cell will not change volume. This is equilibrium.

When a cell is in a hypertonic solution, the salt concentration is greater on the outside of the cell. This results in a net movement out of the cell to bring the salt concentration back to equilibrium. This causes the cell to shrink or crenate. Too much water volume loss by a cell will cause problems with cellular function and reproduction/growth and cause cell death. (Think of putting salt on a slug.)

Hypotonic solutions are caused when the ion concentration is higher inside the cell than outside. This causes a net water movement to go inside the cell to bring the concentrations to equilibrium. If too much water enters the cell, then the pressure pushing outward on the cell membrane will be too much and cause the cell to lyse. This will spill the cell contents and cause cell death.

A study conducted by Sadettin S. Ozturk and Bernhard O. Palsson called "Effect of Medium Osmolarity on Hybridoma Growth, Metabolism, and Antibody Production" recorded data that described the different effects that osmolarity has on hybridoma mammalian cells, as exemplified in Figures B-5a through B-5f. The following osmolarity figures are based on the data. Some important osmolarities to note:

- Ideal osmolarity for mammalian cells: 290 mOsm/kg
- Acceptable range for mammalian cells: 260 to 320 mOsm/kg
- Above 580 mOsm/kg causes cell death/no growth

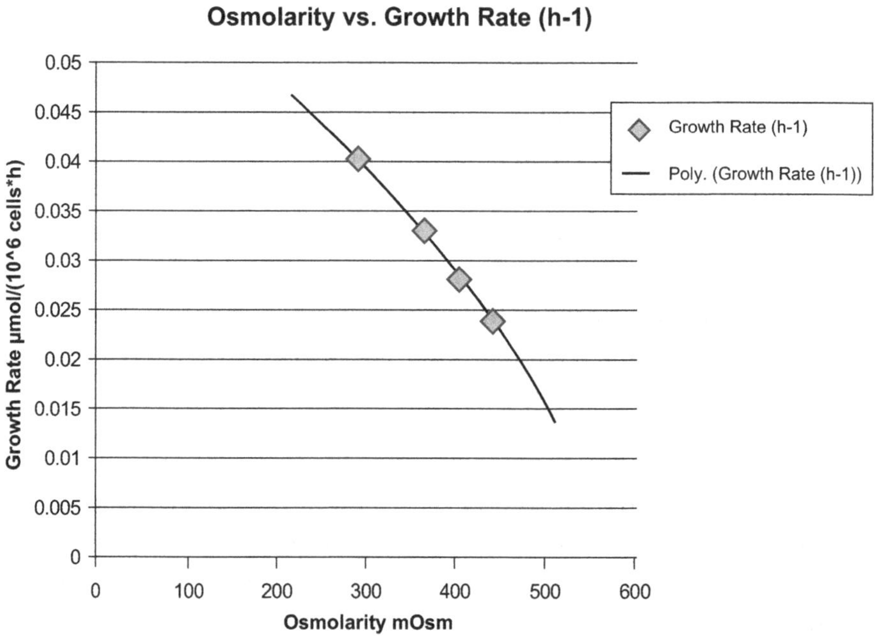

Figure B-5a. Effect of osmolarity on the growth rate.

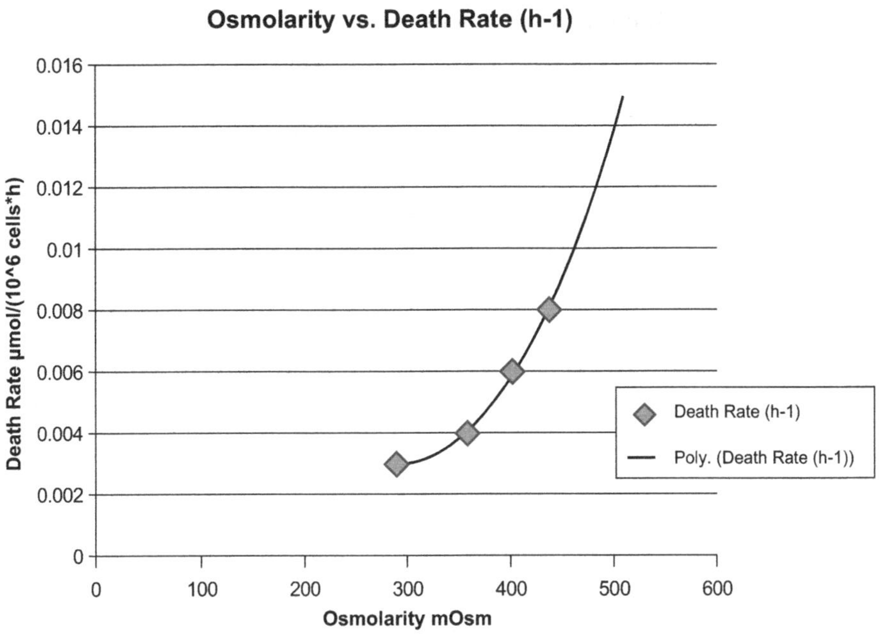

Figure B-5b. Effect of osmolarity on the death rate.

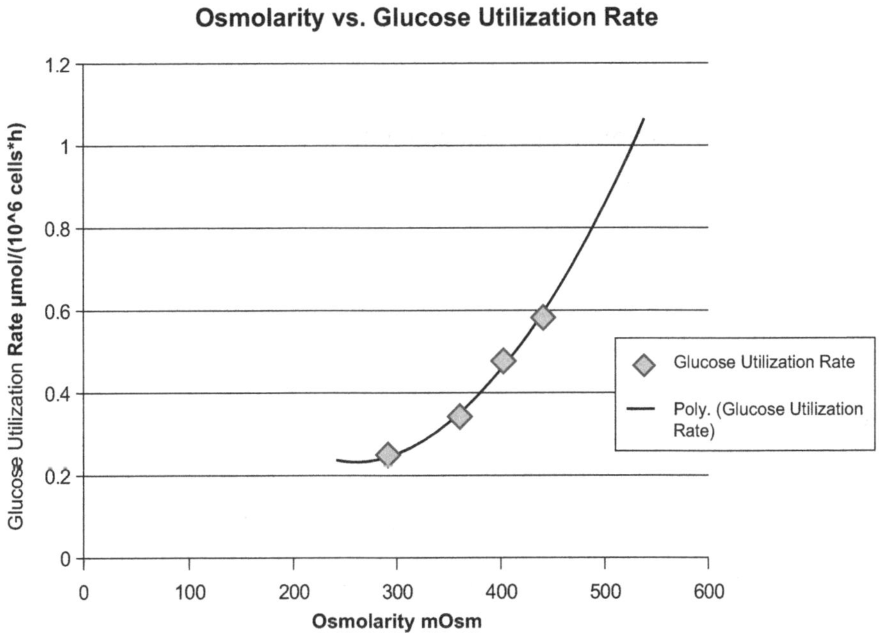

Figure B-5c. Effect of osmolarity on the glucose utilization rate.

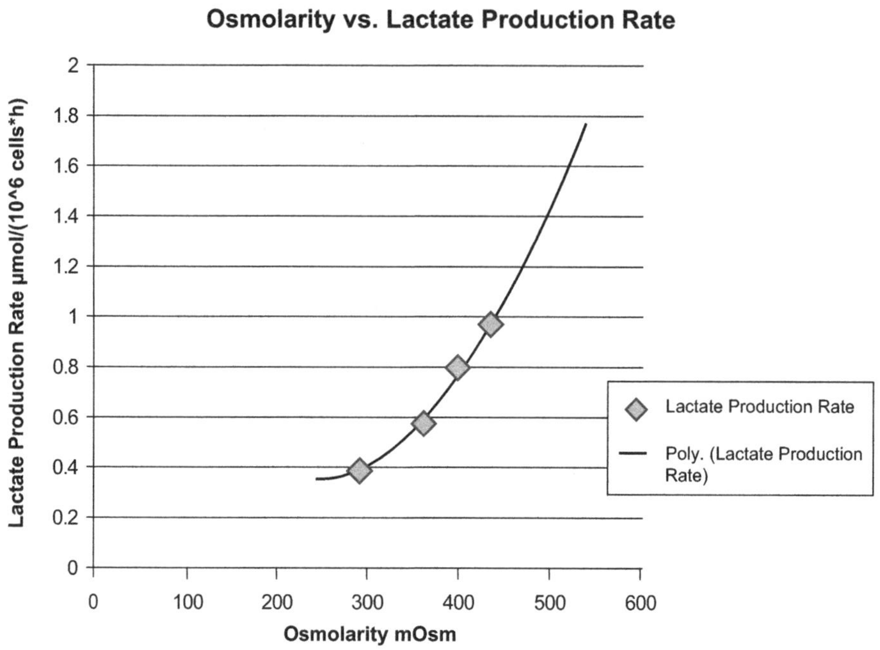

Figure B-5d. Effect of osmolarity on the lactate production rate.

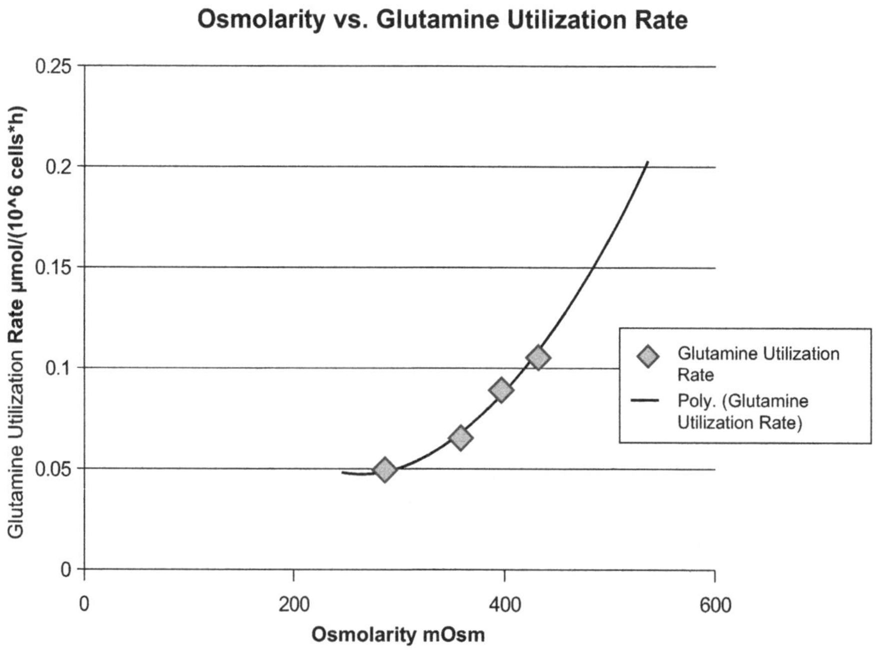

Figure B-5e. Effect of osmolarity on the glutamine utilization rate.

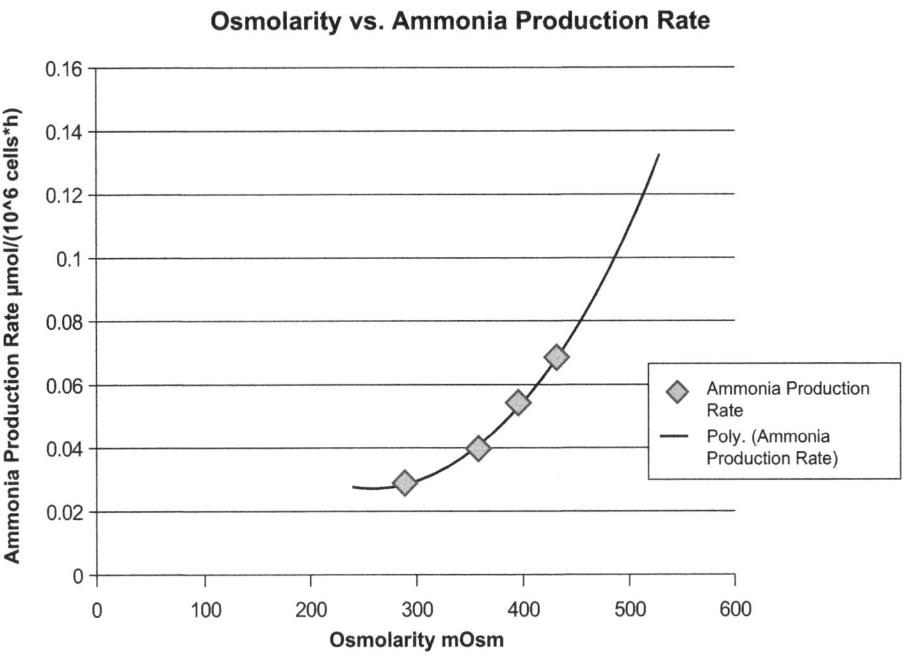

Figure B-5f. Effect of osmolarity on ammonia production rate.

Appendix C: Enhanced PID Controller for Wireless and Analyzer Applications

C.1 Introduction and Overview

It is widely recognized that wireless measurements considerably reduce installation and maintenance costs. However, it is less recognized that the portability of wireless offers the ability to optimize the measurement location for the fastest, most representative process response and to demonstrate/prototype process control improvement for finding and justifying permanent solutions.

The wireless update time, termed *default update rate* (refresh time), is the time interval for periodic reporting. The wireless update sensitivity, the *trigger level,* is the minimum change in the measurement value for exception reporting. The transmitter and its antenna are always turned "on" to enable the mesh network to function, but the electronics go to sleep and wake up periodically to check if the change in sensor value from the last value transmitted is larger than the trigger level. When the change exceeds the trigger level or the time since the last communication exceeds the default update rate, then the signal value is transmitted. The time interval between periodic checks of the sensor is the *triggered update rate* (wake-up time). If only periodic reporting is available, this time interval is the default update rate. Increases in the default update rate and trigger level settings reduce the number of transmissions, which increases battery life. The enhanced proportional-integral-derivative (PID) control eliminates the ramps, limit cycles, and spikes from large values of these settings facilitating an increase in battery life.

For slow processes where a significant process response takes minutes to hours to develop, such as composition, level, and temperature in large volumes, wireless measurements are fast enough for closed-loop control using traditional PID controller. The primary question for these applications is what is the potential detrimental effect of a loss of update and poor sensitivity? The enhanced PID controller protects against communication, sensor, valve, and analyzer failures and eliminates the cycling from wireless exception reporting and sensor, valve, or analyzer sensitivity limits for these and other loops. Thus, the enhanced PID controller improves the reliability and reduces the variability for all loops with and without wireless measurements. Additionally, the enhanced PID controller extends the applicability of wireless measurements and analyzers to processes with a response time faster than the update time [1, 2, 3].

While the enhanced PID controller offers advantages for all loops, wireless measurements should not be used where the process can cause product degradation, shutdowns, equipment damage, or an unsafe condition faster than the triggered update rate (exception reporting) or default update rate (periodic reporting) of the wireless transmitter. Examples of loops that can get into trouble too fast for even the fastest wireless update time are sheet, compressor, and turbine speed control and incinerator and pipeline pressure control. Slow wireless default update rates, such as 30 and 60 s, can be a problem for vessel or header pressure control because the pressure may approach alarm, trip, or relief settings within the default update rate.

In a traditional PID controller, the integral and derivative modes are computed with each execution of the PID block. The PID algorithm uses the execution time in the integral and derivative mode calculations. Thus, the reset and derivative contribution calculated by traditional PID control may not be appropriate when used with a wireless measurement where the default update rate is significant compared to the process response time. In such cases, the traditional PID controller will ramp the controller output through continual integral action even though the actual measurement has not changed. The PID is acting on old information. The detrimental effect of integral mode acting on old information is greatest if the process response has largely responded before the next update, which occurs when the process response is faster than the update time or the measurement value has not been updated because of device or communication failure. The update time of wireless measurements on fast processes, such as flow and pipeline temperature and pH loops, and the update time for analyzers in many applications are slower than the process response. An update failure occurs for loss of transmission, sensor failure, a stuck valve, and a sample or multiplexer system failure. Furthermore, when the wireless measurement is communicated using exception reporting, the ramp of the controller output between updates due to a sensitivity limit results in a perpetual equal amplitude oscillation called a *limit cycle*, even though there are no load disturbances.

When there is an update, the traditional PID controller considers the entire change in the measurement value occurred within the PID execution time for the derivative mode calculation. When the PID is executed faster than the wireless default update rate, the result is a spike in controller output whose size increases with the movement of the process between updates, which is greatest for a fast process or for recovery from an update failure.

The ramp, limit cycle, and spike from a traditional PID controller inflict a disturbance on the loop and possibly other loops. Thus, a traditional PID controller in automatic mode can cause process variability even if there are no disturbances.

The enhanced PID controller computes the integral and derivative mode contributions to the controller output when there is a measurement update and uses the elapsed time between updates in its calculations. Thus, the enhanced PID controller only acts on new information and considers the observed change in the measurement to have occurred not in just the last PID execution time but over the elapsed time. The key functional features of the enhanced PID controller are:

- PID control responds immediately to set-point, feedforward, and manual/remote output changes.
- PID control responds immediately to process variable changes.
- PID action is suspended if there are no changes (integral action stops).
- PID action uses elapsed time instead of PID execution time in derivative calculation.
- PID action computes positive feedback filter exponential response using elapsed time.

If the dead time from a discontinuous signal, whether a digital device or an at-line or online analyzer, is much greater than the open-loop time constant, the enhanced PID controller will see the full process response in the next update. For these cases, the controller gain can be set equal to the inverse of the process gain to provide considerably tighter control. Variation in the update time does not affect the tuning. Consequently, the enhanced PID controller is particularly valuable for closed-loop control using offline analyzers where the time between lab results is large and extremely variable.

When there is a loss of communication with the measurement or final control element (e.g., valve, damper, or variable speed drive), the enhanced PID controller provides no further reset action. The enhanced PID controller waits for new information, whereas the traditional PID controller output ramps to an output limit. When communication is restored, the enhanced PID controller acts on the new information only. The traditional PID controller sees the effect of the ramp to the output limit that occurred during the loss of communication or loss of valve travel. Additionally, the traditional PID controller considers the entire observed change to have occurred within the last execution of the PID output instead of the time duration of the failure. The result is a bump from gain action and spike from rate action in the traditional PID output.

The enhanced PID controller executes as fast as a traditional PID controller independent of the wireless default update rate or analyzer cycle and sample time. As a result, the enhanced PID controller is ready to immediately act on changes in the set point, mode, remote output, feedforward, and tuning. For a proportional gain that is the inverse of the process gain, the set-point response is particularly impressive in that the controller output goes to the value needed to reach the set point within the PID execution time. For a fast loop, the process is already at the set point before the next update.

Secondary loops are designed to be faster than primary loops for cascade control. If a wireless measurement is used in a secondary loop, the enhanced PID controller can prevent the secondary loop response from becoming too slow. The fast load response of the enhanced PID controller helps the secondary loop reject secondary disturbances. The fast set point response of the enhanced PID controller helps the primary loop reach its set point faster and reject primary loop disturbances.

Analyzers with sample systems offer a higher level of control. Ultimately, what you want to know and control is the composition in a process stream. However, the use of analyzers for closed-loop control is rather limited due to the long update time, excessive interruptions and extraneous values, and low sensitivity. Analyzer sensitivity is generally not as good as flow, pH, pressure, and temperature measurement sensitivity. The analyzer sensitivity limit is the result of interferences, analysis method, and sensor sensitivity. Analyzer update time, poor sensitivity, and noise result in a step response that precludes the use rate action even if the process time constant is large. The enhanced PID controller eliminates most of these concerns and enables more analyzers to be used for closed-loop control [1, 2, 3].

If there is a resolution instead of a sensitivity limit, where the response is quantized so that the increment or decrement in the measurement is an integer multiple of the resolution limit, the steps in the response never put the measurement at the set point. Analyzers with sample systems frequently have resolution limits due to analysis methodology and minimum dilution and reagent dose sizes. The enhanced PID controller will eliminate the cycling from reset action and the spikes from rate action for both resolution and sensitivity limits in the measurement. The enhanced PID controller can also eliminate the limit cycles from valve stick-slip and backlash (valve sensitivity limits).

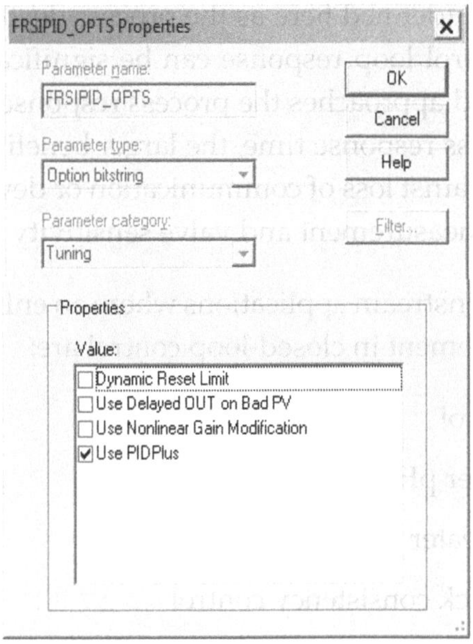

Figure C-1. Enhanced PID control enabled in FRSIPID_OPTS for the PID block.

The enhanced PID controller is available starting in DeltaV v11. The enhanced PID option can be enabled under the FRSIPID_OPTS for the PID block in control studio (Use PIDPlus) as shown in Figure C-1. This PIDPlus option turns on the Dynamic Reset Limit option.

C.2 Target Applications

While the enhanced PID controller can be used in any application to reduce process variability and provide protection against measurement and valve failures, the greatest benefits are seen for loops where:

- Update time is larger than the process response time.
- Update time is significant and variable.
- Update sensitivity limit is larger than the desired process precision.
- Update failures can occur (e.g., communication and electrode failures).
- Measurement resolution or sensitivity is poor, causing limit cycles.
- Valve exhibits significant stick-slip and backlash, causing limit cycles.
- Variable speed drive setup has significant deadband, causing limit cycles.

Because the enhanced PID controller takes into account the effect of update time, the benefit is greatest where the wireless or analyzer update time is much larger than the process response time (defined here as the process dead time plus time constant). The improvement in control loop response can be significant but diminishes as the update time decreases and approaches the process response time. For an update time that is less than the process response time, the larger benefit of an enhanced PID controller is the protection against loss of communication or device failure or the elimination of limit cycles from measurement and valve sensitivity limits.

Some examples of mainstream applications where an enhanced PID controller provides the greatest improvement in closed-loop control are:

- Wireless flow control
- Wireless static mixer pH control
- Wireless desuperheater temperature control
- Pulp and paper stock consistency control
- Cross directional (CD) control of sheet thickness
- At-line analyzer control of pipeline composition
- At-line analyzer control of plug flow reactor product composition
- Off-line analyzer control of almost any unit operation

Examples of mainstream applications with less but still significant improvement in closed-loop control are:

- Wireless vessel pressure control
- Wireless extruder temperature control
- Wireless jacket temperature control
- Wireless heat exchanger temperature control

C.3 Test Setup

A DeltaV v11 module with 1 s execution time was set up with two PID blocks. The first PID block (PID1) had the enhanced PID option enabled in the FRSIPID_OPTS parameter, whereas the second PID block (PID2) did not. The PID analog inputs were provided by SAMPLE_DATA02 composite blocks that simulated the update time and sensitivity limits for wireless transmitters and analyzers (TRANSMITTER1 for PID1 and TRANSMITTER2 for PID2). A composite block (PERFORMANCE02) for performance metrics tallies the number of communications and computes the integrated absolute error (IAE) and peak error for load disturbances and the rise time, settling time, overshoot, and undershoot for set-point changes. The configuration of the module is shown in Figure C-3.

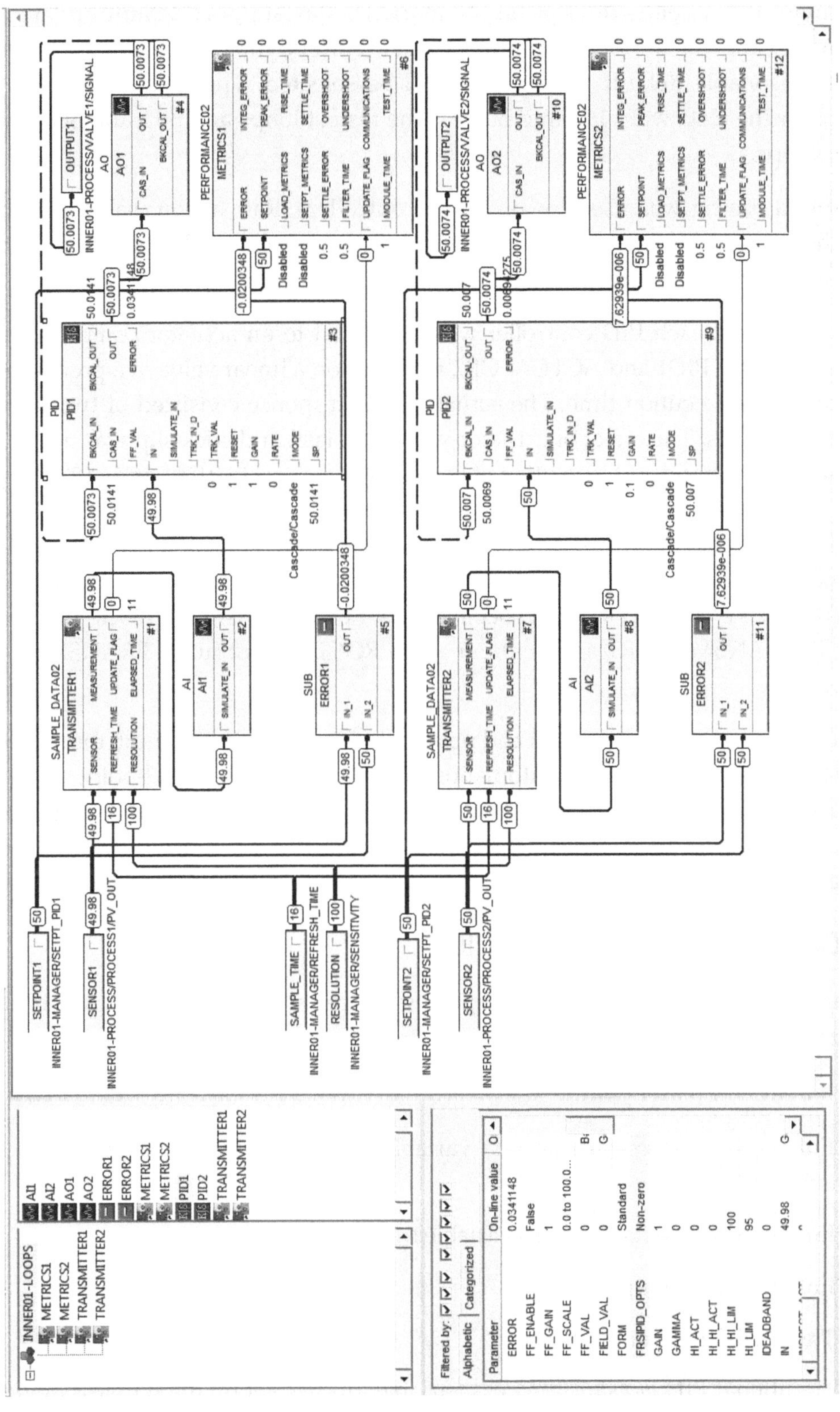

Figure C-3. Digital twin test setup for comparison of enhanced and traditional PID controllers.

Many of the wireless devices on the market today support default update rates of 8, 16, 30, and 60 s. However, to maximize battery life, it is assumed that the longest reasonable default update rate will be selected. The default update rates selected for the tests were 16 s and 30 s for the fastest loops and 30 s and 60 s for the slowest loops.

Lambda tuning was used for both PID controllers. Rate action was not used. In all of the cases studied except vessel temperature control, the PID2 gain had to be decreased to prevent excessive oscillations.

The output of each PID controller was the signal to an actuator composite block (ACTUATOR1 for PID1 and ACTUATOR2 for PID2) on a linear valve in a process module with a 0.2 s execution time. The actuator time response consisted of two 1 s time constants in series. The effect of actuator-valve backlash and stick-slip was not part of study, so the deadband and resolution of the actuator was set to 0.02% and 0.01%, respectively. The actuator block simulated the stroke and computed valve travel. The output of the valve for each loop was an input to a composite template block that simulated the dynamic response of a self-regulating and integrating process. The output of each process block is connected to the respective sensor of the transmitter block (PROCESS1 output is TRANSMITTER1 sensor input and PROCESS2 output is TRANSMITTER2 sensor input).

The setup of process dynamics and update time and sensitivity limit; the computation of tuning settings; and the automation of tests for load steps, set-point steps, and measurement failure were supplied by a test manager module with an execution time of 2 s. The manager module also turned on the performance metrics and the valve travel accumulation during the tests for load upsets and set-point changes.

C.4 Discussion of Test Results

The variables in the trend charts in Figures C-5a, b, and c and Figures C-6a, b, and c are as follows:

- Enhanced PID set point
- Enhanced PID *measured* process variable (update set by the wireless device or analyzer)
- Enhanced PID *actual* process variable (instantaneous update)
- Enhanced PID output
- Traditional PID set point
- Traditional PID *measured* process variable (update set by the wireless device or analyzer)

- Traditional PID *actual* process variable (instantaneous update)
- Traditional PID output

The size of the step change in load or set point for each loop was 10%, except for vessel pressure where the step size was reduced to 5% due to the small vessel size. The actual process response included two 1 s actuator lags besides in addition to the stated process delay (dead time), process lag (time constant), and process gain. Tables C-4a and C-4b show the set-point and load performance metrics, respectively, for all of the test loops.

C.5 Flow Loop

Flow loops are the most common loop in the process industry. The process dynamics are fast, with a response time less than 2 s. The use of a wireless measurement with a traditional PID controller will oscillate severely unless the loop is detuned. The approach taken in the tests was to severely reduce the controller gain approaching an integral-only type of control. The reset time was kept at its original value to help eliminate offsets for ratio control.

The enhanced PID tuning could be made more aggressive than the tuning for a flow loop with a wired measurement. The controller gain could be increased to be the inverse of the process gain. For step set-point and load changes (as shown in Figures C-5a and C-5b), this enabled the enhanced PID controller to make a single almost exact correction. For a set-point change, the correction occurs within one module execution

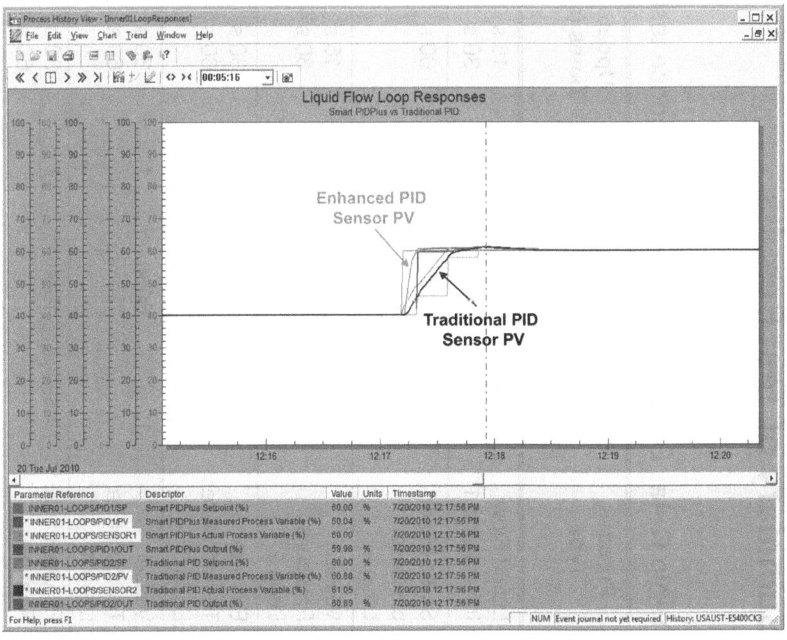

Figure C-5a. Flow set-point response of enhanced and traditional PID (update time = 16 s) controllers.

Summary of Test Results

Table C-4a. Summary of set-point response test metrics.

Process Type	Process Delay (s)	Process Lag (s)	Process Gain	Update Times (s)	Overshoot PIDPlus, PID (%)	Rise Time PIDPlus, PID (s)	Settling Time PIDPlus, PID (%)	Lambda PIDPlus, PID (s)	Gain Setting PIDPlus, PID	Reset Setting PIDPlus, PID (s)
Liquid Flow Inner Loop 1	0.5 s	1 s	1	16 s 30 s	0.5, 1.1 0.0, 2.3	9, 29 10, 23	25, 57 10, 98	2, 2	1.0, 0.1	1, 1
Static Mixer pH Outer Loop 1	6 s	6 s	1	60 s	0.0, 4.0	38, 94	38, 244	8.5, 8.5	1.0, 0.1	6, 6
Vessel Pressure Loop 2	1 s	—	0.1 s^{-1}	16 s 30 s	6.6, 4.9 11.0, 8.2	16, 26 16, 27	86, 156 233, 276	20, 20	0.8, 0.4	42.5, 42.5
Extruder Temperature Loop 3	24 s	6 s	1	30 s 60 s	2.2, 3.9 0.0, 7.9	74, 91 181, 85	167, 286 181, 293	25.5, 25.5	0.5, 0.1	6, 6
Vessel Temperature Loop 4	12 s	60 s	1	30 s 60 s	6.0, 8.7 7.3, 12.1	50, 42 51, 39	199, 272 276, 288	13.5, 13.5	2.2, 2.2	60, 60
Polymer Line Analyzer Loop 5	1 s	10 s	1	30 s 60 s	2.5, 0.0 0.4, 1.9	25, 128 39, 81	63, 128 39, 147	2.5, 2.5	1, 0.2	10, 10

Table C-4b. Summary of load response test metrics.

Process Type	Process Delay (s)	Process Lag (s)	Process Gain	Update Times (s)	IAE PIDPlus, PID (%* min)	Peak Error PIDPlus, PID (%)	Valve Travel PIDPlus, PID (%)	Lambda PIDPlus, PID (s)	Gain Setting PIDPlus, PID	Reset Setting PIDPlus, PID (s)
Liquid Flow Inner Loop 1	0.5 s	1 s	1	16 s 30 s	6.5, 11.9 9.2, 19.2	20, 20 20, 20	30, 33 30, 65	2, 2	1.0, 0.1	1, 1
Static Mixer pH Outer Loop 1	6 s	6 s	1	60 s	45.3, 75.5	20.0, 19.7	—	8.5, 8.5	1.0, 0.1	6, 6
Vessel Pressure Loop 2	1 s	—	$0.1\ s^{-1}$	16 s 30 s	15.3, 32.0 34.8, 53.2	15.1, 21.5 31.5, 32.0	35, 40 82, 79	20, 20	0.8, 0.4	42.5, 42.5
Extruder Temperature Loop 3	24 s	6 s	1	30 s 60 s	34.9, 42.5 69.6, 67.9	20, 20 19.6, 18.8	33, 46 28, 60	25.5, 25.5	0.5, 0.1	6, 6
Vessel Temperature Loop 4	12 s	60 s	1	30 s 60 s	17.2, 16.6 27.4, 40.6	8.9, 8.0 13.9, 14.3	50, 97 59, 212	13.5, 13.5	2.2, 2.2	60, 60
Polymer Line Analyzer Loop 5	1 s	10 s	1	30 s 60 s	13.4, 27.0 23.1, 33.0	11.8, 15.7 11.9, 15.9	31, 33 30, 40	2.5, 2.5	1, 0.2	10, 10

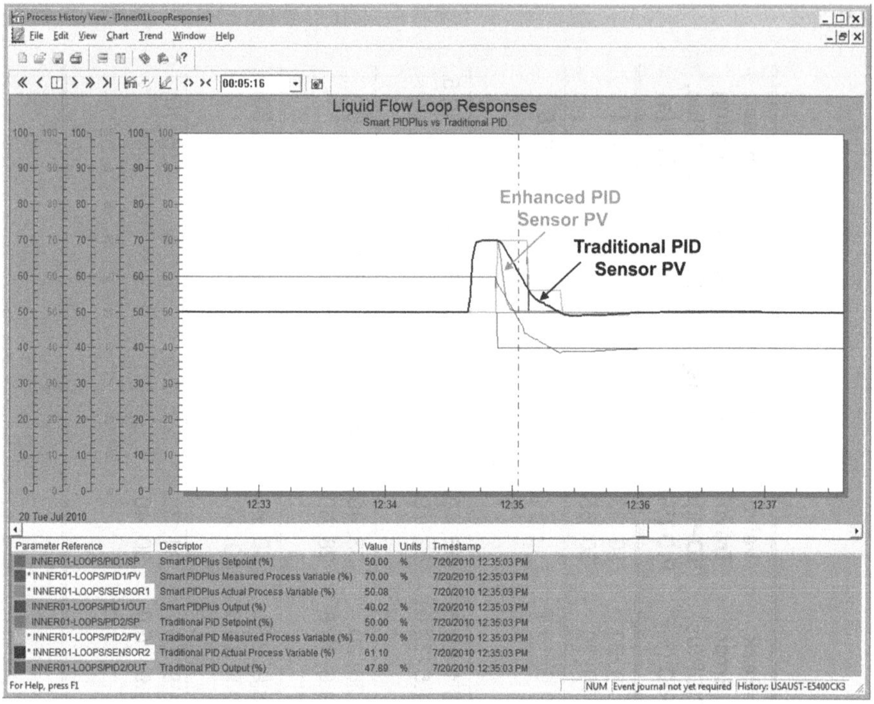

Figure C-5b. Flow load response of enhanced and traditional PID (update time = 16 s) controllers.

time. How soon the response fully appears in the actual flow measurement is limited only by the valve response time (two 1 s lags in series), the process dead time (0.5 s), and the process time constant (1 s). However, this incredibly fast response is not apparent to the user because the measurement update is much longer. In the trend charts, the actual process measured process variable is plotted. The enhanced PID load response is much faster than a traditional PID load response even though the correction is delayed on the average by about half of the update time after the upset. The exact delay in the recognition of the disturbance depends on when the disturbance arrives in the update time interval.

For a 16 s pH update time and load upsets, the enhanced PID controller reduced the integrated absolute error (IAE) by 45%. Because the flow loop has a loop dead time that is not significantly less than the process time constant, the peak error of the enhanced PID controller and the traditional loop for a load disturbance is the full open-loop error (error with loop in manual). For a 16 s pH update time and set-point changes, the enhanced PID controller reduced the rise time (time to reach the set-point) by 69%, the settling time (time to settle at the set point) by 56%, and the overshoot by 55%.

If there is an update failure, the traditional PID controller will ramp toward an output limit. When the measurement is restored, the traditional PID controller sees the huge self-inflicted upset and overshoots the set point in its recovery. The enhanced PID controller

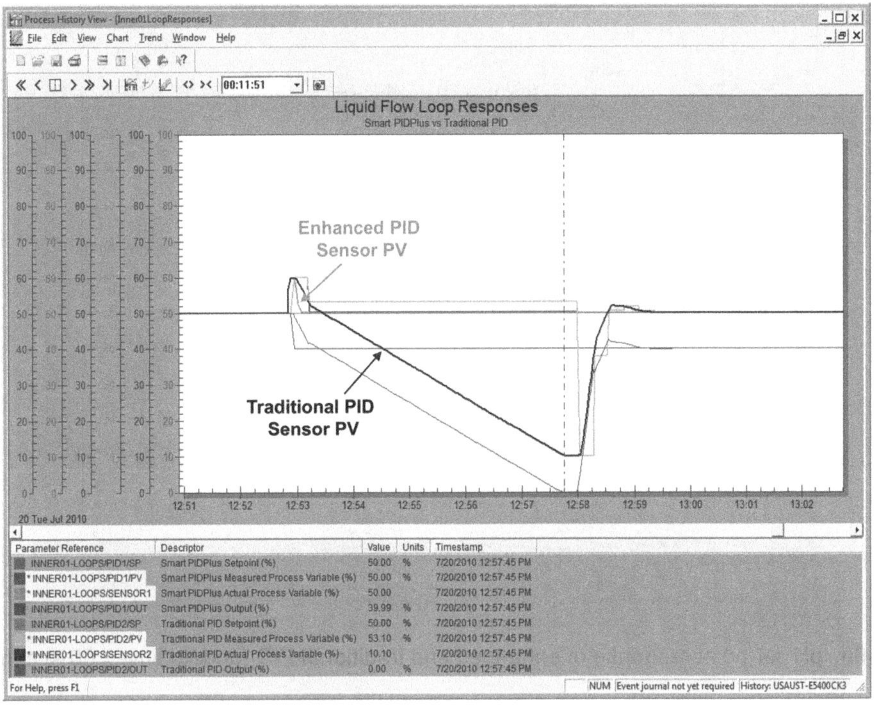

Figure C-5c. Flow failure response of enhanced and traditional PID (update time = 16 s) controllers.

corrects for upsets just before the failure and waits for the next update. The bump to the process is minimal, as shown in Figure C-5c, if there are no upsets during the failure.

C.6 Static Mixer pH Loop

The enhanced PID controller dramatically improves the response of cascade loops by enabling an immediate response of the secondary loop to changes in its set point. Because the static mixer process dead time (6 s due to injection delay) and process time constant (6 s due to electrode lag) is much less than the default update rate (60 s), the enhanced PID controller gain can be increased. With both the secondary flow and the primary pH loop controller gain equal to the inverse of the process gain, the response to pH set-point changes and load disturbances is impressive, as shown in Figures C-6a and C-6b. In contrast, the traditional PID controller gains for both loops were a factor of 10 lower to prevent oscillations. If the default update rate of the secondary loop is not about four times faster than the primary loop, the enhanced PID controller gain must be detuned to stabilize the loops because the oscillations from the secondary loop can be amplified by the primary loop. For a primary pH loop default update rate of 60 s, the secondary flow default update rate must be 16 s or faster.

For a 60 s pH update time and load upsets, the enhanced PID controller reduced the IAE by 40%. Because both the flow and pH loop have a process dead time that is not significantly less than the process time constant, the peak error of the enhanced PID

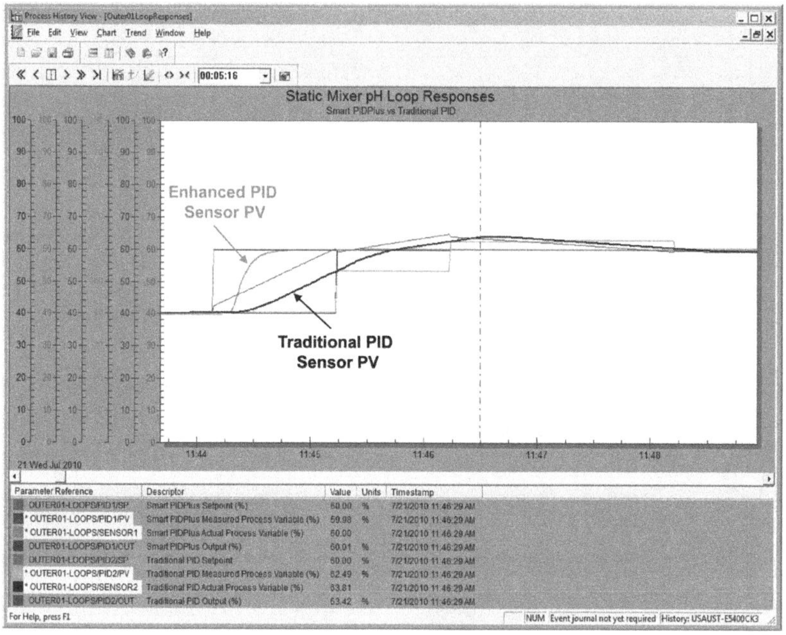

Figure C-6a. pH set-point response of enhanced and traditional PID (update time = 60 s) controllers.

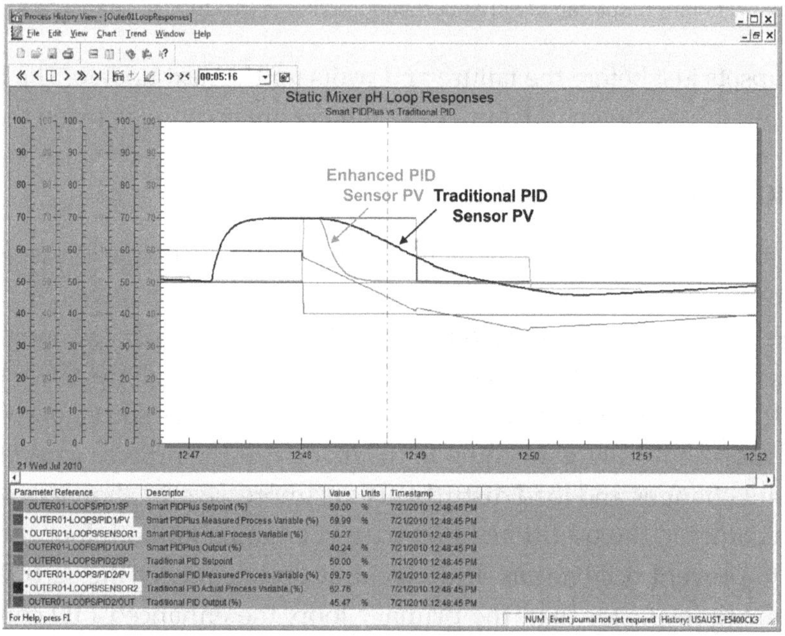

Figure C-6b. pH load response of enhanced and traditional PID (update time = 60 s) controllers.

controller and traditional loop for a load disturbance is the full open-loop error (error with loop in manual). For a 60 s pH update time and set-point changes, the enhanced PID controller reduced the rise time by 60%, reduced the settling time by 84%, and eliminated the overshoot that was 40% of the set-point change. For a failure to update, the bump to the process is minimal, as shown in Figure C-6c, if there are no upsets during the failure.

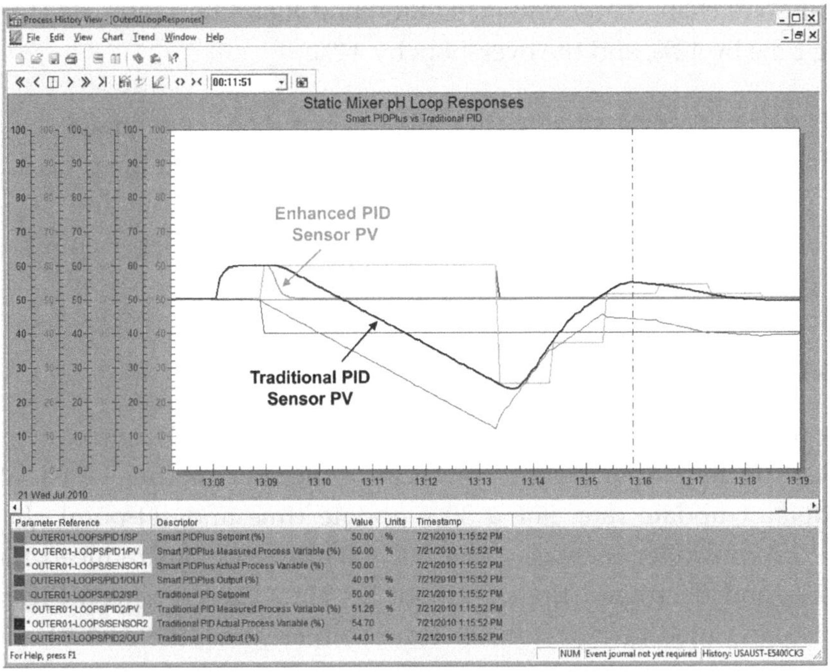

Figure C-6c. pH failure response of enhanced and traditional PID (update time = 60 s) controllers.

C.7 Vessel Pressure Loop

The vessel pressure loop has an integrating response with a short process dead time (1 s) and a relatively fast integrating process gain (0.1 %/sec/%) due to a small vessel volume or narrow measurement span. Even with a relatively fast default update rate of 16 s, control is challenging. The set point and load change for this test was 5%, whereas for all other tests it was 10%.

For a 16 s update time and load upsets, the enhanced PID controller reduced the IAE by 52% and the peak error by 30%. The reduction in peak error is important to prevent activation of relief devices and safety instrumented systems (SISs). For a 16 s update time and set-point changes, the enhanced PID controller reduced the rise time by 39% and the settling time by 45%. However, the overshoot of the enhanced PID controller was 35% larger. Set-point rate of change limits could be used to reduce the overshoot, but this would increase the rise time.

C.8 Extruder Temperature Loop

For a 30 s update time and load upsets, the enhanced PID controller reduced the IAE by 18%. Because the extruder has a process dead time larger than the process time constant, the peak error of the enhanced PID controller and traditional loop for a load disturbance is the full open-loop error (error with loop in manual). For a 30 s update time and set-point changes, the enhanced PID controller reduced the rise time by 19%, the settling time by 42%, and the overshoot by 44%.

An extruder may have some radial mixing, but there is negligible axial mixing (back mixing). As a result, the process dead time is larger than the process time constant, and process fluctuations with time are not smoothed out and appear in the product. Consequently, the ability of the enhanced PID controller to eliminate process variability directly translates to improved product quality.

C.9 Vessel Temperature Loop

For a 30 s update time and load upsets, the enhanced PID controller offered no improvement in the IAE or peak error because the process time constant is larger than the default update rate. For a 30 s update time and set-point changes, the enhanced PID controller increased the rise time by 19% but reduced the settling time by 27%, and decreased the overshoot by 31%. The enhanced PID controller could have been tuned for a faster rise time by increasing the controller gain. The vessel studied was exceptionally small. The performance of temperature loops is excellent on the larger vessel sizes typically seen in industrial plants if noise, sensitivity, and resolution limits are not significant and the controller is tuned with the high gain permitted by the extremely large process time constant. Performance improvement is normally not needed to enable the use of wireless temperature transmitters on large volumes.

C.10 Polymer Line Analyzer Loop

For a 60 s update time and load upsets, the enhanced PID controller reduced the IAE by 30% and the peak error by 25%. For a 60 s update time and set-point changes, the enhanced PID controller reduced the rise time by 52%, the settling time by 74%, and the overshoot by 79%.

A polymer line analyzer may have some radial mixing, but there is negligible axial mixing (back mixing). Thus, the enhanced PID controller offers an exceptional improvement in polymer and sheet line product quality by reducing variability.

C.11 Sensitivity Limit

In the future, most wireless devices will support exception reporting. When exception reporting is selected, the sensitivity limit is the minimum change that is reported. Once the change exceeds the trigger level (sensitivity limit) or the time since the last communication exceeds the default update rate, the full change and exact change are reported. For a traditional PID controller, a sensitivity or resolution limit, whether due to wireless exception reporting, sensor, valve, or analyzer sample and processing system, will cause a continuous oscillation known as a *limit cycle*. The amplitude of the square wave in the measurement is the sensitivity or resolution limit. Because the valve or measurement did not respond, traditional PID reset action continues to drive the output even though there is no process disturbance. When there is an update, the jump in the measurement goes beyond the set point and the ramp goes in the opposite direction. This ramping of the traditional PID output up and down continues indefinitely. Load disturbances and set-point changes will temporarily disrupt the pattern. Because the enhanced PID integral mode only makes a calculation when there is an update, there is no ramping of the controller output. If there are no set-point changes or load disturbances, the measurement draws a straight line. The actual process variable is closer to the set point than indicated by the measurement. The benefits are less process variability and less valve travel (longer valve packing life).

The enhanced PID controller will also inherently eliminate limit cycles from control valve stiction and backlash and variable-speed-drive resolution limits, as shown in Figure C-11. When the final control element is not responding, there is no process variable update unless there is a disturbance, and consequently the ramping action of the

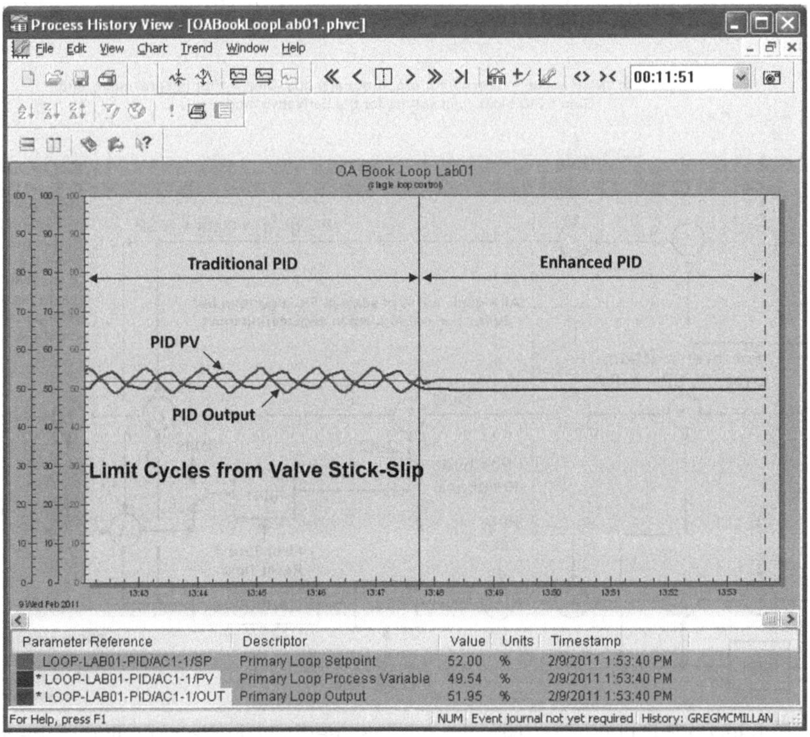

Figure C-11. Enhanced PID control eliminates the limit cycle that persists in traditional PID controller.

integral mode is suspended. To ensure measurement noise does not trigger an update, any signal filtering needed should be judiciously done by transmitter damping. The wireless update trigger level should also be increased to screen out noise. To absolutely prevent an unnecessary update that would trigger PID action, a simple small sensitivity limit should be added to the process variable used by the PID block.

C.12 Block Diagram

A key part of the enhanced PID controller uses the positive feedback implementation of the integral mode. When the enhanced PID capability is enabled, the output of the filter block in the positive feedback path of Figure C-12 is computed as a first-order exponential response for the elapsed time. The derivative calculation uses elapsed time instead of the module execution time. The enabling of external-reset feedback (e.g., dynamic reset limit) by the enhanced PID controller protects against a burst of oscillations when the primary PID output tries to change faster than a secondary PID controller, control valve, or variable speed drive can respond. The external-reset feedback also allows the use of directional move suppression by setting up and down set-point rate limits without having to retune the enhanced PID controller. The directional move suppression is useful for providing a slow approach to an optimum and a fast getaway for an abnormal condition and for slowing down movement to the split-range point reducing unnecessary crossings of the split range point and the resulting

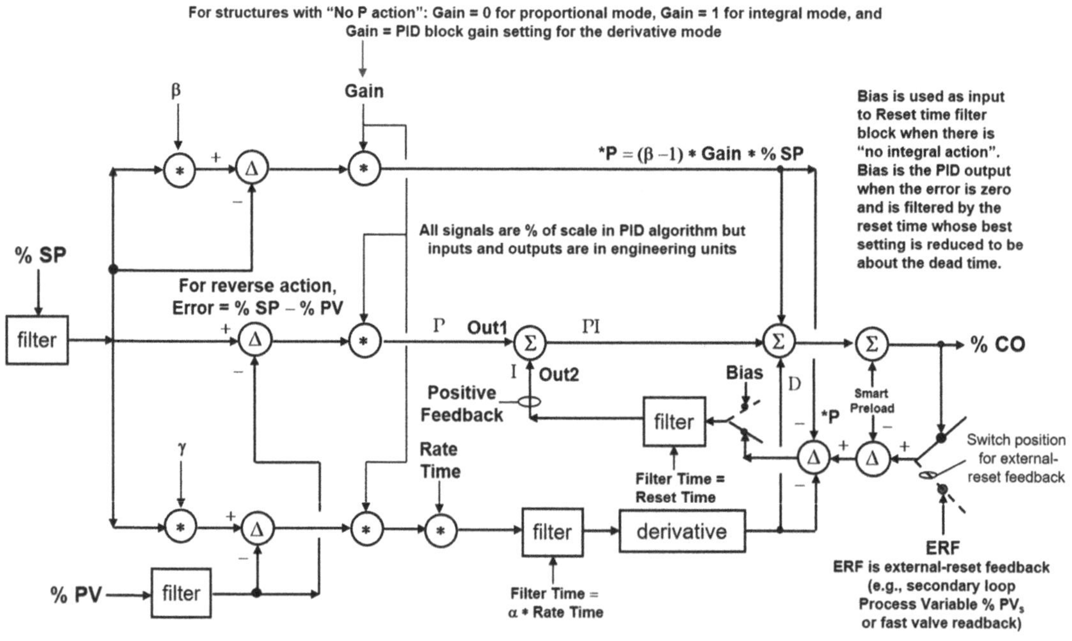

Figure C-12. Enhanced PID controller uses the positive feedback implementation of the integral mode as shown here for the standard form.

cross-neutralization of reagents or cycling of heating and cooling. The insertion of a dead time block in the external-reset feedback enables dead time compensation similar to a Smith Predictor but without the disruption to the operator interface and without the need to set a model gain and time constant. If the block dead time is updated so that it is the total loop dead time with an error less than 10% despite changes in the production rate, the PID reset setting can be greatly decreased and the PID gain setting increased. There is no appreciable improvement until the PID tuning is made much more aggressive. Contrary to public opinion, dead-time compensators offer a greater improvement for lag dominant rather than dead-time dominant processes, and the compensator performance deteriorates much more quickly for a model dead time greater instead of smaller than the actual total loop dead time [1, 2, 3].

References

1. Gregory K. McMillan, "Wireless: Overcoming Challenges in PID Control & Analyzer Applications," *InTech* (July–August 2010).

2. Gregory K. McMillan and P. Hunter Vegas, *Industrial/Process Instruments and Controls Handbook*, 6th ed. (New York: McGraw-Hill Education, 2019).

3. Gregory K. McMillan and P. Hunter Vegas, "Best Practices for PID," *Control* (December 2018).

Appendix D: Modern Myths

Myth 1 – You need an advanced degree to do advanced control. Not so anymore. The digital twin, advanced control tools, and simulation software can be used to explore, develop, and test innovations in a digital twin with first-principle dynamic models to automate much of the required expertise and eliminate the need for special interfaces. The user can now focus mostly on the application and the goal.

Myth 2 – Dynamic simulations and model-based control are only applicable to continuous processes. Because most of the applications are in the continuous industry, this is a common misconception. Although it is true that a steady-state simulation is not valid because there is by definition no steady state in batch, dynamic simulation can follow a batch as long as the software can handle zero flows and empty vessels. Model-predictive control (MPC), which looks at trajectories, is suitable for the real-time optimization (RTO) of fed-batch processes by the use of controlled variables that are the slopes of key batch profiles, such as cell growth rate and product formation rate. The opportunities MPC provides to improve a process's efficiency add up to about 25% for batch compared to 5% for continuous operations.

Myth 3 – You need consultants to maintain dynamic simulations and model-based control systems. No longer true. The new software's ease of use enables the user to get much more involved, which is critical to ensuring that the plant gets the most value out of the models. Previously, the benefits started to diminish as soon as the consultant left the jobsite. Now, the user should be able to tune, troubleshoot, and update the models.

Myth 4 – You do not need good operator displays and training if you have well-designed RTO and MPC systems. The operators are the biggest constraint in most plants. Even if the models used for real-time optimization and model-based control are perfect, operators will take these systems off-line if they do not understand them. The new guy in town is always suspect, so the first time an operational problem occurs and no one is around to answer questions, the RTO and MPC systems are turned off—even if they are doing the right thing. Training sessions using the digital twin described in Chapters 5 and 9 and displays should focus on showing the individual contribution of the trajectories for each controlled, disturbance, and constraint variable in relation to the observed changes in the manipulated and optimization variables.

Myth 5 – Simple step (bump) tests are never enough. You must do a pseudo random binary sequence (PRBS) test. A complete PRBS test may take too long. The plant may have moved to an entirely different state, tripped, or—in the case of a batch operation—finished before a PRBS test is complete. As a minimum, one step in each direction should be held to steady state. The old rule is true: if you can see the model from a trend, it is there. Sometimes, the brain can estimate the process gain, time delay, and lag better than a software package.

Myth 6 – You need to know your process before you start an RTO or MPC application. This would be nice, but often the benefits provided by a model stem from the knowledge discovered during the systematic building and identification procedures. Frequently, the understanding gained by developing models leads to immediate benefits in terms of better set points and instruments. The commissioning of the RTO and MPC is the icing on the cake and locks in benefits for varying plant conditions.

Myth 7 – Optimization by pushing constraints will decrease on-stream time. Not true. MPC and RTO recognize future violations of constraints for unforeseen problems and will back off from the edge to increase on-stream time.

Appendix E: Enzyme Inactivity Decreased by Controlling the pH with a Family of Bézier Curves

This appendix is an excerpt from the article, "Enzyme Decreased by Controlling the pH with a Family of Bézier Curves," published in the May 2006 issue of *Pharmaceutical* magazine.

Michael J. Needham, Validation Engineer ROVISYS

Aurora, Ohio

A client asked if it was possible to better control the inactivity of enzymes that were pH sensitive. The question arose because the client needed a stable, or possibly increasing, microbial growth environment that would prevent or shift the biosynthesis capabilities of the process harvest. The biosynthesis required an essential active enzyme concentration in a cell-culture growth medium. Reduced biosynthesis capabilities, caused by an inactive enzyme, posed an economic risk for the client.

The biosynthesis of pharmaceutical drug products produced through biotechnical manufacturing processes begins in a laboratory. The informational components essential to this development include charts showing the production capability of the drug of interest compared with pH, time, substrate concentration, enzyme concentration, and the like. These informational components are the basis and starting point for the innovative enzyme controller (IEC) to be developed. To understand how these laboratory charts fundamentally improve the approach to process control, this appendix begins with a brief review of the methods used by physical biochemists to analyze these types of reactions.

The biological cellular growth mediums that make up enzyme-catalyzed reactions are difficult to work with analytically for the following key reasons [1]:

- To achieve lower activation energies for a reaction, enzymes require extremely low concentrations compared to their substrate. These low concentrations as well as the complexity of enzyme-substrate binding make it difficult to measure what the enzyme is actually doing.
- The kinetic constants of complex enzymes cannot be determined by steady-state experiments. In a step reaction sequence to produce a particular drug, there are often at least six kinetic constants and more are possible (see Figure E-1).
- The molecular dissociation in these biological mediums is difficult to determine.
- Conventional steady-state equilibrium methods are not linearly applicable to enzymes.
- The activities of many enzymes vary with pH in the same way that simple acids and bases ionize. Enzymes involved in the sequential reactions associated with a microorganism's biosynthetic pathway growth kinetics are very complex, diverse, and huge (molecular weights can be as high as 1,000,000 Daltons).
- The activity of enzyme reaction velocities can be inhibited by reactive competition from other components in the medium or even the substrate [2].

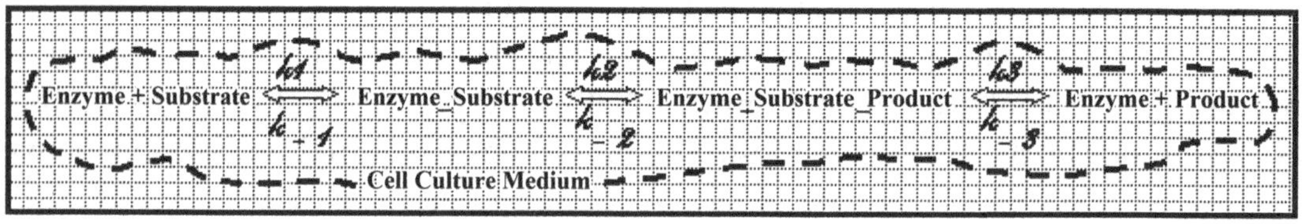

Figure E-1. Reaction mechanism.

Recently, physical biochemists have derived a standard state where the activities and concentrations normally used to analyze reaction kinetics are not required. Hammes developed an analytical method called "Relaxation Methods" to study fast chemical reaction components at equilibrium. The Hammes method can be briefly described as follows:

- A chemical system at equilibrium is suddenly perturbed in such a way that it finds itself out of equilibrium under the new conditions. It then relaxes to a state of equilibrium under the new conditions, and the rate of relaxation is governed by the rate constants and mechanism of the process.

In this appendix, we will use relaxation method analysis and steady-state theory to describe the physical model of an enzyme-catalyzed reaction as shown in Figure E-1. Our precepts are as follows:

- At equilibrium, the rates of forward and back reactions are the same:

$$k1 = k - 1$$

$$k2 = k - 2$$

$$k3 = k - 3$$

- Individual rate constants cannot be determined, only their ratio.
- Many reaction properties undergo constant and rapid fluctuations around some average or median state. Any long-term measurements yield the same result if the system is truly at equilibrium.
- Relaxation kinetics give simple first-order exponential time regardless of how many molecules are involved as either reactants or products.
- After the measurement is analyzed, the concentrations charted over time should describe a straight line.
- Steady state is an approximation since the substrate is gradually being depleted. However, this approximation will be very good as long as the rate measurements are restricted to a short time interval during which the substrate's concentration does not greatly change.

Departing from conventional control strategies, the principle object of the IEC will be:

- Maintaining the equilibrium by controlling the reaction perturbation as defined by the laboratory charts, which is technically supported by the Relaxation Method's physical model for the enzyme-catalyzed reaction of interest.

To divide the laboratory charts into different reactant intermediate states, we will use the transition state theory [3] (see Figure E-2). This theory is supported by the Hammond postulate for Gibbs free energy of mathematical maximums and minimums for reaction growth diagrams. Stated briefly, the transition state theory for enzyme-catalyzed chemical reactions is as follows:

- Transition states occur at mathematical maximums in reaction growth diagrams for a reaction, and intermediates occur at mathematical minimums.
- At mathematical maximums on reaction growth diagrams, chemical bonds are in the process of being made and broken.
- At mathematical minimums on reaction growth diagrams, intermediates have fully formed chemical bonds.
- The reaction growth diagram plots the energy of the reagents (we use both growth and pH) as the reaction proceeds.

Briefly stated, the portion of the Hammond postulate of interest is as follows:

- If there is an unstable intermediate in the chemical reaction pathway, the reaction's transition state will resemble the structure of the intermediate. The reasoning is that the unstable intermediate will be a mathematical maximum in the reaction coordinate diagram.
- Gibbs free energy curves for a substrate and product intersect on the reaction coordinate at the position of the transition state. A change of structure away

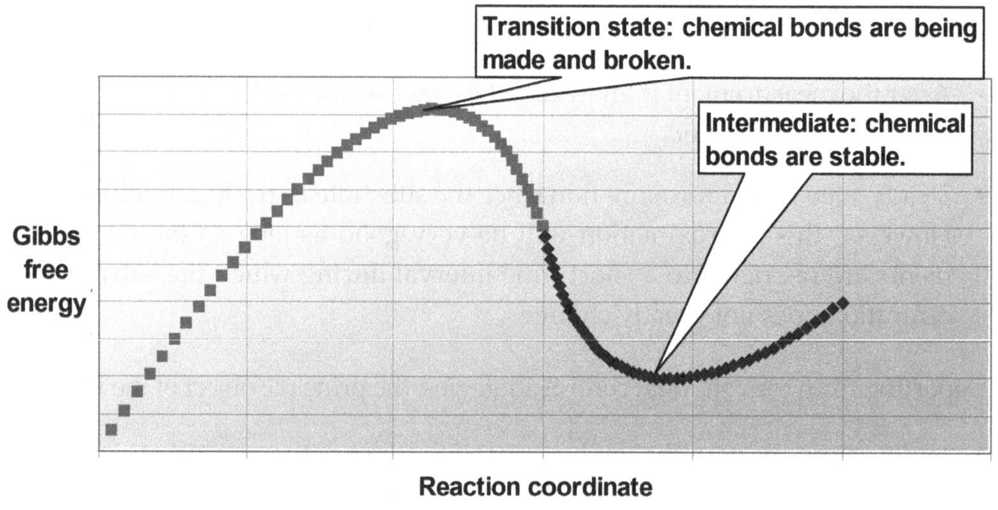

Figure E-2. Transition state theory.

from the seat of reaction destabilizes it, displacing its energy curve higher. The point of intersection moves closer to the substrate.

The importance of transition state theory is that it relates the rate of a reaction to the difference in Gibbs energy between the transition state and the ground state.

We approached the selection of the nonlinear control strategy conceptually [4]. As Antje Kann has noted [5], "models are often developed...with knowledge of theory and algorithms, but who may not be intimately familiar with the users' situations. Modeling choices made ... are often guided by practical concerns such as solvability, run-time issues, and ease of implementation in code. However, these choices may impair the ease with which a model can be applied to solve the users' problems." Following this train of thought, we took a fresh look. The laboratory charts would be the best starting point for the bases of the controller development. Well-documented analysis would be available for all the process dynamics, measured variables, responses, and also unanswerable process questions for the drug of interest. Classical multivariable, nonlinear control strategies also represented well-documented information. Also, a good curve-generating engine would help turn data from the laboratory charts into information.

Parametric equations allow curves to have the unusual ability to move straight up or even curve back and cross themselves (see Figure E-3). Cubic Bézier curves, which belong to the family of parametric curve equations, were selected as the first set of parametric curves to use because they do not need a tangent vector. Bézier curves are characterized by defining four coordinate points (origin, destination, values of X, and

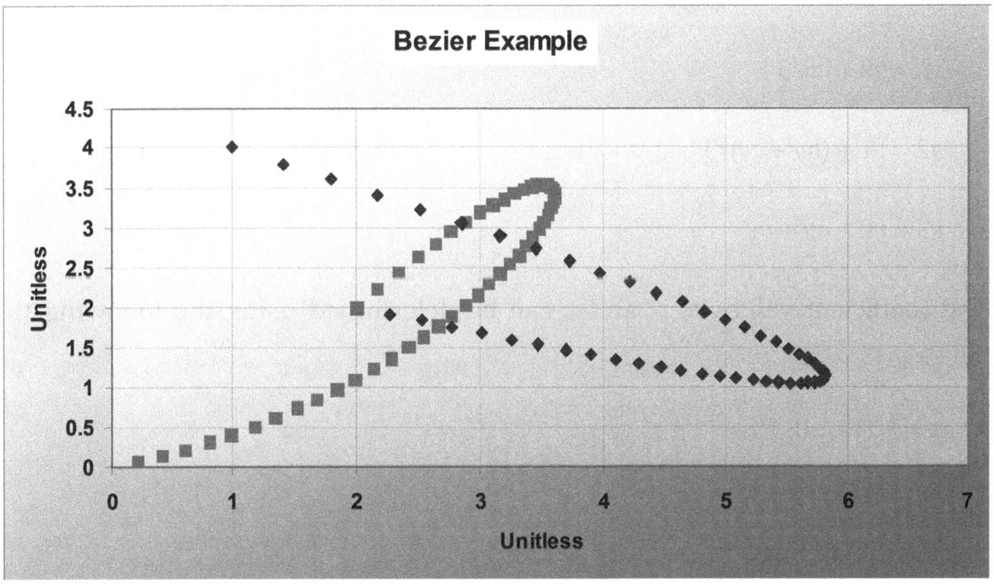

Figure E-3. Example of a parametric curve crossing itself twice.

values of Y), one linear variable coordinate point (t), and two polynomial equations. The two polynomial equations are used to calculate the values of X and Y between an increasing third variable (t).

The following is a simple example illustrating the development of a cubic Bézier curve [6]:

Two polynomial equations define the coordinate points on the curve. Both are evaluated for an arbitrary number of values of (t). In our example, we will increment as follows:

- $\boldsymbol{t} = \boldsymbol{t} + 0.02$, $\boldsymbol{t} \Rightarrow 1$ (that is, 50 values of $\boldsymbol{t}$)

As increasing values for (t) are supplied to the equations, the point defined by x (t), y (t) moves from the origin to the destination:

- $x(t) = axt^3 + bxt^2 + cxt + x0$

  $x1 = x0 + cx/3$

  $x2 = x1 + (bx + cx)/3$

  $x3 = x0 + ax + bx + cx$

- $y(t) = ayt^3 + byt^2 + cyt + y0$

  $y1 = y0 + cy/3$

  $y2 = y1 + (by + cy)/3$

  $y3 = y0 + ay + by + cy$

The coefficient values a, b, and c can be determined using the following three equations:

- $cx = 3(x1 - x0)$
- $bx = 3(x2 - x1) - cx$
- $ax = x3 - x0 - cx - bx$

First, let us select four control cubic coordinates at random:

- origin: xp0 = 3.5 and yp0 = 0
- control point 1: xp1 = 3.5 and yp1 = 1
- control point 2: xp2 = 5 and yp2 = 6
- destination: xp3 = 6 and yp3 = 3

These four coordinates can be evaluated as coefficients:

- cx = 0
- cy = 3
- bx = 4.5
- by = 12
- ax = –2
- ay = –12

Then, calculating the first coordinate pair at $t = 0.02$:

- x(0.02) = 3.50
- y(0.02) = 0.06

For $t = 0.04$:

- x(0.02) = 3.50
- y(0.02) = 0.14

For $t = 0.06$, and so on.

Figure E-4 graphically demonstrates the dramatically different results obtained from two selections for control coordinates.

We used loosely coupled data from the laboratory charts and borrowed concepts from classical controllers with multivariable, [7] nonlinear capabilities that are generally used for pH regulation. This approach produced the following filtered list of controller types:

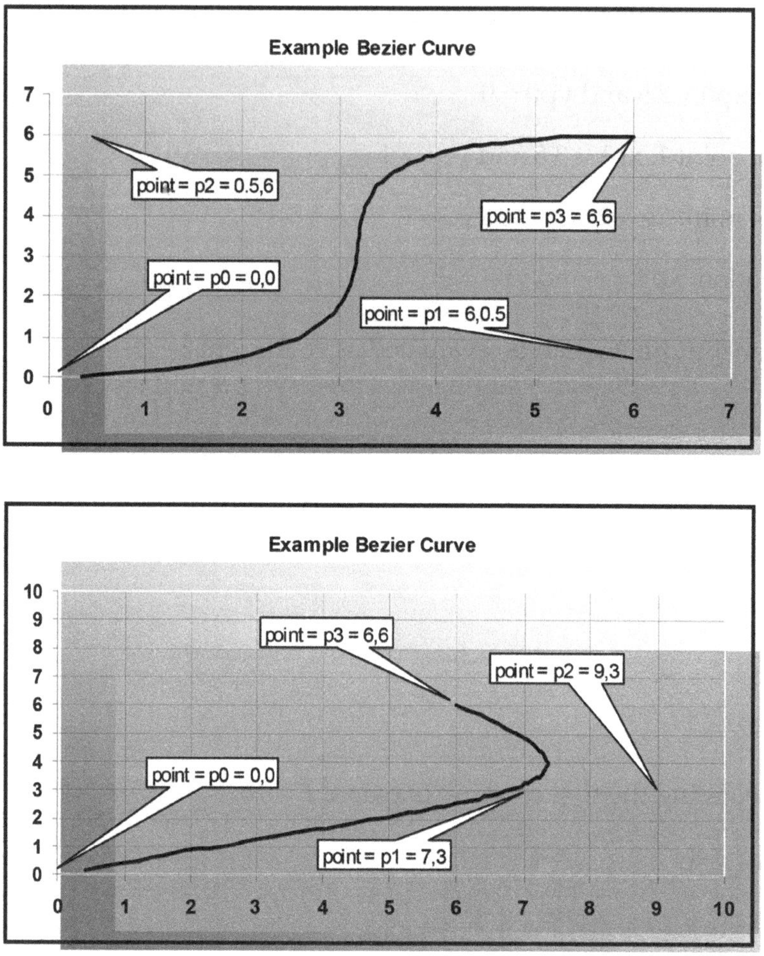

Figure E-4. Examples of cubic Bézier curves.

- Endpoint Control (EpC)
  - Strengths: Stoichiometrics.
  - How IEC uses this concept: The stoichiometrics associated with enzyme-substrate reactions that occur in cell culture growth mediums are much more complex, both physically and chemically. We will use measured variables as an indication value and as associated other parameters before we determine the value of the final control element.
- Gain Compensation Control (GCC)
  - Strengths: Can function with nonlinear control measurements in real time.
  - How IEC uses this concept: We reduced or in some cases eliminated controller response by giving the laboratory charts for pH, time, and the product

growth rate's Bézier capability and by adjusting the known process nonlinearities to be predictable.

- Adaptive Control (AC)
  - Strengths: Controller mode values change with the dynamics of the measured variables.
  - How IEC uses this concept: The controller and process will be self-regulating if you give the laboratory charts for pH, time, and product growth rate's Bézier capability and adjust the controller output to known measurement dynamics.
- Model Predictive Control (MPC)
  - Strengths: Able to predict near future variable values based on the model developed by the pseudo random process variation (PRPV) studies.
  - How IEC uses this concept: A very definitive model is available if you use the laboratory charts as the bases for the controller algorithm,

The IEC can now be fully developed using the following properties, theories, methods, control strategies, and states for the drug of interest:

- Laboratory informational charts developed for the drug of interest
- Physical biochemical model precepts by Hammes' Relaxation Method
- Transition state theory with the Hammond postulate
- Steady-state dynamics
- Pseudo steady state
- Endpoint control (EpC) concept of stoichiometrics
- Gain compensation control (GCC) concept for nonlinear measurements
- Adaptive control (AC) concept of mode changes with measured dynamics
- Model predictable control (MPC) concept of near-future predictions for process variables

The following detail requirement serves as the basis for the controller:

- Three laboratory-determined sets of data for product growth rate versus pH, growth rate versus time, and growth rate per pH versus time will be experimentally developed.

- The three sets of laboratory data for product growth rate will be formulated using cubic Bézier control.
- EpC has control components that control the proportional feeding (stoichiometric) of reactive agents in either correct or excess proportions. The IEC will calculate the standard deviation of the laboratory's experimental data and determine if the batch has either a correct or excess proportion of reactive agents.
- GCC has a control component that varies the loop-gain compensation with nonlinear compositions. The IEC will calculate the required loop gain based on the time interval from the state of the batch, pH, and estimated growth rate.
- AC has control components that change controllers' control modes based on the IEC's active control mode. The IEC will change the slave PID controller modes and values based on the following:
    - Set-point limits based on the standard deviation of the active control mode
    - The integral set by the active median value
    - Derivative set the active estimated growth rate
    - AUTOMATIC/MANUAL/REMOTECASCADE based by operator selection
- MPC has a control horizon component that uses the model developed for the various coupled control variables to estimate a future value. The growth rate curve determined by experimental methods in the laboratory will be the model.
- Laboratory-determined data point coordinates would be formulated using cubic Bézier control.
- The following measurements will be made and their values compared with the desired values:
    - Time batch started
    - Real-time clock reset to zero at time batch started
    - pH
    - Gain (pH desired at time x/pH actual at time x or growth rate at pH actual at time x/growth rate estimated at time x)

The functional specifications for the IEC that are to be developed are as follows:

- pH measured and displayed on a screen with historical capability

- pH measured data given cubic Bézier capability and displayed on a screen with historical capability
- Chemical transition states, each displayed on a screen
- Chemical transition states data given cubic Bézier capability
- Chemical intermediates, each displayed on a screen
- Chemical intermediates data given cubic Bézier capability
- Real-time growth curve for the microorganism of interest displayed on two screens, each with historical capability
- Growth rate curve data will be linked to either a chemical transition state or chemical intermediate
- Estimated growth rate curve screens will have a real-time clock displayed linearly
- pH measured data will be linked to each growth rate curve screen
- pH measured data gain will be linked to each growth rate curve screen
- Cubic Bézier control capability for all data sets with historical capability

References

1. Ignacio Tinoco Jr., Kenneth Sauer, and James C. Wang, *Physical Chemistry Principles and Applications in Biological Sciences*, 2nd ed. (Englewood Cliffs, NJ: Prentice-Hall, 1985).

2. Pamela C. Champe, Richard A. Harvey, and Denise R. Ferrier, *Biochemistry*, 3rd ed. (Philadelphia: Lippincott Williams & Wilkins, 2005).

3. Alan Fersht, *Structure and Mechanism in Protein Science* (New York: W. H. Freeman, 2000).

4. *Standards Library for Automation and Control: Guidelines for Quality, Safety, & Productivity*, 2nd ed., CD (Research Triangle Park, NC: ISA [International Society of Automation], 2006).

5. Conversations with Antje Kann, President, Analytics Interactive.

6. Bézier Curves, http://www.math.ubc.ca/~ cass/gfx/bezier.html.

7. Gregory K. McMillan, *Tuning and Control Loop Performance*, 2nd ed. (Research Triangle Park, NC: ISA [International Society of Automation], 1990).

Appendix F: First-Principle Process Relationships

F.1 General Implications

First-principle relationships can define process cause and effects that can lead to improved controller tuning and performance by the selection of better tuning rules and process variables for scheduling of tuning settings. They also affect the choice of control valve trim and the feedforward design. The understanding of these relationships does not require a degree in chemical engineering but presumes some understanding of common terms (e.g., heat transfer coefficient and area), relationships (e.g., ideal gas law), and physical concepts (e.g., conservation of mass and energy).

Equations have been developed from first-principle relationships for the process gains, dead times, and time constants of volumes with various degrees of mixing. The results show that for well-mixed volumes with negligible injection delays, the effect of flow cancels out for the controller gain if one of the following methods is used: lambda self-regulating rule where lambda is set equal to the dead time, or the reaction curve method. The effect of flow also cancels out for the reset time in addition to the controller gain if the process is treated as a "near-integrator" and the lambda integrating tuning rule is used. This is because the flow rate cancels out in the computation of the ratio of process gain to time constant that is the "near-integrator" gain. This ratio and "near-integrator gain" are inversely proportional to the process holdup mass (e.g., liquid mass). However, for temperature control the effect of changes in liquid mass cancels out because a change in level increases the heat transfer surface area covered. Several authors have mistakenly tried to schedule controller tuning based on liquid level for reactor temperature control. One author has reported being bewildered by its failure. This is not the case for gas pressure control. The equations show that liquid level has a profound effect on the process integrating gain for vessel pressure control because it changes the vapor space volume without any competing effect. To summarize, the integrator gain for composition and gas pressure is inversely proportional to liquid level (liquid mass). For temperature, the effect of level cancels out unless the level is above or below the heat transfer surface area, which is unusual but can occur at the beginning or end of a batch when coils instead of a jacket are used for heat transfer. For temperature, the integrator gain is nearly always proportional to the overall heat transfer coefficient that is a function of mixing, process composition, and fouling or frosting.

The equations also show that if the transport delay for flow injection is large compared to the time constant, which does occur for reagent injection in dip tubes for pH control, then the controller gain will be proportional to flow. Note that pH control is a class of concentration control.

For the control of temperature and concentration in a pipe, the process dead time and process gain are both inversely proportional to flow and the process time constant is essentially zero, which makes the actuator, sensor, transmitter, or signal filter time lag the largest time constant in the loop. Thus, the largest automation system lag determines the dead-time to time-constant ratio. For a static mixer, there is some mixing, and the process time constant is inversely proportional to flow but is usually quite small compared to other lags in the loop. The controller gain is generally proportional to flow for both cases.

Finally, the above has implications as to whether a flow feedforward multiplier or summer should be used and whether a linear or equal percentage trim should be used. A flow feedforward multiplier and equal percentage trim, which both have a gain proportional to flow, can help compensate for a process gain that is inversely proportional to flow provided the process time constant is not also inversely proportional to flow. This is generally the case for temperature and concentration control of plug flow volumes (pipelines, static mixers, and heat exchangers). For well-mixed volumes, feedforward summers and an installed linear characteristic for valves are generally best. For control valves, this corresponds to a linear trim when the available pressure drop is much larger than the system pressure drop or critical pressure drop so the installed flow characteristic is close to the inherent flow characteristic.

The results are also useful for determining the dead-time to time-constant ratio, which has a profound effect on the tuning factors used and the performance of dead time compensation, which has been discussed on the *Control* magazine Control Talk Blog (https://www.controlglobal.com/blogs/controltalkblog/) and the Control Talk column "Voices" (https://www.controlglobal.com/voices/) sites.

F.2 Batch Implications

Plug flow volumes can always be considered as continuous because the volume is completely full and anything entering will be discharged after a transportation delay.

A back mixed volume is partially full. If the liquid discharge flow is zero, this volume can be considered to be in the batch mode rather than the continuous mode. If the flows are all sequenced and charged based on time and/or totals, the vessel operation can be considered to be pure batch. If reactor flows are ratioed and manipulated by a control loop, the vessel operation can be classified as fed-batch.

Level has an integrating response whether in the batch or continuous mode. In the batch mode, it is a zero load integrator in that all the feed flows must be zero for the level to stop rising. Level has a one-sided integrating response in the batch mode because the level can only rise and not drop. This type of response causes overshoot for any controller with reset action. Proportional plus derivative (PD) controllers with a zero or negative bias can be tuned for zero overshoot.

The temperature response of a back mixed volume remains self-regulating even for zero liquid discharge flow unless the liquid level is above or below the heat transfer surfaces. However, the temperature response does lose some self-regulation and behaves more like a "near-integrator."

The concentration response of a back mixed volume becomes integrating for a zero discharge flow. This is not obvious because the discharge flow cancels out of the differential equation from the application of the multiplicative rule of integration in the transition of the derivative from the rate of accumulation of component mass to the rate of accumulation of component concentration. The effect of zero discharge flow is more recognizable if we consider the case of a zero reaction rate so that the process time constant is simply the residence time (liquid mass divided by the liquid feed flow rate). The increase in mass for a fixed feed rate over the residence time is simply the feed rate multiplied by the residence time. The result is an increase in mass equal to the existing mass. This doubling of mass doubles the residence time and therefore the process time constant. Consequently, the process never reaches a steady state because the process time constant is constantly increasing as the level is rising for a zero discharge flow.

For the case of zero reaction, the integrator is a zero load integrator because the feed of the component must be zero for the concentration of the component to stop rising. The concentration here has a one-sided integrating response in that the concentration can only increase and not decrease. This would also be the case for reaction products where there is only a forward reaction (no reverse or side reactions). As with level, overshoot is a problem unless PD controllers are used. Alternately, the controlled variable can be translated to a rate of change of concentration as noted in application of model-predictive control (MPC) for bioreactor biomass and product concentration.

Gas pressure is an integrator regardless of liquid discharge flow as long as the pressure in the vessel has a negligible effect on vent flow, which is the case for large or critical pressure drops. If this is not the case, the gas pressure response becomes self-regulating, but for large volumes and small vent flows it behaves like a "near-integrator."

F.3 Results

These equations are developed in the "Back Mixed Volumes" and "Plug Flow Volumes" sections in this appendix. They are numbered according to the point of derivation so the user can better locate and understand their origination. The nomenclature with engineering units is summarized at the end of the appendix.

The integrating process gain (K_{ip}) for the control of liquid level by the manipulation of a flow:

$$K_{ip} = 1/(\rho_o \bullet A_o) \tag{F-4d}$$

The integrating process gain (K_{ip}) for the control of pressure per gas law with a specific gas constant (R_g) by the manipulation of a flow:

$$K_{ip} = \left[(R_g \bullet T_g)/ V_g \right] \tag{F-5d}$$

For the manipulation of jacket temperature to control outlet temperature, the main process time constant (τ_p) is (positive feedback if heat of feed and reaction exceeds the product of heat transfer coefficient and area):

$$\tau_p = (C_p \bullet M_o)/ \left[C_p \bullet F_f - \Delta Q_r / \Delta T_o + U \bullet A\right] \tag{F-6g}$$

For the manipulation of jacket temperature to control outlet temperature, the integrating process gain (K_p) is:

$$K_p = (U \bullet A)/\left[C_p \bullet F_f - \Delta Q_r / \Delta T_o + U \bullet A\right] \tag{F-6h}$$

For the manipulation of jacket temperature to control outlet temperature, the near-integrating process gain (K_{nip}) is:

$$K_{nip} = (U \bullet A) / (C_p \bullet M_o) \quad \text{(F-6i)}$$

For the manipulation of feed temperature to control outlet temperature, the process gain (K_p) is:

$$K_p = (C_p \bullet F_f) / \left[C_p \bullet F_f - \Delta Q_r / \Delta T_o + U \bullet A\right] \quad \text{(F-6j)}$$

For the manipulation of feed flow to control outlet temperature, the process gain (K_p) is:

$$K_p = (C_p \bullet T_f) / \left[C_p \bullet F_f - \Delta Q_r / \Delta T_o + U \bullet A\right] \quad \text{(F-6k)}$$

For manipulation of jacket temperature, the additional small secondary process time constant associated with the heat capacity and mass of the jacket wall is:

$$\tau_{p2} = (C_w \bullet M_w) / \left[U \bullet A\right] \quad \text{(F-6l)}$$

The process dead time (θ_p) from the turnover time for temperature and concentration control in a well-mixed volume is:

$$\theta_p = (M_o / \rho_o) / \left[(F_f + F_a + F_r) / \rho_o + F_v / \rho_v\right] \quad \text{(F-6m)}$$

The process dead time (θ_p) from injection delay for concentration control is:

$$\theta_p = V_1 / (F_1 / \rho_1) \quad \text{(F-6n)}$$

For the manipulation of feed flow to control reactant concentration (X_{Ao}), the main process time constant (τ_p) is:

$$\tau_p = M_o / (R_x + F_f) \quad \text{(F-7g)}$$

For the manipulation of feed flow to control reactant concentration (X_{Ao}), the process gain (K_p) is:

$$K_p = X_{Af} / (R_x + F_f) \quad \text{(F-7h)}$$

For the manipulation of feed flow to control reactant concentration, the near-integrating process gain (K_{nip}) is:

$$K_{nip} = X_{Af} / M_o \quad \text{(F-7i)}$$

For the manipulation of feed concentration to control reactant concentration, the process gain (K_p) is:

$$K_p = F_f / (R_x + F_f) \tag{F-7j}$$

For plug flow volumes where different streams are being combined, the process gain (K_p) for controlling the temperature of the mixture (T_f) by the manipulation of flow is:

$$K_p = dT_f / dF_1 = T_1 / \Sigma F_i \tag{F-9a}$$

For plug flow volumes where different streams are being combined, the process gain (K_p) for controlling the composition of component A (X_{Af}) in the mixture by the manipulation of flow is:

$$K_p = dX_{Af} / dF_1 = X_{A1} / \Sigma F_i \tag{F-9b}$$

For plug flow volumes where different streams are being combined, the process dead time for controlling the temperature or composition of the mixture by the manipulation of flow is:

$$\theta_p = V_1 / (F_1 / \rho_1) + V_p / \Sigma(F_i / \rho_i) \tag{F-9c}$$

The process time constant is essentially zero for true plug flow. For a static mixer there is some back mixing; the residence time in Equation F-9c is split between a dead time and time constant per Equations F-9d and F-9e.

$$\theta_p = V_1 / (F_1 / \rho_1) + x \bullet V_p / \Sigma(F_i / \rho_i) \tag{F-9d}$$

$$\tau_p = (1 - x) \bullet V_p / \Sigma(F_i / \rho_i) \tag{F-9e}$$

F.4 Derivations

There are three types of processes: self-regulating, integrating, and runaway as shown in Figures F-4a, F-4b, and F-4c, respectively. A self-regulating process will decelerate to a new steady state. An integrating process will continually ramp. A runaway process will accelerate until hitting a relief or interlock setting.

More than 90% of the processes are self-regulating. However, many of the continuous and fed-batch processes in the chemical industry with the greatest direct economic benefits behave and can be best treated as "near-integrating" processes. The classic integrating process is a pure batch or level process. Fewer than 1% of the processes are

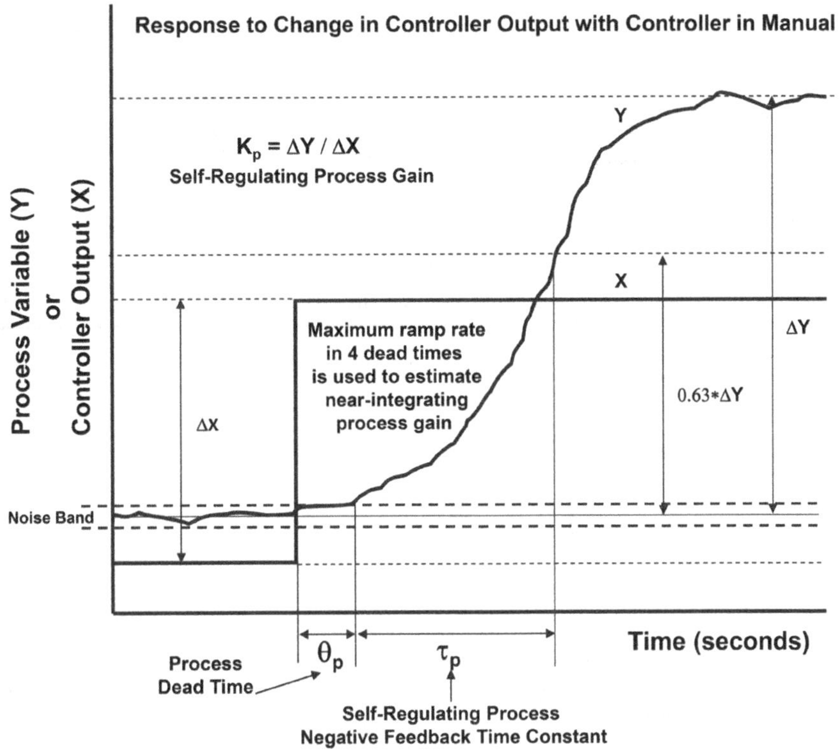

Figure F-4a. Self-regulating (negative feedback) process.

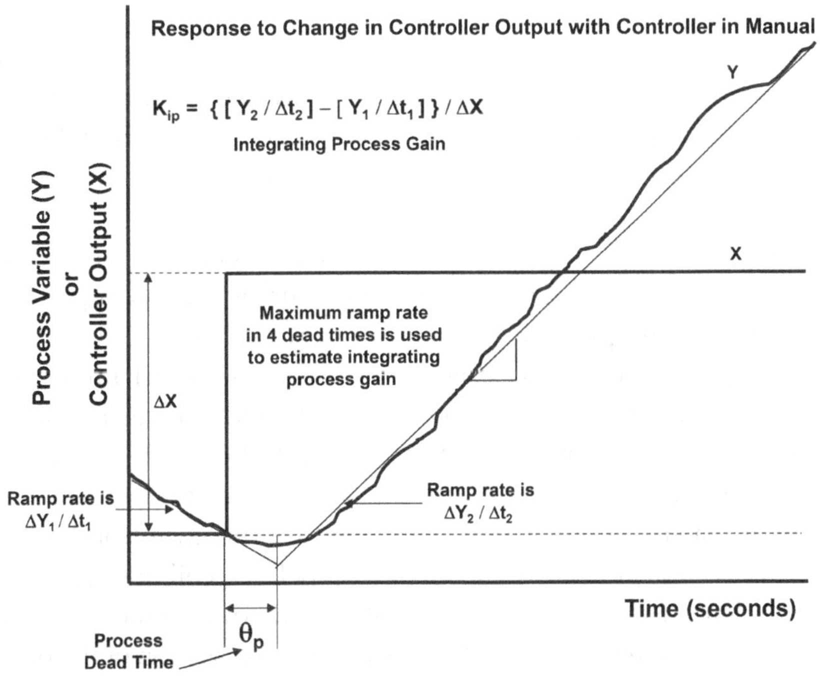

Figure F-4b. Integrating (zero feedback) process.

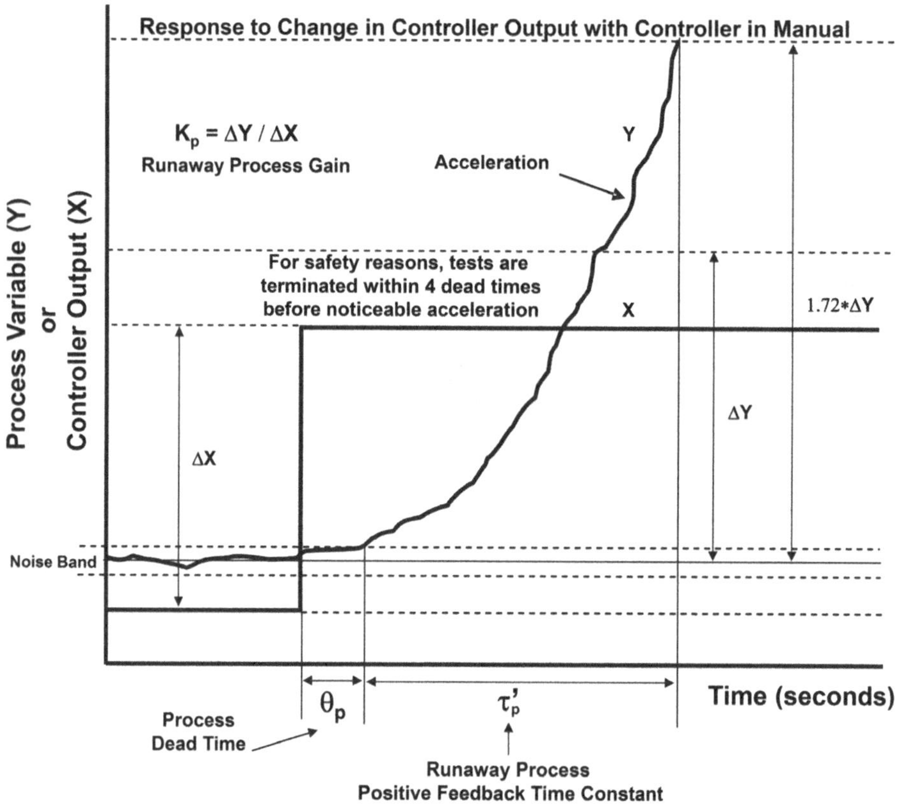

Figure F-4c. Runaway (positive feedback) process.

runaway. When these exist, understanding the runaway response is critical in terms of safety and control because of the propensity to accelerate and reach a point of no return. Runaway responses are almost exclusively associated with highly exothermic reactors used in plastics and specialty chemical production.

This appendix develops the equations for process dynamics for back mixed volumes and plug flow volumes. The back mixed volumes section applies to volumes whenever an agitator pumping rate, an eductor or recirculation liquid flow rate, or a gas evolution or sparge rate produces enough turbulence and back mixing to make the mixture more uniform in the axial in addition to the radial direction of the volume. Gas volumes generally have enough turbulence and fast enough gas dispersion rates to be treated as a back mixed volume. The equations therefore hold relatively well for an evaporator and a single distillation stage due to turbulence from the vapor flow. The plug volume section applies to static mixers, pipelines, coil inlets, and jacket inlets where the turbulence from pipe fittings or internal mixing elements creates enough radial mixing to make the mixture uniform over the cross section of the pipe or nozzle inlet but provides little axial mixing.

F.5 Back Mixed Volumes

For a back mixed volume, the process gains and time constants can be readily identified if the ordinary differential equations for the rate of accumulation of energy or material in the volume are set up so that the process output of interest (Y) is on right side with a unity coefficient. From this simple generic form, we can identify the process time constant (τ_p) as the coefficient of the derivative of the process output (dY/dt) and the process gain (K_p) as the coefficient of the process input (X). The process output (Y) and net input (X) can be viewed as the controlled variable and manipulated variable biased by a load, respectively. Many other terms can exist, but these are not shown in the following equations. These missing terms can be categorized as disturbances.

If the sign of the unity coefficient of the process output on the right side is negative (Equation F-1a), the process has negative feedback. As the process output changes, the negative feedback slows down and eventually halts the excursion of the process output at its new steady state when it balances out the effect of the process input and the disturbances.

$$\tau_p \bullet dY / dt = K_p \bullet X - Y \tag{F-1a}$$

The integration of this equation provides the time response of a change in the process output (Δ_Y) for a step change in the process input (ΔX). The step occurs at $t = 0$.

$$\Delta Y = K_p \bullet \left(1 - e^{-t/\tau_p}\right) \bullet \Delta X \tag{F-1b}$$

If the process output does not appear on the right side (Equation F-2a), there is no process feedback. As the process output changes, there is no feedback to slow it down or speed it up so it continues to ramp. There is no steady state. The ramping will stop only when X is zero, which occurs when the manipulated variable is equal to the load.

$$dY/dt = K_{ip} \bullet X \tag{F-2a}$$

$$\Delta Y = K_{ip} \bullet t \bullet \Delta X \tag{F-2b}$$

Often in the more important loops for concentration, pressure, and temperature control of large volumes, the time constant in Equation F-1a is so large that the time to reach steady state is beyond the time frame of interest. Because these loops with small dead-time to time-constant ratios should be tuned with integrating process-tuning rules as discussed in Chapter 3, the controller only sees the first part of the excursion before the inflection point and deceleration by negative process feedback. In this case, the response is best characterized by a near-integrating process gain calculated per Equation F-2c.

$$K_{nip} = K_p / \tau_p \tag{F-2c}$$

If the sign of the unity coefficient of the process output on the right side is positive (Equation F-3a), the process has positive feedback. As the process output changes, the positive feedback speeds up the excursion unless disturbances counteract the effect of the process input and output.

$$\tau_p' \bullet dY/dt = K_p \bullet X + Y \tag{F-3a}$$

$$\Delta Y = K_p \bullet (e^{t/\tau_p'} - 1) \bullet \Delta X \tag{F-3b}$$

Consider a mixed volume with a jacket and vapor space. There are liquid reactant feeds, gas feeds (sparged through the liquid and added directly to the vapor space), an outlet liquid flow, a vent gas flow, and a jacket coolant flow. Usually, there are multiple components of interest. For example, consider liquid or gas acid and base reagent or reactant components (a, b) to produce primary and secondary liquid or gas products (c, d, e). Consider also that there are typically water and nitrogen gas components (w, n).

The ordinary differential equation for the accumulation of liquid mass as shown in Equation F-4a includes inlet flows added directly to the liquid volume (ΣF_i), vapor flow rates from evaporation and vaporization (ΣF_v), and an outlet liquid flow rate (F_o). The liquid level depends on the density and cross-section area of the liquid. Equation F-4a can then be reformulated to Equation F-4b to include the process variable of interest, liquid level (L_l), in the derivative.

$$dM_o/dt = \Sigma F_i - \Sigma F_v - F_o \tag{F-4a}$$

$$d(\rho_o \bullet A_o \bullet L_o) / dt = \Sigma F_i - \Sigma F_v - F_o \tag{F-4b}$$

If we consider the density (ρ_o) to be a weak function of composition and therefore constant like the cross-sectional area (A_o), we can take these terms outside the derivative and divide through to get an equation for level (L_o) in the form of Equation F-2a. Now it is evident that the integrating process gain (K_{ip}) for manipulation of flows in or flow out is simply the inverse of the product of the liquid density and cross-section area (Equation F-4d).

$$dL_o/dt = [1 / (\rho_o \bullet A_o)] \bullet [\Sigma F_i - \Sigma F_v - F_o] \tag{F-4c}$$

$$K_{ip} = 1 / (\rho_o \bullet A_o) \tag{F-4d}$$

The ordinary differential equation for the accumulation of gas mass as shown in Equation F-5a includes inlet flows added directly to the gas volume (ΣF_i); vapor flow rates from gas sparging, evolution, and vaporization (ΣF_v); and an exit gas flow rate (F_g). Equations of state such as the ideal gas law can be used to express this relationship for a given composition using a specific gas constant (R_g) that includes a compressibility factor. Equation F-5a can then be reformulated to Equation F-5b to include the process variable of interest, gas pressure (P_g), in the derivative.

$$dM_g/dt = \Sigma F_i + \Sigma F_v - F_g \tag{F-5a}$$

$$d\left[(P_g \bullet V_g)/(R_g \bullet T_g)\right]/dt = \Sigma F_i + \Sigma F_v - F_g \tag{F-5b}$$

If we consider changes in the gas volume (V_g) and gas temperature (T_g) to be much slower than changes in the gas pressure (P_g) and therefore relatively constant during the integration step, we can take these terms outside the derivative and divide through to get an equation for pressure in the form of Equation F-2a. Now it is evident that the integrating process gain (K_{ip}) for manipulation of flows in or flow out is simply the product of the specific gas coefficient (R_g) and the absolute gas temperature divided by the gas volume (Equation F-5d). This assumes a change in pressure does not significantly change the gas flow out of the volume, which is usually the case for a pressure drop across the vent valve that is large or critical.

$$dP_g/dt = \left[(R_g \bullet T_g)/V_g\right] \bullet \left[\Sigma F_i + \Sigma F_v - F_g\right] \tag{F-5c}$$

$$K_{ip} = \left[(R_g \bullet T_g)/V_g\right] \tag{F-5d}$$

The ordinary differential equation (ODE) for the accumulation of energy as shown in Equation F-6a includes the effects of feed temperature, heat of reaction as a function of temperature, heat of vaporization, and heat transfer to the jacket. If we consider the specific heat capacity to be relatively constant and use the multiplicative rule of integration, we can express the differential equation in the generic form of Equation F-1a in terms of temperature to show the process feedback in Equation F-6f, the final form of the ODE where the discharge flow cancels out. The relative magnitude of the terms in the denominator of Equation F-6g determines the feedback sign.

$$\begin{aligned} dQ_o/dt = {} & C_p \bullet \Sigma(F_i \bullet T_i) - C_p \bullet F_o \bullet T_o + (\Delta Q_r/\Delta T_o) \bullet T_o \\ & + H_x \bullet R_x - H_v \bullet F_v - U \bullet A \bullet (T_o - T_j) \end{aligned} \tag{F-6a}$$

$$dQ_o/dt = d(C_p \bullet M_o \bullet T_o)/dt = C_p \bullet (dM_o/dt) \bullet T_o + C_p \bullet M_o \bullet (dT_o/dt) \tag{F-6b}$$

$$F_f = \Sigma F_i \tag{F-6c}$$

$$T_f = \Sigma(F_i \bullet T_i) / \Sigma F_i \tag{F-6d}$$

$$C_p \bullet (dM_o/dt) \bullet T_o = C_p \bullet (F_f - F_o) \bullet T_o \tag{F-6e}$$

$$C_p \bullet M_o \bullet (dT_o/dt) = C_p \bullet F_f \bullet T_f + H_x \bullet R_x - H_v \bullet F_v + U \bullet A \bullet T_j - \left[C_p \bullet F_f - \Delta Q_r / \Delta T_o + U \bullet A\right] \bullet T_o \tag{F-6f}$$

For the manipulation of jacket temperature to control outlet temperature, the main process time constant (τ_p) is (positive feedback if heat of feed and reaction exceeds the product of heat transfer coefficient and area):

$$\tau_p = (C_p \bullet M_o) / \left[C_p \bullet F_f - \Delta Q_r / \Delta T_o + U \bullet A\right] \tag{F-6g}$$

For the manipulation of jacket temperature to control outlet temperature, the process gain (K_p) is:

$$K_p = (U \bullet A) / \left[C_p \bullet F_f - \Delta Q_r / \Delta T_o + U \bullet A\right] \tag{F-6h}$$

For the manipulation of jacket temperature to control outlet temperature, the near-integrating process gain (K_{nip}) is:

$$K_{nip} = (U \bullet A) / (C_p \bullet M_o) \tag{F-6i}$$

For the manipulation of feed temperature to control outlet temperature, the process gain (K_p) is:

$$K_p = (C_p \bullet F_f) / \left[C_p \bullet F_f - \Delta Q_r / \Delta T_o + U \bullet A\right] \tag{F-6j}$$

For the manipulation of feed flow to control outlet temperature, the process gain (K_p) is:

$$K_p = (C_p \bullet T_f) / \left[C_p \bullet F_f - \Delta Q_r / \Delta T_o + U \bullet A\right] \tag{F-6k}$$

For manipulation of jacket temperature, the additional small secondary process time constant associated with the heat capacity and mass of the jacket wall is:

$$\tau_p = (C_w \bullet M_w) / \left[U \bullet A\right] \tag{F-6l}$$

Any change in the temperature at the heat transfer surfaces or the feed inlet must be dispersed and back mixed into the volume. This process dead time (θ_p) is the turnover time that can be approximated as the liquid inventory divided by the summation of the feed flow rate (F_f), agitator pumping rate (F_a), recirculation flow rate (F_r), and vapor evolution rate or vapor bubble rate (F_v). Because this turnover time is computed in terms of volumetric flow rates, the liquid mass and the mass flow rates are divided by their respective densities as shown in Equation F-6m.

$$\theta_p = (M_o/\rho_o) / \left[(F_f + F_a + F_r)/\rho_o + F_v/\rho_v\right] \tag{F-6m}$$

If there is an injector (dip tube or sparger ring) volume, a change in composition at the nozzle must propagate by plug flow to the discharge points of the dip tube or sparger ring. The dead time for a feed flow (F_1) is the injector volume (V_1) divided by the injection mass flow (F_1) divided by its respective density (ρ_1).

$$\theta_p = V_1 / (F_1 / \rho_1) \tag{F-6n}$$

The ODE for the accumulation of liquid reactant mass (M_A) as shown in Equation F-7a includes the effects of feeds (F_i) with a reactant mass fraction (X_{Ai}), reaction rate (R_x), and outlet flow (F_o). The feeds can be from raw material, intermediate products, recycle streams, or multistage reactors. If we use the multiplicative rule of integration, we can express the differential equation in the generic form of Equation F-1a in terms of concentration to show the process feedback in Equation F-7f, the final form of the ODE.

$$dM_A/dt = \Sigma(F_i \bullet X_{Ai}) - (R_x + F_o) \bullet X_{Ao} \tag{F-7a}$$

$$dM_A/dt = d(M_o \bullet X_{Ao})/dt = (dM_o/dt) \bullet X_{Ao} + M_o \bullet (dX_{Ao}/dt) \tag{F-7b}$$

$$F_f = \Sigma F_i \tag{F-7c}$$

$$X_{Af} = \Sigma(F_i \bullet X_{Ai}) / \Sigma F_i \tag{F-7d}$$

$$(dM_o/dt) \bullet X_{Ao} = (F_f - F_o) \bullet X_{Ao} \tag{F-7e}$$

$$M_o \bullet (dX_{Ao}/dt) = F_f \bullet X_{Af} - (R_x + F_f) \bullet X_{Ao} \tag{F-7f}$$

For the manipulation of feed flow to control reactant concentration (X_{Ao}), the main process time constant (τ_p) is:

$$\tau_p = M_o/(R_x + F_f) \tag{F-7g}$$

For the manipulation of feed flow to control reactant concentration (X_{Ao}), the process gain (K_p) is:

$$K_p = X_{Af}/(R_x + F_f) \tag{F-7h}$$

For the manipulation of feed flow to control reactant concentration, the near-integrator gain (K_{nip}) is:

$$K_{nip} = X_{Af} / M_o \tag{F-7i}$$

For the manipulation of feed concentration to control reactant concentration, the process gain (K_p) is:

$$K_p = F_f / (R_x + F_f) \tag{F-7j}$$

The process dead times from turnover time and from feed injection are the same as computed in the section for temperature control (Equations F-6m and F-6n).

F.6 Plug Flow Volumes

For plug flow volumes where different streams are being combined, the process gain for controlling the temperature (T_f) or composition (X_{Af}) of the mixture (often a feed to downstream equipment) can be computed by taking the derivative of Equations F-6d and F-7d with respect to the manipulated flow for stream 1 (F_1) to give Equations F-9a and F-9b, respectively. In both cases, the process gain is inversely proportional to total flow (ΣF_i).

$$K_p = dT_f/dF_1 = T_1 / \Sigma F_i \tag{F-9a}$$

$$K_p = dX_{Af}/dF_1 = X_{A1} / \Sigma F_i \tag{F-9b}$$

The process dead time for the manipulation of a flow for stream 1 (F_1) is the summation of the injection delay for stream 1 and the piping delay from the point of injection to the point of temperature or composition measurement. For plug flow the residence time, which is the second expression in Equation F-9c, completely becomes dead time.

$$\theta_p = V_1 / (F_1/\rho_1) + V_p / \Sigma(F_i/\rho_i) \tag{F-9c}$$

The process time constant is essentially zero for true plug flow. For a static mixer there is some back mixing; the residence time in Equation F-9c is split between a dead time and a time constant per Equations F-9d and F-9e.

$$\theta_p = V_1/(F_1/\rho_1) + x \bullet V_p / \Sigma(F_i/\rho_i) \tag{F-9d}$$

$$\tau_p = (1 - x) \bullet V_p / \Sigma(F_i/\rho_i) \tag{F-9e}$$

It is obvious from the above that both the process gain and dead time are inversely proportional to total flow.

F.7 Controller Tuning

The implication of the results can be best seen if lambda is set equal to the total loop dead time (θ_o), resulting in Equation F-10a for the proportional-integral-derivative (PID) gain. If the open-loop time constant (τ_o) is large compared to the dead time, the process response is termed *near-integrating* and the ratio of the open-loop self-regulating process gain (K_o) to the time constant is used to approximate the open-loop integrating process gain (K_i) in Equation 5.10b. This PID gain is about ½ of the PID gain estimated by the Ziegler-Nichols reaction curve method.

$$K_c = 0.5 \bullet \frac{\tau_o}{K_o \bullet \theta_o} \tag{F-10a}$$

$$K_c = 0.5 \bullet \frac{1}{K_i \bullet \theta_o} \tag{F-10b}$$

$$K_o = K_v \bullet K_p \bullet K_m \tag{F-10c}$$

The open-loop time constant (τ_o) in the numerator of Equation F-10a is the largest time constant in the loop wherever it occurs. Hopefully, the process is mixed well enough and the instrumentation is fast enough that the largest time constant is in the process ($\tau_o = \tau_P$) and not the automation system. A large time constant in the process slows down the disturbance and is desirable. A large time constant in the measurement and final element is detrimental because it slows down the ability of the controller to see and react to disturbance, respectively.

The open-loop self-regulating process gain (K_o) in the denominator is dimensionless (Δ%PV/Δ%OUT). The open-loop gain is the product of the manipulated variable gain (K_v); the process gain (K_p); the gain of nonlinear process variables, such as pH (slope of the titration curve); and the measurement gain (K_m). For a loop that manipulates a control valve or variable speed drive, the manipulated variable gain is the slope of the valve's or variable speed drive's installed flow characteristic. For the primary loop of a cascade control system, the manipulated variable gain is the secondary loop set-point span divided by 100%. The measurement gain is 100% divided by the process variable measurement span. Thus, changes in calibration span affect the computed controller gain. A decrease in measurement span may increase measurement accuracy, but controller gain may need to be decreased to ensure robustness and a less oscillatory response.

Finally, the dead time (θ_o) in the denominator is really the total loop dead time, which is summation of the process dead time (θ_p) plus all the small time lags and delays in the loop. While the names *open-loop time constant* (τ_o), *open-loop self-regulating process gain* (K_o), and *total loop dead time* (θ_o) for the parameters in Equation F-10a are more definitive, most of the control literature uses the terms *process time constant, process gain,* and *process dead time* indiscriminately.

F.8 Nomenclature

Process Parameters

A_o = cross-sectional area of liquid level (m^2)

A = heat transfer surface area (m^2)

C_p = heat capacity of process (kJ/kg • °C)

C_w = heat capacity of wall of heat transfer surface (kJ/kg • °C)

F_a = agitator pumping rate (kg/s)

F_f = total feed flow (kg/s)

F_g = gas flow (kg/s)

F_i = feed stream *i* flow (kg/s)

F_o = vessel outlet flow (kg/s)

F_r = recirculation flow (kg/s)

F_v = vaporization rate (kg/s)

H_v = heat of vaporization (kJ/kg)

H_x = heat of reaction (kJ/kg)

L_o = liquid level (m)

M_A = component A mass (kg)

M_g = gas mass (kg)

M_o = liquid mass (kg)

M_w = mass of wall of heat transfer surface (kg)

P_g = gas pressure (kPa)

T_f = total feed temperature (°C)

T_g = gas temperature (°K)

T_i = feed stream i temperature (°C)

T_o = vessel outlet temperature (°C)

t = time (s)

Q_o = total heat of liquid (kJ)

Q_r = heat from reaction (kJ)

R_x = reaction rate (kg/s)

R_g = specific gas constant including compressibility factor ((kPa • m^3)/(°K • kg per mole))

ρ_g = gas density (kg/m^3)

ρ_i = stream i density (kg/m^3)

ρ_o = liquid density (kg/m^3)

ρ_v = density of vapor (kg/m^3)

U = overall heat transfer coefficient (kJ/m^2 • °C)

V_g = gas volume (m^3)

V_i = injection (e.g., dip tube or sparger ring) volume (m^3)

V_p = piping volume (m^3)

x = fraction of volume that is plug flow

X_{Af} = total feed component A concentration (mass fraction)

X_{Ai} = feed stream i component A concentration (mass fraction)

X_{Ao} = vessel outlet component A concentration (mass fraction)

Generic Terms

X = process input (manipulated variable) (eu)

Y = process output (controlled variable) (eu)

Dynamic Parameters

K_c = PID controller gain (dimensionless)

K_i = open-loop integrating process gain (PV percent per second per CO percent) (1/seconds)

K_{ip} = integrating process gain (PV engineering units/MV engineering units)

K_m = measurement gain (PV percent/PV engineering units)

K_o = open-loop self-regulating process gain (PV percent/CO percent) (dimensionless)

K_p = process gain (PV engineering units/MV engineering units)

K_{nip} = near-integrating process gain (PV engineering units per second/MV engineering units)

K_v = valve or VSD gain (slope of installed flow characteristic) (flow engineering units/CO percent)

τ_p = negative feedback process time constant (seconds)

τ_p' = positive feedback process time constant (seconds)

τ_o = open-loop time constant (primary time constant) (seconds)

θ_p = process dead time (seconds)

θ_o = total loop dead time (seconds)

Appendix G: Gas Pressure Dynamics

The interactive time constants of multiple gas volumes (e.g., vessels, columns, and headers) in series separated by flow resistances (e.g., valves and fittings) can be estimated by Equations G-1 through G-4 for turbulent and sonic flow [1].

$$\tau_i = \frac{C_i}{\sum_i \frac{1}{R_i}} \tag{G-1}$$

$$C_i = \frac{V_i}{P_a} \tag{G-2}$$

For turbulent flow:

$$R_i = \frac{2 \bullet \Delta P_i}{F_i} \tag{G-3}$$

For sonic flow:

$$R_i = \frac{P_i}{F_i} \tag{G-4}$$

where

C_i = capacitance of the gas volume i (scf/psi)
F_i = gas flow through resistance i (scfm)
P_a = atmospheric pressure at resistance i (psi)
P_i = inlet pressure at resistance i (psi)
ΔP_i = pressure drop across resistance i (psi)
R_i = resistance to the gas flow into or out of volume i (psi/scfm)
V_i = gas volume i (scf)
τ_i = interactive time constant of gas volume i (min)

If the intermediate flow resistance pressure drops for turbulent nonsonic flow between volumes in a gas system are less than 2% of the system inlet and outlet flow resistance pressure drops, then all the individual volumes can be summed and substituted into Equation G-2 to give a single process time constant for the gas system. The universal gas sizing equation can be used to estimate the C_v of the fitting or valve necessary to prevent the creation of multiple time constants. If the pressure drops between volumes are not negligible (flow coefficient of internal resistance does not satisfy Equation G-5), then the system should be partitioned into two major volumes and Appendix I used in conjunction with the equations here to estimate the equivalent two noninteractive time constants for two volumes in series. The sum of the noninteractive time constants should be then used to estimate the process dead time.

$$C_v > \frac{F}{P \bullet \sqrt{\frac{0.02 \bullet \Delta P}{P}}} \bullet \sqrt{\frac{G \bullet T}{520}} \qquad \text{(G-5)}$$

where

C_v = gas flow coefficient of internal resistance (scfm/psi)
F = gas flow through the system (scfm)
G = gas specific gravity in the system (dimensionless)
P = pressure at the system inlet (psi)
ΔP = pressure drop at the system outlet (psi)
T = absolute temperature in the system (°R)

Reference

1. Peter Harriott, *Process Control* (New York: McGraw-Hill, 1964).

Appendix H: Charge Balance to Model pH

H.1 Introduction

Aqueous systems are composed of ions in various degrees of dissociation in water. These contributions come from acids, bases, and salts.

Salts

$$B_{\varepsilon}^{+\lambda}A_{\beta}^{-\omega} \leftrightarrow \varepsilon B^{+\lambda} + \beta A^{-\omega} \tag{H-1a}$$

Acids

$$H_{\omega}A^{-\omega} \underset{pK_{a1}}{\leftrightarrow} H^{+} + H_{\omega-1}A^{-1} \underset{pK_{a2}}{\leftrightarrow} H^{+} + H_{\omega-2}A^{-2} \underset{pK_{a3}}{\leftrightarrow} \ldots \tag{H-1b}$$

Bases

$$B^{+\lambda}(OH)_{\lambda} \underset{pK_{b1}}{\leftrightarrow} OH^{-1} + B(OH)_{\lambda-1}^{+1} \underset{pK_{b2}}{\leftrightarrow} OH^{-1} + B(OH)_{\lambda-2}^{+2} \underset{pK_{b3}}{\leftrightarrow} \ldots \tag{H-1c}$$

H.2 Definitions

pH

$$pH = -\log_{10}[H^{+}] \tag{H-2a}$$

pK_a

$$pK_a = -\log_{10} K_a \tag{H-2b}$$

K_w

$$H_2O \underset{K_w}{\leftrightarrow} H^+ + OH^- \quad and \quad K_w = K_b \bullet K_a = [OH^-][H^+] \tag{H-2c}$$

For a system composed of these species, there are three fundamental relationships available:

- Principle of electroneutrality in a charge balance
- Conservation of mass in a species balance
- Equilibrium relationships

The way these relationships come together will depend on the valence of the ions considered. The derivations for monovalent through trivalent ions follow.

H.3 Monovalent Ion

The monovalent ion can enter the system either through an acid (if an anion), base (if a cation), or a salt. Consider the dissociation of an anion in an acid:

$$[HA] \overset{K_{a1}}{\leftrightarrow} [A^{-1}] + [H^{+1}] \quad and \quad K_{a_1} = \frac{[H^+][A^-]}{[HA]} \tag{H-3a}$$

Additionally, there may be any number of salts contributing the anion to the system as:

$$B_{\varepsilon}^{+\lambda} A_{\beta}^{-1} \rightarrow \varepsilon B^{+\lambda} + \beta A^{-1} \tag{H-3b}$$

We can then write a species balance about the system at a constant volume as:

$$[HA] + [A^{-1}] = X_a + \sum_{j=1}^{n} \beta_j \bullet X_j = E \tag{H-3c}$$

where X_a and X_j are initial concentrations of acid and salt, respectively. The β_j is the stoichiometric coefficient associated with the anion for the jth salt. The right side of this equation is a constant and can be noted as E.

$$[HA]+[A^{-1}]=E \tag{H-3d}$$

Taking the equilibrium expression and rearranging:

$$\frac{[H^{+1}][A^{-1}]}{K_{a_1}}=[HA] \tag{H-3e}$$

Substituting Equation H-3d into Equation H-3e:

$$\frac{[H^{+1}][A^{-1}]}{K_{a_1}}+[A^{-1}]=E \tag{H-3f}$$

Solving for $[A^{-1}]$:

$$[A^{-1}]=\frac{E}{\left(\frac{[H^{+}]}{K_{a_1}}+1\right)} \tag{H-3g}$$

Using the definition of pK_a and pH:

$$[A^{-}]=\frac{E}{(10^{-pH}\bullet 10^{pK_{a1}}+1)}=\frac{E}{(10^{pK_{a1}-pH}+1)} \tag{H-3h}$$

For a base, K_b can be used instead:

$$B^{+1}(OH)_1 \underset{pK_{b1}}{\leftrightarrow} OH^{-1}+B^{+} \tag{H-3i}$$

$$[B(OH)]+[B^{+1}]=X_b+\sum_{j=1}^{n}\varepsilon_j \bullet X_j=E \tag{H-3j}$$

And:

$$K_b=\frac{[OH^{-}][B^{+}]}{B(OH)} \tag{H-3k}$$

$$[B^{+}]=\frac{E}{\left(\frac{[OH^{-}]}{K_{b_1}}+1\right)} \tag{H-3l}$$

For $\frac{[OH^-]}{K_{b_1}}$:

$$[OH^-]\,[H^+] = K_w \rightarrow [OH^-] = \frac{K_w}{[H^+]} \tag{H-3m}$$

And with:

$$K_w = K_a \bullet K_b \rightarrow K_b = \frac{K_w}{K_a} \tag{H-3n}$$

Then:

$$\frac{[OH^-]}{K_{b_1}} = \underbrace{\frac{K_w}{[H^+] \bullet K_{b_1}}}_{Substituting\ Eq.\ \text{H-3n}} = \frac{\overbrace{K_w}}{[H^+] \bullet \frac{K_w}{K_a}} = \underbrace{\frac{K_a}{[H^+]}}_{Simplifying} = 10^{pH - pK_{a1}} \tag{H-3o}$$

By introduction of s, which is positive for bases and negative for acids, we can define P_1:

$$P_1 = 10^{s \bullet (pH - pK_{a1})} \tag{H-3p}$$

The charge balance now becomes (with respect to the acid form with no base and salts):

$$0 = \sum_{i=1}^{n} \pm X_i \bullet [B_i^{\pm X_i}] \tag{H-3q}$$

$$0 = -[A^-] - [OH^-] + \sum_{j=1}^{l} \lambda_j \bullet \beta_j \bullet [B^{+\lambda}] + [H^+] \tag{H-3r}$$

$$0 = -\underbrace{\frac{E}{[P_1 + 1]}}_{M_1} + [OH^-] + \underbrace{\sum_{j=1}^{l} \lambda_j \bullet \beta_j \bullet [B^{+\lambda}]}_{C_S} + [H^+] \tag{H-3s}$$

$$0 = -M_1 - [OH^-] + C_S + [H^+] \tag{H-3t}$$

Using Equation H-3m:

$$0 = -M_1 - \frac{K_w}{[H^+]} + C_S + [H^+] \tag{H-3u}$$

Substituting definitions for pK_w and pH:

$$0 = -M_1 - 10^{pH - pK_w} + 10^{-pH} + C_s \tag{H-3v}$$

If instead a base is used, the sign of M_1 will change, giving:

$$M_1 = s \bullet \frac{X_a + \sum_{j=1}^{n} \beta_j \bullet X_j}{P_1 + 1} = s \bullet \frac{E}{P_1 + 1} \tag{H-3w}$$

H.4 Divalent Ion

As before, the divalent ion can enter the system through an acid (if an anion), a base (if a cation), or a salt. For a divalent acid:

$$[H_2A] \overset{K_{a1}}{\leftrightarrow} [HA^{-1}] + [H^{+1}] \quad and \quad K_{a_1} = \frac{[H^+][HA^-]}{[H_2A]} \tag{H-4a}$$

$$[HA^{-1}] \overset{K_{a2}}{\leftrightarrow} [H^+] + [A^{-2}] \quad and \quad K_{a_2} = \frac{[H^+][A^{-2}]}{[HA^-]} \tag{H-4b}$$

And for a salt that completely dissociates:

$$B_{\varepsilon}^{+\lambda} A_{\beta}^{-2} \rightarrow \varepsilon B^{+\lambda} + \beta A^{-2} \tag{H-4c}$$

The species balance becomes:

$$[H_2A] + [HA^{-1}] + [A^{-2}] = X_a + \sum_{j=1}^{n} \beta_j \bullet X_j = E \tag{H-4d}$$

Substituting equilibrium expressions:

$$\frac{[H^+][HA^-]}{K_{a_1}} + \frac{[H^+][A^{-2}]}{K_{a_2}} + [A^{-2}] = E \tag{H-4e}$$

$$\frac{[H^+]^2[A^{-2}]}{K_{a_1}K_{a_2}}+\frac{[H^+][A^{-2}]}{K_{a2}}+[A^{-2}]=E \tag{H-4f}$$

Solving for A^{-2}:

$$[A^{-2}]=\frac{E}{1+\frac{[H^+]}{K_{a2}}\left(1+\frac{[H^+]}{K_{a_1}}\right)} \tag{H-4g}$$

The charge balance becomes:

$$0=C_S-[HA^-]-2[A^{-2}]+[H^+]-[OH^-] \tag{H-4h}$$

With:

$$[HA^-]=\frac{[H^+][A^{-2}]}{K_{a_2}} \tag{H-4i}$$

The charge balance becomes:

$$0=C_S-\frac{[H^+][A^{-2}]}{K_{a_2}}-2[A^{-2}]+[H^+]-[OH^-] \tag{H-4j}$$

Isolating $[A^{-2}]$:

$$0=C_S-[A^{-2}]\bullet\left(2+\frac{[H^+]}{K_{a2}}\right)+[H^+]-[OH^-] \tag{H-4k}$$

Substituting for $[A^{-2}]$:

$$0=C_S-\left(2+\frac{[H^+]}{K_{a_2}}\right)\frac{E}{1+\frac{[H^+]}{K_{a2}}\left(1+\frac{[H^+]}{K_{a_1}}\right)}+[H^+]-[OH^-] \tag{H-4l}$$

Casting into terms of P_1 and P_2:

$$0=C_s-\underbrace{(2+P_2)\frac{E}{1+P_2(1+P_1)}}_{M_2}+[H^+]-[OH^-] \tag{H-4m}$$

Note that a similar approach from the monovalent case can be used to show the derivation for bases where the sign of M_2 will change, leading to the final expression:

$$M_2 = s \cdot (2+P_2) \cdot \frac{\left(X_a + \sum_{j=1}^{n} \beta_j \cdot X_j\right)}{1+P_2(1+P_1)} = s \cdot (2+P_2) \cdot \frac{E}{1+P_2(1+P_1)} \quad \text{(H-4n)}$$

H.5 Trivalent Ion

For the trivalent acid:

$$[H_3A] \overset{K_{a1}}{\leftrightarrow} [H_2A^{-1}] + [H^{+1}] \overset{K_{a2}}{\leftrightarrow} [H^{+1}] + [HA^{-2}] \overset{K_{a3}}{\leftrightarrow} [H^{+1}] + [A^{-3}] \quad \text{(H-5a)}$$

$$K_{a_1} = \frac{[H_2A^-][H^+]}{[H_3A]} \quad K_{a_2} = \frac{[HA^{-2}][H^+]}{[H_2A^-]} \quad K_{a_3} = \frac{[A^{-3}][H^+]}{[HA^{-2}]} \quad \text{(H-5b)}$$

The species balance becomes:

$$[H_3A] + [H_2A^-] + [HA^{-2}] + [A^{-3}] = X_a + \sum_{j=1}^{n} \beta_j \cdot X_j = E \quad \text{(H-5c)}$$

Substituting for equilibrium as before:

$$\frac{[A^{-3}][H^+]^3}{K_{a_1}K_{a_2}K_{a_3}} + \frac{[A^{-3}][H^+]^2}{K_{a_2}K_{a_3}} + \frac{[A^{-3}][H^+]}{K_{a_3}} + [A^{-3}] = E \quad \text{(H-5d)}$$

Solving for $[A^{-3}]$:

$$[A^{-3}] = \frac{E}{\left(\frac{[H^+]^3}{K_{a_1}K_{a_2}K_{a_3}} + \frac{[H^+]^2}{K_{a_2}K_{a_3}} + \frac{[H^+]}{K_{a_3}} + 1\right)} \quad \text{(H-5e)}$$

The charge balance in this instance becomes:

$$0 = C_S + [H^+] - [OH^-] - [H_2A^-] - 2[HA^{-2}] - 3[A^{-3}] \quad \text{(H-5f)}$$

Substituting for equilibrium expressions:

$$0 = C_S + [H^+] - [OH^-] - \frac{[A^{-3}][H^+]^2}{K_{a_2}K_{a_3}} - 2\frac{[A^{-3}][H^+]}{K_{a_3}} - 3[A^{-3}] \quad \text{(H-5g)}$$

Isolating $[A^{-3}]$:

$$0 = C_S + [H^+] - [OH^-] - [A^{-3}] \bullet \left(\frac{[H^+]^2}{K_{a_2} K_{a_3}} + 2\frac{[H^+]}{K_{a_3}} + 3 \right) \quad \text{(H-5h)}$$

Substituting for $[A^{-3}]$:

$$0 = C_S + [H^+] - [OH^-] - \frac{E}{\left(\frac{[H^+]^3}{K_{a_1} K_{a_2} K_{a_3}} + \frac{[H^+]^2}{K_{a_2} K_{a_3}} + \frac{[H^+]}{K_{a_3}} + 1 \right)} \bullet \left(\frac{[H^+]^2}{K_{a_2} K_{a_3}} + 2\frac{[H^+]}{K_{a_3}} + 3 \right)$$

(H-5i)

Substituting for P_1, P_2, and P_3:

$$0 = C_S + [H^+] - [OH^-] - \frac{3 + P_3 \bullet (2 + P_2)}{1 + P_3 \bullet (1 + P_2 \bullet (1 + P_1))} \bullet E \quad \text{(H-5j)}$$

Note that a similar approach from the monovalent case can be used to show the derivation for bases where the sign of M_3 will change, leading to the final expression:

$$M_3 = s \bullet \left(X_a + \sum_{j=1}^{n} \beta_j \bullet X_j \right) \bullet \frac{3 + P_3 \bullet (2 + P_2)}{1 + P_3 \bullet (1 + P_2 \bullet (1 + P_1))} = s \bullet E \frac{3 + P_3 \bullet (2 + P_2)}{1 + P_3 \bullet (1 + P_2 \bullet (1 + P_1))}$$

(H-5k)

H.6 Temperature

Note that equilibrium constants (pK_a values, including K_w) will be functions of temperature. Many sources in the literature have a variety of data collected at various temperatures. K_w is also a function of temperature.

H.7 Activity

As the concentrations of ionic species in the system increase, the ideality of the system will diminish. For a general equilibrium equation, activity coefficients can be included:

$$K_a' = \gamma_{H^+} \bullet [H^+] \bullet \gamma_{A^-} \bullet \frac{[A^-]}{[HA] \bullet \gamma_{HA}} = K_a \bullet \frac{\gamma_{H^+} \bullet \gamma_{A^-}}{\gamma_{HA}} \quad \text{(H-7a)}$$

However, for relatively dilute systems, activity can be used to correct for nonidealities using the Debye-Hückle equation:

$$log(\gamma_{H^+}) = -0.5 \bullet \frac{I^{0.5}}{1+I^3} \quad \text{(H-7b)}$$

where I is the ionic strength. Ionic strength can be calculated as:

$$I = 0.5 \bullet \sum m_i \bullet Z_i^2 \quad \text{(H-7c)}$$

where m_i is molal concentration, Z_i is the ion's concentration, and i is an index over the ionic species.[1]

H.8 Nomenclature

β = stoichiometric coefficient associated with an anion
ε = stoichiometric coefficient associated with a cation
λ = positive charge associated with a cation
ω = negative charge associated with an anion
B = cation (such as Ca^{2+})
A = anion (such as Cl^-)
C_S = contribution of salts to a charge balance
E = molar concentration of an acid or base and its common ions to a system
I = ionic strength (see Equation H-7c)
K_a = acid dissociation constant
P_i = pH factor of a particular dissociation i
s = sign of a particular contribution; defined as –1 for acids and +1 for bases
pH = negative of the base 10 log of hydronium ion concentration
pK_a = negative of the base 10 log of the acid dissociation constant
pK_w = negative of the base 10 log of the dissociation constant of water
M_i = overall contribution of the ith component to the charge balance
m_i = molal concentration of the ith species (moles/(kg of solvent))
X = molar concentration of acid, base, or salt in solution (mol/L)

1 Gregory K. McMillan and Robert A. Cameron, *Advanced pH Measurement and Control*, 3rd ed. (Research Triangle Park, NC: ISA [International Society of Automation]), 2005.

Appendix I: Interactive to Noninteractive Time Constant Conversion

Interactive time constants can be partitioned into two subsystems in series with their capacitance and resistance components to calculate the equivalent noninteractive time constants. The quadratic equation can then be used find the two noninteractive time constants for two capacitances that have an inlet and outlet resistance and are separated by a resistance [1]. Interactive time constants exist when the flow or transfer of mass or heat into, within, and out of a system depend on the operating conditions as driving forces (e.g., pressure drop for flow, concentration differences for mass transfer, and temperature difference for heat transfer). For gas pressure systems, the flow resistance of valves and fittings into, within, and out of the volumes are important. For thermal systems such as vessel coils and jackets, heat exchanger tubes, and thermowells, the resistance of convective heat transfer of the inside and outside surfaces and the resistance of conductive heat transfer within the surface are important. The capacitances offer the retention of energy or material in a subsystem, and the resistances oppose the transfer of material and energy between the subsystems.

I.1 Equivalent Noninteractive Time Constants

$$\tau_{n1} = \frac{2 \bullet A}{B - \sqrt{B^2 - 4 \bullet A \bullet C}} \tag{I-1a}$$

$$\tau_{n2} = \frac{2 \bullet A}{B + \sqrt{B^2 - 4 \bullet A \bullet C}} \tag{I-1b}$$

$$A = R_1 \bullet C_1 \bullet R_2 \bullet C_2 \bullet R_3 \tag{I-1c}$$

$$B = R_1 \bullet C_1 \bullet R_3 + R_1 \bullet C_1 \bullet R_2 + R_2 \bullet C_2 \bullet R_3 + R_1 \bullet C_2 \bullet R_3 \tag{I-1d}$$

$$C = R_1 + R_2 + R_3 \tag{I-1e}$$

where

A = first term of quadratic equation
B = second term of quadratic equation
C = third term of quadratic equation
C_1 = capacitance of section 1
C_2 = capacitance of section 2
R_1 = resistance at inlet to section 1
R_2 = resistance between sections 1 and 2
R_3 = resistance at outlet of section 2
τ_{n1} = noninteractive time constant 1 (minutes)
τ_{n2} = noninteractive time constant 2 (minutes)

I.2 Original Interactive Time Constants

The following equations can be used to compute the interactive time constants for a system partitioned into two sections with an inlet resistance (R_1) and outlet resistance (R_3) separated by a resistance (R_2). The capacitances and resistances for gas pressure systems, including engineering units, are defined in Appendix G.

$$\tau_{i1} = \frac{C_1}{\frac{1}{R_1} + \frac{1}{R_2}} \tag{I-2a}$$

$$\tau_{i2} = \frac{C_2}{\frac{1}{R_2} + \frac{1}{R_3}} \tag{I-2b}$$

where

C_1 = capacitance of section 1
C_2 = capacitance of section 2
R_1 = resistance at inlet to section 1
R_2 = resistance between sections 1 and 2
R_3 = resistance at outlet of section 2
τ_{i1} = interactive time constant 1 (minutes)
τ_{i2} = interactive time constant 2 (minutes)

Reference

1. Peter Harriot, *Process Control* (New York: McGraw-Hill, 1964).

Appendix J: Jacket and Coil Temperature Control

The heating and cooling system is critical for reactors because temperature plays such a huge role in determining reaction rate and selectivity (formation of desired product). ***The limit to what a valve position controller (VPC) can do to increase the production rate depends on the capability of the heating and cooling system. The system can also be the source of discontinuities and nonlinearities. Most reactor temperature control problems beyond tuning can be traced back to deficiencies in the reactor coil or jacket temperature system design.***

Heating is required for endothermic reactions (reactions that consume energy), liquids that are being vaporized, or bringing a reactor up to operating temperature. Cooling is needed for exothermic reactions (reactions that produce heat), vapors that are being condensed, and bringing a reactor down to operating temperature. Often both heating and cooling are required. Split-range control is used to go back and forth between heating and cooling. In split-range control for an exothermic reactor with a fail-open cooling valve, as the temperature controller output increases from 0% to the split-range point, the coolant valve goes from wide open to completely closed. As the temperature controller output increases from the split-range point to 100%, the heating valve goes from closed to wide open. The split-range point is traditionally set at 50%

but would be better set so that the change in temperature for a change in flow is about the same for each valve. Although cooling and heating may be done by coils as well as jackets, many of the control considerations are the same. Jacketed reactors will be used for purposes of discussion and illustration. Cascade temperature control is the most prevalent strategy in which the primary reactor temperature proportional-integral-derivative (PID) controller provides a set point to a secondary jacket temperature PID for the throttling of hot and cold fluids (Figures J-1a and J-1b). For highly exothermic reactors, boiler feedwater is added under level control and the reactor temperature PID output is the jacket outlet steam pressure PID set point. These jacket control schemes are suitable for batch as well as continuous operation.

The use of cascade control, where the reactor temperature controller output is the set point of a coil or jacket temperature loop, offers considerable performance improvements. The jacket temperature controller can correct for disturbances to the jacket before they affect the reactor temperature. The jacket temperature loop also isolates process and valve gain nonlinearity from the reactor loop. For a negligible

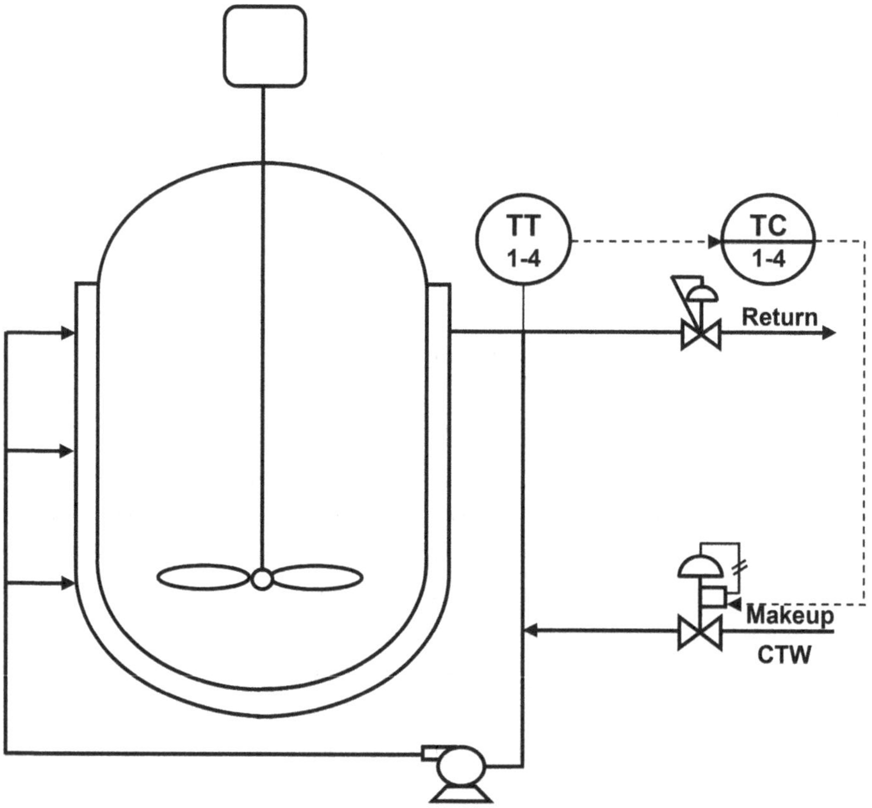

Figure J-1a. Jacket-outlet temperature control offers moderation of mixing and phase discontinuities from the attenuation by jacket volume.

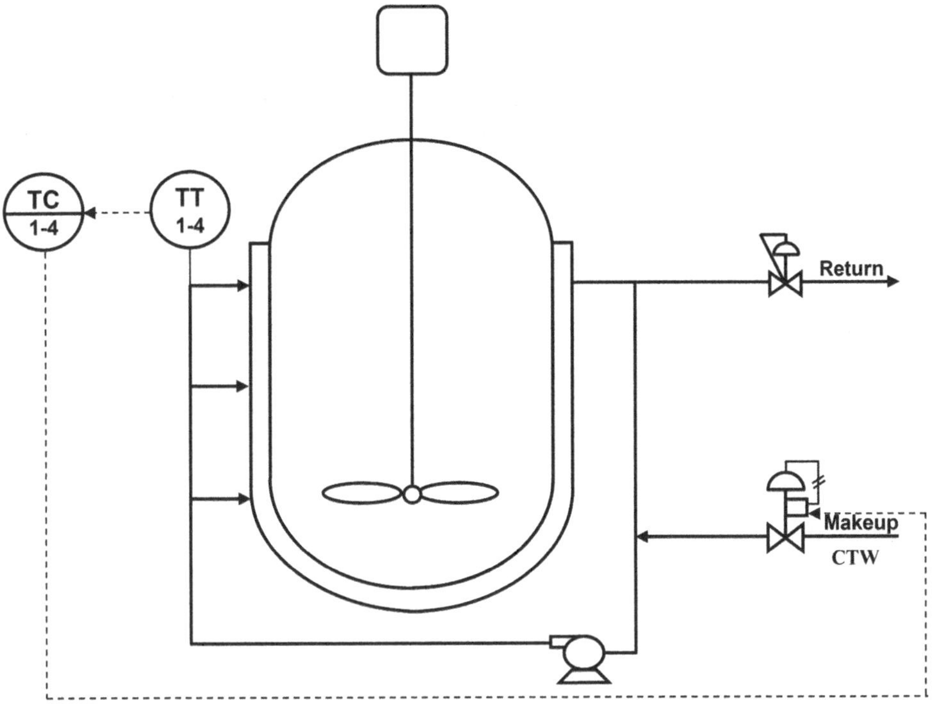

Figure J-1b. Jacket-inlet offers faster correction of jacket inlet disturbances such as coolant temperature.

increase in the heat of reaction compared to heat transfer capability, the process gain for the reactor temperature loop approaches unity.

Coils generally offer a faster temperature response than a jacket by a decrease in the volume and an increase in the velocity. Both of these work to decrease the process dead time, which is the coil volume divided by the utility flow rate. The increase in velocity increases the heat transfer coefficient, but this is partially offset by a decrease in the surface area. An increase in the product of the heat transfer coefficient and surface area (UA) will decrease the secondary process lag in the thermal response. The transition in split-range operation is faster, which is useful for a valid transition between hot and cold utility streams but can be problematic for inadvertent transitions from stick-slip and an integrating response in the process or controller.

Whether the secondary loop uses coil or jacket inlet or outlet temperature is often a matter of tradition for a particular company or process industry. The dynamic response of the cascade control system to reactor disturbances such as feed and reaction rate are the same for jacket inlet and outlet temperature control. ***The coil or jacket inlet temperature control will correct for changes in cooling or heating utility supply temperature and pressure sooner than the transportation delay through the jacket.*** The jacket loop process dead time is also less by the amount of this delay, enabling a faster reset time setting and faster correction of valve nonlinearities. ***At the coil or jacket inlet, the mixing of the recirculation with the hot or cold makeup flow may be incomplete and the discontinuity in the transition from hot to cold more abrupt.*** The location of the temperature sensor on the jacket outlet offers time for mixing and volume for smoothing transitions. Less measurement temperature noise can translate to a higher controller gain and less overreaction to the discontinuity at a split-range transition.

The difference between the reactor and the jacket outlet temperature (approach temperature) can provide an inferential measurement of the heat transfer coefficient (U) for a constant jacket circulation flow, a given production rate, and the heat transfer area (A). The approach temperature increases as UA decreases. For residence time control, the increase in level will increase the heat transfer area covered by reactants, offsetting the increase in heat release with production rate, and eliminating the need for production rate and level correction. For fed-batch reactors and continuous reactors without residence time control, a correction for level is needed to compute U from the UA.

For the coil or jacket temperature control to provide rapid adjustments of cooling and heating for disturbances to the jacket and set-point changes from the reactor temperature control, the coil or jacket PID response must be fast and achieved by a fast sensor and fast tuning.

A tight-fitting sensor bottomed in a thermowell with a high thermal conductivity metal and tapered tip provides a fast measurement. Spring loading can ensure the sensor sheath is bottomed. The clearance between the sheath outside diameter and the thermowell inside diameter must be minimized and the fluid velocity maximized. While grounded thermocouple sensors are a few seconds faster than resistance temperature detector (RTD) sensors, the difference is insignificant compared to the effect of thermowell design and fluid velocity. The thermowell design details are also important for reactor temperature. The greater sensitivity and lower drift of the RTD is important for jacket and reactor temperature measurements. The lower drift reduces maintenance, and the higher sensitivity provides faster recognition. The use of RTDs also facilitates more accurate online heat transfer computation for process diagnostics and inferential measurements of reaction rate. Thermocouples are preferred for temperatures above 400°C where RTD insulation resistance and sensor integrity become problematic.

A high controller gain and a small reset time provide fast tuning. The jacket temperature controller is a self-regulating process with a maximum PID gain that is about half of the open-loop time constant divided by the product of the open-loop gain and dead time as detailed in the section in Chapter 3 on controller tuning. The minimum reset setting is about four times the loop dead time for the jacket temperature PID controller.

Most temperature loops tend to develop an oscillation across the split-range point in making the transition between heating and cooling. The transition creates a discontinuity from the change in utility fluid and the increase in friction and loss of sensitivity in the valve seating or sealing. The worst case is often the transition between steam and coolant due to the huge difference in temperatures and the creation of bubbles in the coolant or water droplets in the steam. Steam valves tend to have higher seating and sealing friction due to the higher temperatures and pressures. Hot and cold liquids and tempered water systems offer a much smoother transition than steam and coolant, and they reduce nonlinearities.

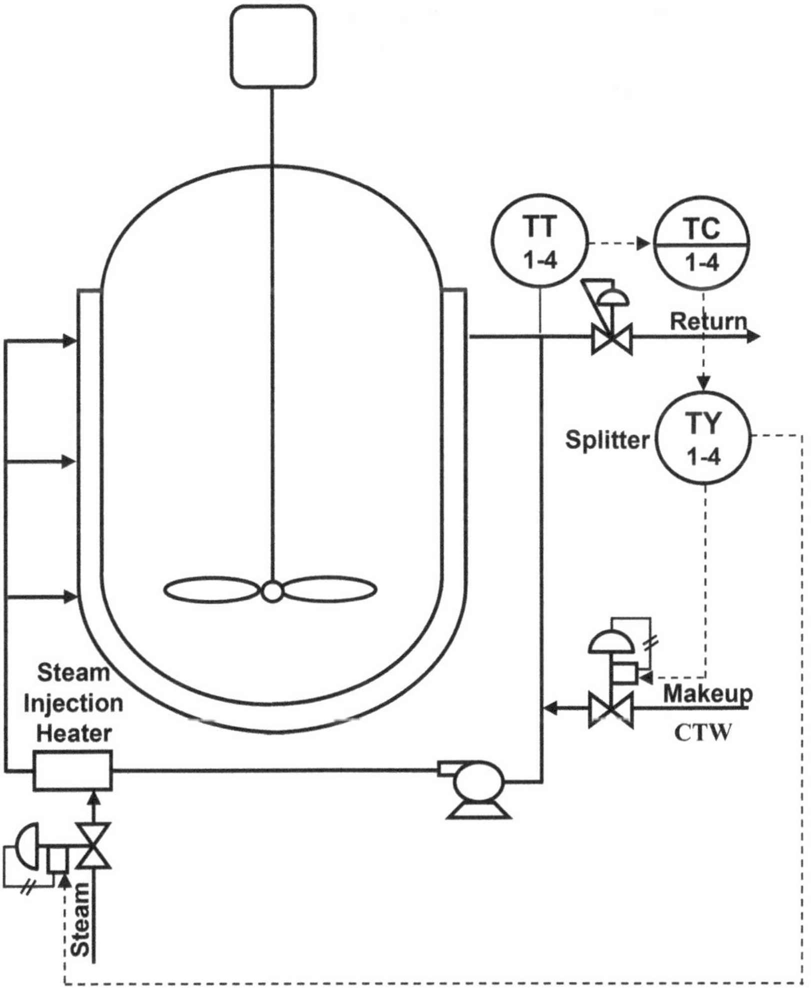

Figure J-2. Using steam injection heaters to create hot water offers rapid heating and tight control of coil or jacket temperature with smoother split-ranged transitions.

Steam does not provide uniform heating in coils or a jacket. Steam typically collects in the top of the jacket and condensate collects in the bottom. Hot spots can develop around inlets. Thermal shock and steam hammer can damage glass-lined vessels. The time required to drive steam completely out of the jacket before cooling water is introduced causes a considerable delay in the control system. Improper trap design or operation can cause condensate buildup.

The addition of hot water instead of steam directly into the jacket provides a more uniform heat distribution, a dramatically smoother transition between heating and cooling, and a more efficient and maintainable system. For rapid heating, the use of direct steam injection heaters and pressurized water can provide hot water temperatures well above 100°C (Figure J-2). If the injection heater has hundreds of small orifices, the bubbles are extremely small and are rapidly and quietly mixed into the water. Variable orifice steam injection heaters are not as quiet, and the mixing is not as complete. The use of jacket outlet temperature reduces the possibility of bubbles hitting the temperature sensor.

Appendix K: PID Forms and Conversion of Tuning Settings

K.1 Conversion of Settings to Correct Units and Form

The equations in this book are based on the Standard Form with specific tuning setting units where the controller gain is dimensionless, the reset time is seconds (seconds per repeat), and the rate time is seconds. Proportional-integral-derivative (PID) tuning settings must first be checked for units and, if necessary, undergo a units conversion before being used.

1. Convert proportional band (%) to controller gain (%/%) (dimensionless).

   Gain = 100 %/proportional band

2. Convert reset setting in repeats per minute to reset time in seconds per repeat.

   Seconds per repeat = 60/repeats per minute

3. Convert rate time in minutes to rate time in seconds.

   Seconds = 60 • minutes

4. After the tuning setting units are verified, Series and Parallel Forms convert tuning settings to the Standard Form tuning settings per the following equations.

The primed tuning settings are for the Series Form, the double-primed tuning settings are for the Parallel Form, and the unprimed tuning settings are for the Standard Form used in this book.

To convert from Series to Standard Form controller gain:

$$K_c = \frac{T_i' + T_d'}{T_i'} \bullet K_c' \tag{K-1a}$$

To convert from Series to Standard Form reset (integral) time:

$$T_i = \frac{T_i' + T_d'}{T_i'} \bullet T_i' = T_i' + T_d' \tag{K-1b}$$

To convert from Series to Standard Form rate time:

$$T_d = \frac{T_i'}{T_i' + T_d'} \bullet T_d' \tag{K-1c}$$

Note that if the rate time is zero, the Series Form (Figure K-1a) and the Standard Form (Figure K-1c) settings are identical. When using the Standard Form, if the rate time is greater than ¼ the reset time, the response can become oscillatory. If the rate time exceeds the reset time, the response can become unstable from a reversal of action from these modes. The Series Form inherently prevents this instability by increasing the effective reset time as the rate time is increased.

We can convert from the Standard Form to the Series Form using the following equations if the reset time is equal to or greater than four times the rate time ($T_i \geq 4 * T_d$).

$$K_c' = \frac{K_c}{2} \bullet \left[1 + \left(1 - 4 \bullet \frac{T_d}{T_i}\right)^{0.5}\right] \tag{K-1d}$$

$$T_i' = \frac{T_i}{2} \bullet \left[1 + \left(1 - 4 \bullet \frac{T_d}{T_i}\right)^{0.5}\right] \tag{K-1e}$$

$$T_d' = \frac{T_i}{2} \bullet \left[1 - \left(1 - 4 \bullet \frac{T_d}{T_i}\right)^{0.5}\right] \tag{K-1f}$$

The Parallel Form (Figure K-1b) is used in some distributed control systems (DCSs) and programmable logic controllers (PLCs) to isolate the proportional mode tuning setting from the other modes. The gain setting does not affect the contribution from

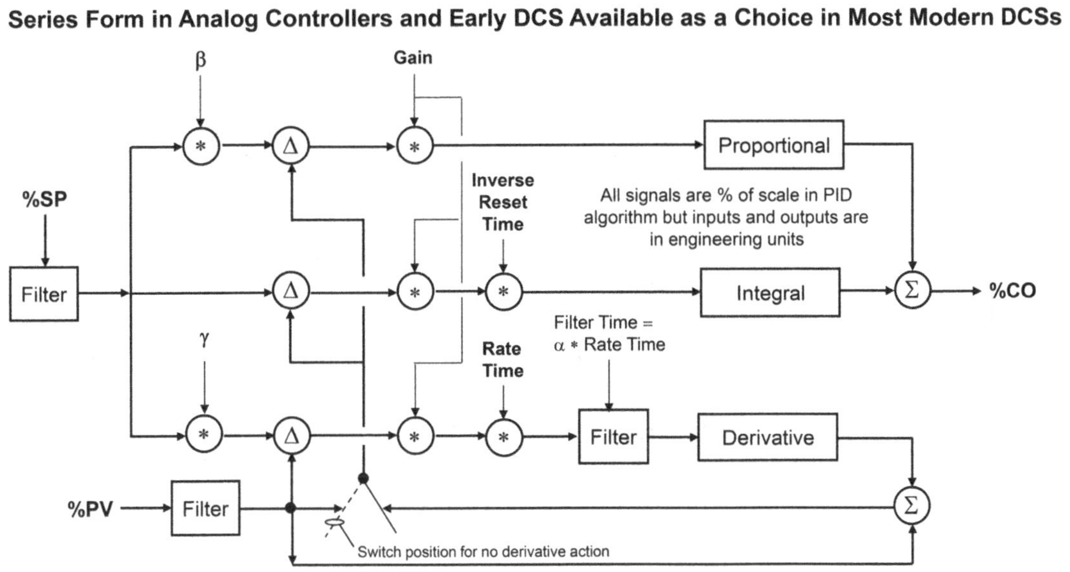

Figure K-1a. Series Form has derivative mode computed in series with other modes.

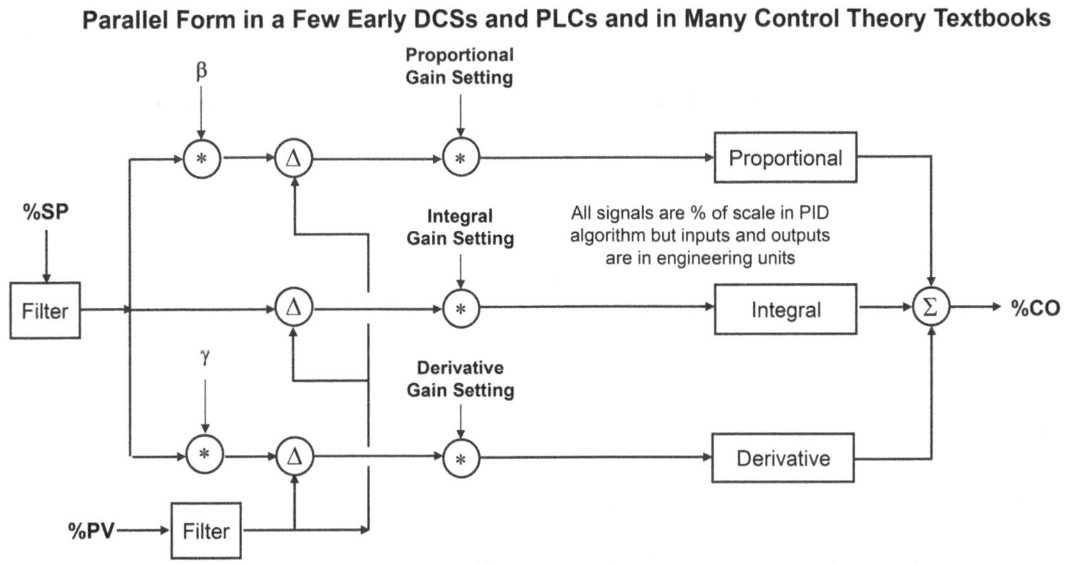

Figure K-1b. Parallel Form has independent mode gain settings.

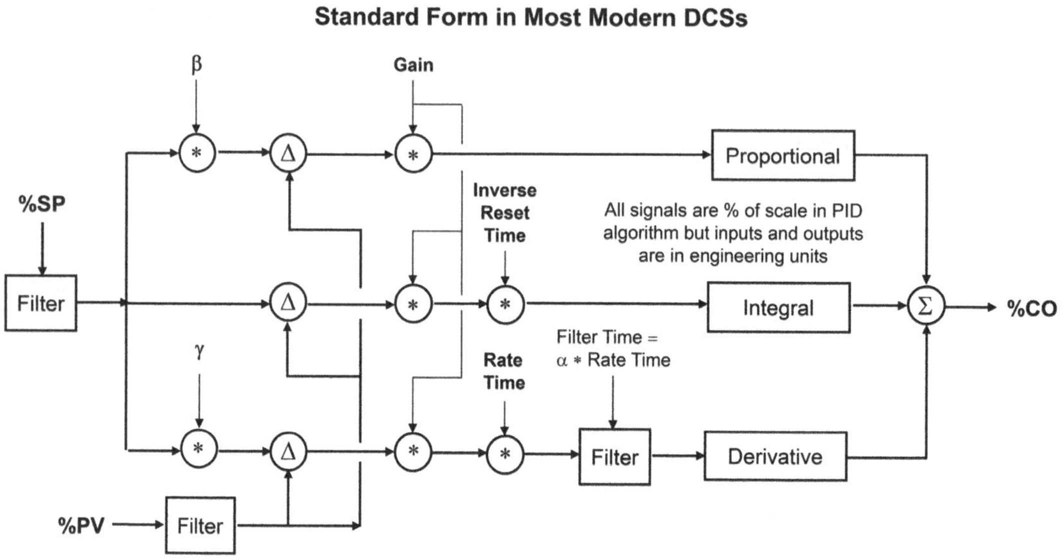

Figure K-1c. Standard Form has modes in parallel with proportional gain factor.

the integral and derivative modes and is sometimes called *noninteracting*. We can convert from the Parallel Form to the Standard Form by using the following equations.

$$K_c = K_c'' \tag{K-1g}$$

$$T_i = K_c'' \bullet T_i'' \tag{K-1h}$$

$$T_d = \frac{T_d''}{K_c''} \tag{K-1i}$$

We can convert from the Standard Form to the Parallel Form by using the following equations:

$$K_c'' = K_c \tag{K-1j}$$

$$T_i'' = \frac{T_i}{K_c} \tag{K-1k}$$

$$T_d'' = K_c \bullet T_d \tag{K-1l}$$

where

K_c = controller gain for *Standard* Form (%/%) (dimensionless)

K_c' = controller gain for *Series* Form (%/%) (dimensionless)

K_c'' = controller gain for *Parallel* Form (%/%) (dimensionless)

T_i = integral time (reset time) for *Standard* Form (seconds)

T_i' = integral time (reset time) for *Series* Form (seconds)

$T_i^{''}$ = integral time (reset time) for *Parallel* Form (seconds)

T_d = derivative time (rate time) for *Standard* Form (reset time) (seconds)

$T_d^{'}$ = derivative time (rate time) for *Series* Form (seconds)

$T_d^{''}$ = derivative time (rate time) for *Parallel* Form (seconds)

K.2 Difference Equations for Standard and Series Form

The difference equations for a reverse-acting Standard Form PID controller with rate limiting where the filter time on the derivative mode is a fraction α of the rate time are as follows:

$$\%CO_n = P_n + I_n + D_n + \%CO_i \tag{K-2a}$$

$$P_n = K_c \bullet (\beta \bullet \%SP_n - \%PV_n) \tag{K-2b}$$

$$I_n = \frac{K_c}{T_i} \bullet (\%SP_n - \%PV_n) \bullet \Delta t_x + I_{n-1} \tag{K-2c}$$

$$D_n = \frac{K_c \bullet T_d \bullet [\gamma \bullet (\%SP_n - \%SP_{n-1}) - (\%PV_n - \%PV_{n-1})] + \alpha \bullet T_d \bullet D_{n-1}}{\alpha \bullet T_d + \Delta t_x} \tag{K-2d}$$

The difference equations for a reverse-acting Series Form PID controller with rate limiting where the filter time on the derivative mode is a fraction α of the rate time are as follows:

$$\%CO_n = PD_n^{'} + ID_n^{'} + \%CO_i \tag{K-2e}$$

$$PD_n^{'} = K_c^{'} \bullet (\beta * \%SP_n - \%PV_n - D_n^{'}) \tag{K-2f}$$

$$ID_n^{'} = \frac{K_c^{'}}{T_i^{'}} \bullet (\%SP_n - \%PV_n - D_n^{'}) \bullet \Delta t_x + ID_{n-1}^{'} \tag{K-2g}$$

$$D_n^{'} = \frac{K_c^{'} \bullet T_d^{'} \bullet [\gamma \bullet (\%SP_n - \%SP_{n-1}) - (\%PV_n - \%PV_{n-1})] + \alpha \bullet T_n^{'} \bullet D_{n-1}^{'}}{\alpha \bullet T_d^{'} + \Delta t_x} \tag{K-2h}$$

If the structure used does not have integral action, the term $\%CO_i$ is an adjustable bias and the integral terms I_n and ID'_n are zero (Equations K-2c and K-2g are not used).

K.3 Positive Feedback Implementation of the Integral Mode

The positive feedback implementation of the integral mode as shown in Figure K-3 effectively yields Equations K-2a through K-2d as shown in the following derivation for a proportional-integral (PI) controller using Laplace transforms. Instead of an integrator, a filter whose input is the controller output and whose output is added to the contribution from the proportional mode is used to provide integral action in the

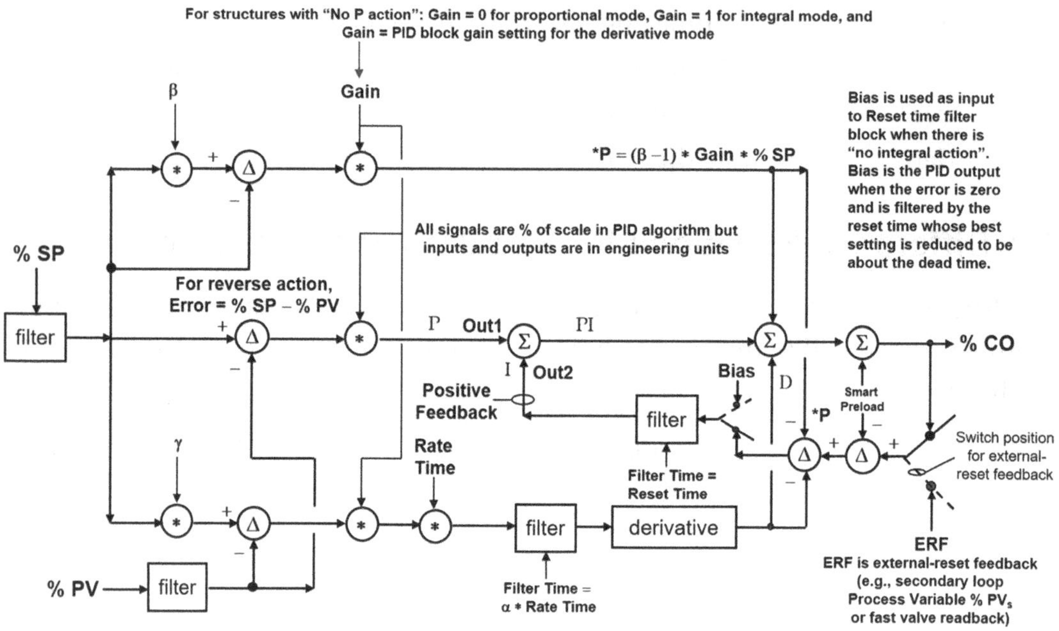

Figure K-3. Standard Form with positive feedback implementation of integral mode and external-reset feedback.

positive feedback implementation. The filter time is the integral time setting. When external-reset feedback is enabled, the input to the filter is switched from the controller output to the external-reset feedback signal.

The use of the positive feedback implementation of the integral mode is the basis for the enhanced PID controller developed for analyzers and wireless devices (described in Appendix C) that is effective for analyzer and wireless devices especially when the update time is much longer than the process 63% response time.

$$O(s) = K_c \bullet E(s) + \frac{1}{1+T_i \bullet s} \bullet O(s) \tag{K-3a}$$

$$O(s) \bullet \left(1 - \frac{1}{1+T_i \bullet s}\right) = K_c \bullet E(s) \tag{K-3b}$$

$$O(s) \bullet \left(\frac{T_i \bullet s}{1+T_i \bullet s}\right) = K_c \bullet E(s) \tag{K-3c}$$

$$\frac{O(s)}{E(s)} = K_c \bullet \left(\frac{1+T_i \bullet s}{T_i \bullet s}\right) = K_c + \frac{K_c}{T_i \bullet s} \tag{K-3d}$$

where

$\%CO_i$ = controller output at the transition from manual (MAN) or remote output (ROUT) modes (%)
$\%CO_n$ = controller output for current scan n (%)
$\%CO_{n-1}$ = controller output for last scan n–1 (%)
$\%PV_n$ = process variable for current scan n (%)
$\%PV_{n-1}$ = process variable for last scan n–1 (%)
$\%SP_n$ = set point for current scan n (%)
$\%SP_{n-1}$ = set point for last scan n–1 (%)
K_c = controller gain for *Standard* Form (%/%) (dimensionless)
K_c' = controller gain for *Series* Form (%/%) (dimensionless)
T_i = integral time (reset time) for *Standard* Form (seconds)
T_i' = integral time (reset time) for *Series* Form (seconds)
T_d = derivative time (rate time) for *Standard* Form (reset time) (seconds)
T_d' = derivative time (rate time) for *Series* Form (seconds)
P_n = proportional mode contribution for *Standard* Form for current scan n (%)
P_{n-1} = proportional mode contribution for *Standard* Form for last scan n–1 (%)
PD_n' = proportional-derivative mode contribution for *Series* Form for current scan n (%)
PD_{n-1}' = proportional-derivative mode contribution for *Series* Form for last scan n–1 (%)
I_n = integral mode contribution for *Standard* Form for current scan n (%)
I_{n-1} = integral mode contribution for *Standard* Form for last scan n–1 (%)
ID_n' = integral-derivative mode contribution for *Series* Form for current scan n (%)
ID_{n-1}' = integral-derivative mode contribution for *Series* Form for last scan n–1 (%)
Δt_x = execution time of PID block (seconds)
α = alpha factor for derivative mode filter (fraction of rate time) (0–1) (dimensionless)
β = beta set-point weight factor for proportional mode (0–1) (dimensionless)
γ = gamma set-point weight factor for derivative mode (0–1) (dimensionless)
$E(s)$ = Laplace transform of controller error (%)
$O(s)$ = Laplace transform of controller output (%)

Appendix L: Liquid Mixing Dynamics

The dead time and time constant of tanks, static mixers, and pipelines for liquid composition control can be estimated by Equations L-1 and L-2 for an estimated or computed equivalent agitation flow (F_a). For static mixers and pipelines, the agitation flow for back mixing is minimal and rarely exceeds 25% of the feed flow (F_f). These volumes are characterized as plug flow, and most of the residence time (volume divided by feed flow) becomes dead time. The lack of a process time constant translates to no slowing down of a fast disturbance. The saving grace is that the residence time of these systems is relatively short (e.g., 3 s), so the dead time introduced is minimal and may be less than the electrode time constant or signal filter that provides some smoothing and prevents the control loop from becoming dead time dominant.

$$\theta_p = 0.5 \bullet \frac{V}{F_f + F_r + F_a} \tag{L-1}$$

$$\tau_p = \frac{V}{F_f} - \theta_p \tag{L-2}$$

For axial agitation where liquid height is between 50% and 150% of vessel diameter:

$$F_a = 7.48 \bullet N_q \bullet N_s \bullet D_t^3 \tag{L-3a}$$

$$N_q = \frac{0.4}{\left[\frac{D_i}{D_t}\right]^{0.55}} \tag{L-3b}$$

The volume cancels out in the ratio of the process dead time to 63% response time (dead time plus time constant), which is important for the minimum peak error if the recirculation plus agitation flow is much larger than the feed flow.

$$\frac{\theta_p}{\tau_p + \theta_p} = 0.5 \bullet \frac{F_f}{F_f + F_r + F_a} \tag{L-4}$$

where

D_i = impeller diameter for axial (back) mixing (feet)
D_t = tank diameter for axial (back) mixing (feet)
N_s = impeller speed for axial (back) mixing (revolutions per minute)
N_s = impeller discharge coefficient for axial (back) mixing (gallons per minute per revolutions per minute • ft^3)
F_a = agitation flow for axial (back) mixing (gallons per minute)
F_f = total feed flow (gallons per minute)
F_r = recirculation flow (gallons per minute)
V = liquid volume (gallons)
θ_p = process dead time (minutes)
τ_p = process time constant (minutes)

For self-regulating processes, the peak error is proportional to the dead-time to time-constant ratio for a dead time much smaller than the time constant. If the recirculation flow and agitation flow are much greater than the feed flow, the volume cancels out in the ratio of Equation L-2 to Equation L-1, giving Equation L-4 showing that the size of the recirculation and agitation flow relative to feed flow is important. Since the integrated (accumulated) error is proportional to the ratio of the dead time squared to time constant, the volume does not cancel out and will increase the amount of product off-spec. The recirculation flow may tend to provide more radial mixing than axial agitation unless the entry point is vertical instead of side entry. An eductor can magnify the recirculation flow and improve the mixing pattern.

For pipelines, there is no recirculation flow and the agitation flow for back mixing is about 20% of the feed flow. For static mixers and laminar flow, the back mixing is less, resulting in more of the residence time becoming dead time and making the volume more plug flow with only radial mixing (no axial mixing). Plug flow is desirable for reactions because the residence time is constant for each infinitesimal volume of reactants [1]. Plug flow is undesirable for in-line systems where reaction time is not a consideration (e.g., static mixers for blending, stock consistency, and neutralization) because there is no smoothing of axial fluctuations.

Equations L-1 and L-2 lead to some interesting conclusions about the use of well-mixed tanks for composition control. If the tank volume is increased but the feed, recirculation, and agitation flows are kept constant, the peak error is the same but the integrated (accumulated) error is increased in proportion to the increase in tank volume. If the tank volume is made exceptionally large, the dead time may be so large that the minimum reset time setting may cause a reset cycle. If the recirculation and agitation flows were increased in proportion to the tank volume, the peak error would be reduced and the integrated error would be about the same as before the increase in volume and recirculation and agitation flows. The increased tank volume will serve to slow down upstream disturbances and attenuate upstream oscillations for downstream control systems. Thus, exceptionally large tanks improve composition control performance if the loops are upstream or downstream but not on the tank, unless the agitation flow is scaled up with the volume. If the large tank is upstream of the loop, the reagent or coolant usage will be reduced when there are short-term fluctuations in reagent or cooling demand.

Reference

1. T. Bor, "The Static Mixer as a Chemical Reactor," *British Chemical Engineering* 16, no. 7 (1971), 610.

Appendix M: Mammalian Bioreactor Model

A bioreactor model has been demonstrated for a mammalian cell culture producing a modern-day biologic product. The bioreactor batch that normally takes 10 days completes a batch in 15 minutes by running 1000 times real time. This is achieved by using a kinetic speedup factor of 20 and a liquid (media) speedup factor of 50 with the total speedup being the product of the two factors. The bioreactor uses a smart proportional-integral-derivative (PID) control system shown in Figure M-1 that has glucose and glutamine concentration control in addition to split-range pH and dissolved oxygen control. An oxygen uptake rate (OUR) from a mass spectrometer provides a feedforward adjustment of glucose feed that is ratioed to give a feedforward adjustment of glutamine feed. Concentration controllers correct these feed set points by feedforward summers.

The configuration in Figure M-2 shows how the streams are simply connected to the bioreactor advanced modeling object. Figures M-3 and M-4 show the operator view of operation at day 5 and day 10, which occurs within about 8 min and 16 min, respectively for the 1000:1 speedup. Figures M-5 and M-6 show how readily the generic kinetics can be easily tailored to match plant operation by the powerful Michaelis-Menten equations for the effect of media concentrations and the convenient cardinal equations for pH and temperature. The kinetic parameters shown can be readily estimated from plant trend charts that show key process variables, feed rates, and set points. Figure M-7 shows the batch trend chart.

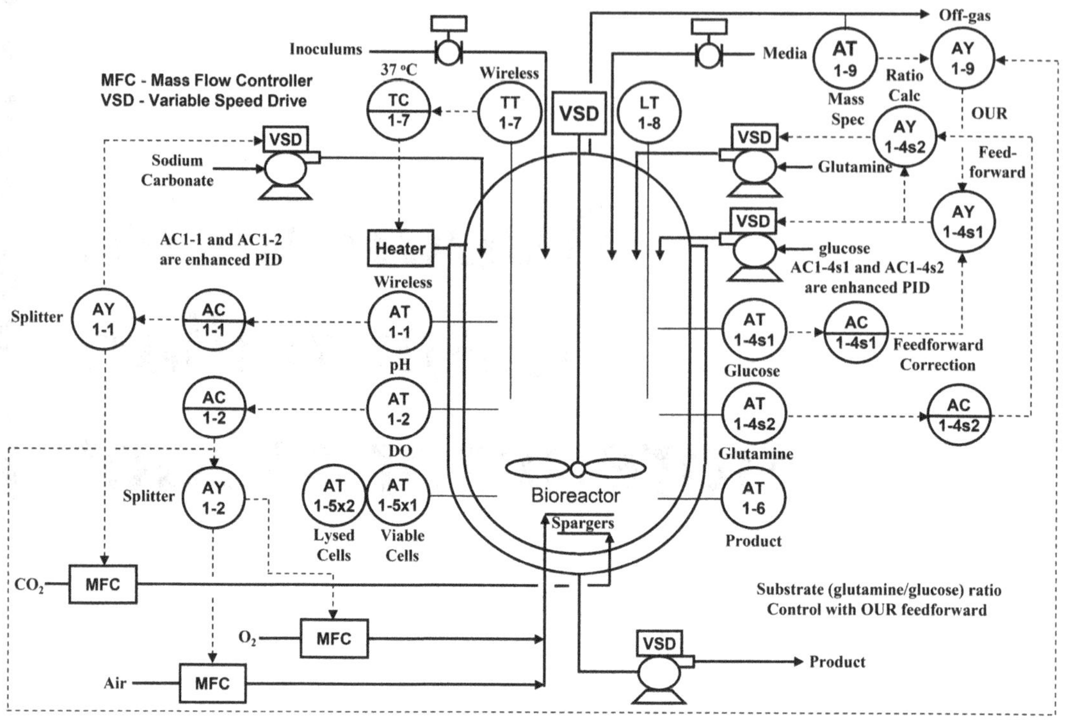

Figure M-1. Mammalian bioreactor smart PID control.

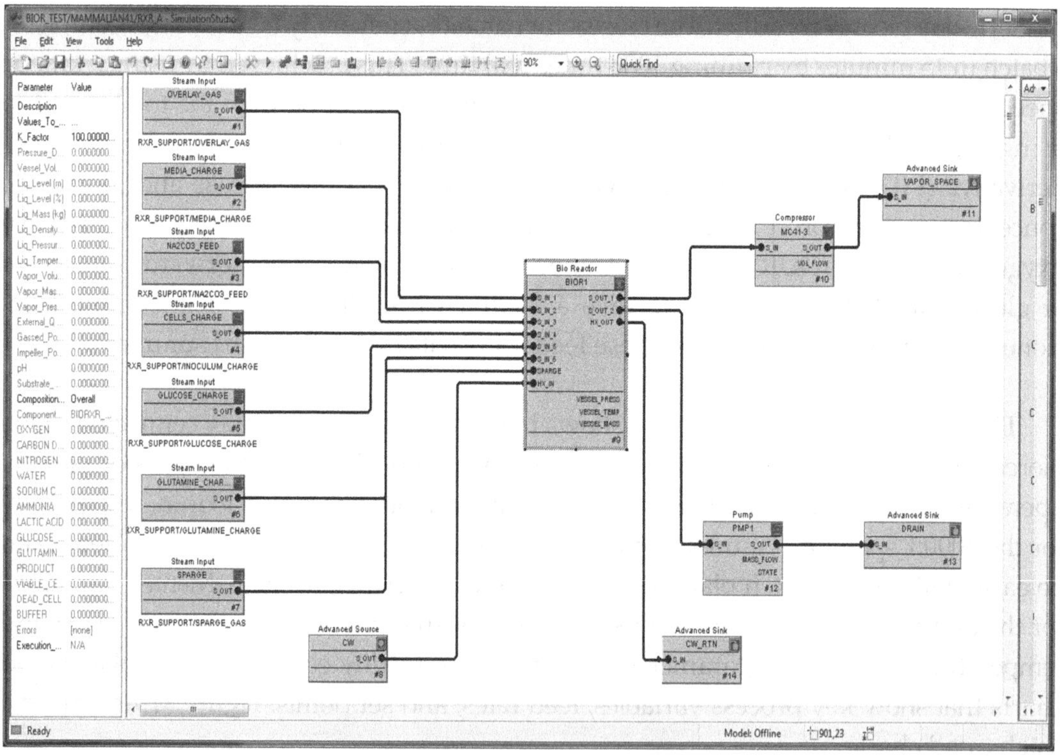

Figure M-2. Mammalian bioreactor advanced modeling object setup.

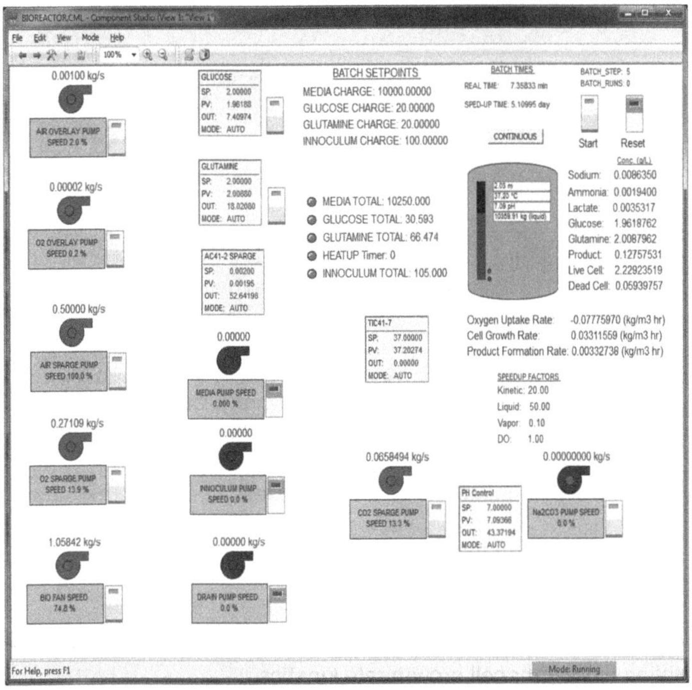

Figure M-3. Mammalian bioreactor day 5 operator view.

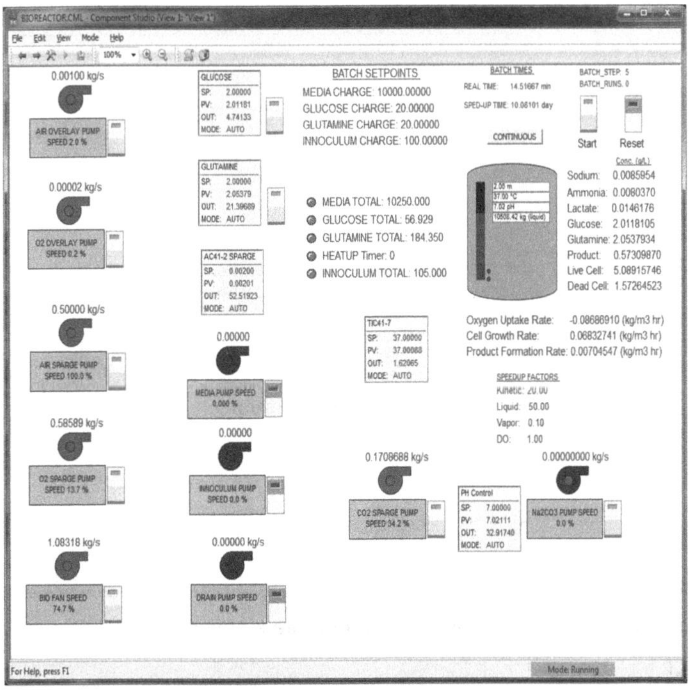

Figure M-4. Mammalian bioreactor day 10 operator view.

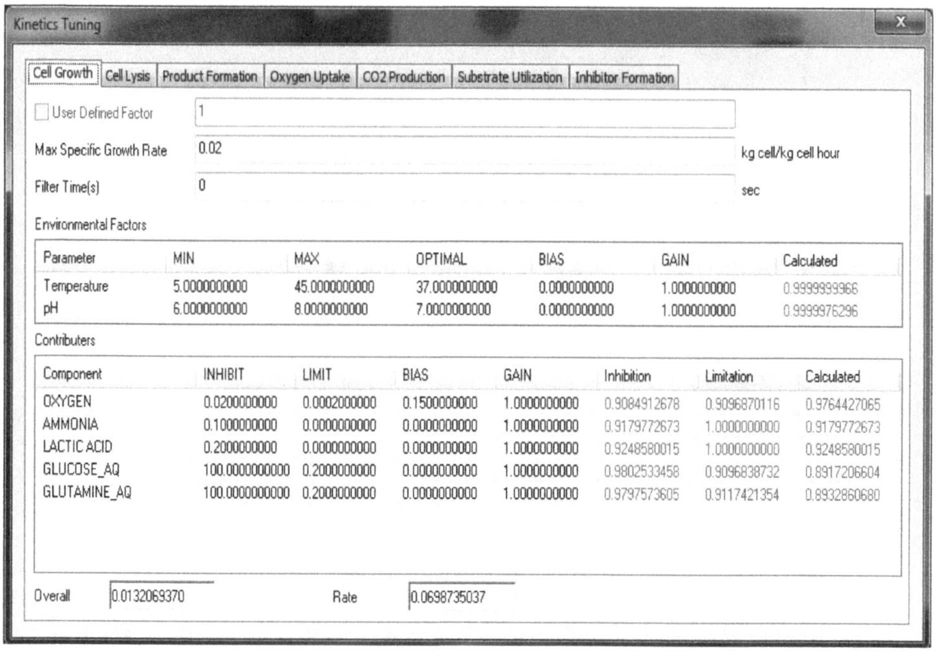

Kinetics Tuning

Cell Growth | Cell Lysis | Product Formation | Oxygen Uptake | CO2 Production | Substrate Utilization | Inhibitor Formation

User Defined Factor: 1

Max Specific Growth Rate: 0.02 kg cell/kg cell hour

Filter Time(s): 0 sec

Environmental Factors

Parameter	MIN	MAX	OPTIMAL	BIAS	GAIN	Calculated
Temperature	5.0000000000	45.0000000000	37.0000000000	0.0000000000	1.0000000000	0.9999999966
pH	6.0000000000	8.0000000000	7.0000000000	0.0000000000	1.0000000000	0.9999976296

Contributers

Component	INHIBIT	LIMIT	BIAS	GAIN	Inhibition	Limitation	Calculated
OXYGEN	0.0200000000	0.0002000000	0.1500000000	1.0000000000	0.9084912678	0.9096870116	0.9764427065
AMMONIA	0.1000000000	0.0000000000	0.0000000000	1.0000000000	0.9179772673	1.0000000000	0.9179772673
LACTIC ACID	0.2000000000	0.0000000000	0.0000000000	1.0000000000	0.9248580015	1.0000000000	0.9248580015
GLUCOSE_AQ	100.0000000000	0.2000000000	0.0000000000	1.0000000000	0.9802533458	0.9096838732	0.8917206604
GLUTAMINE_AQ	100.0000000000	0.2000000000	0.0000000000	1.0000000000	0.9797573605	0.9117421354	0.8932860680

Overall: 0.0132069370 Rate: 0.0698735037

Figure M-5. Mammalian bioreactor cell growth rate kinetics.

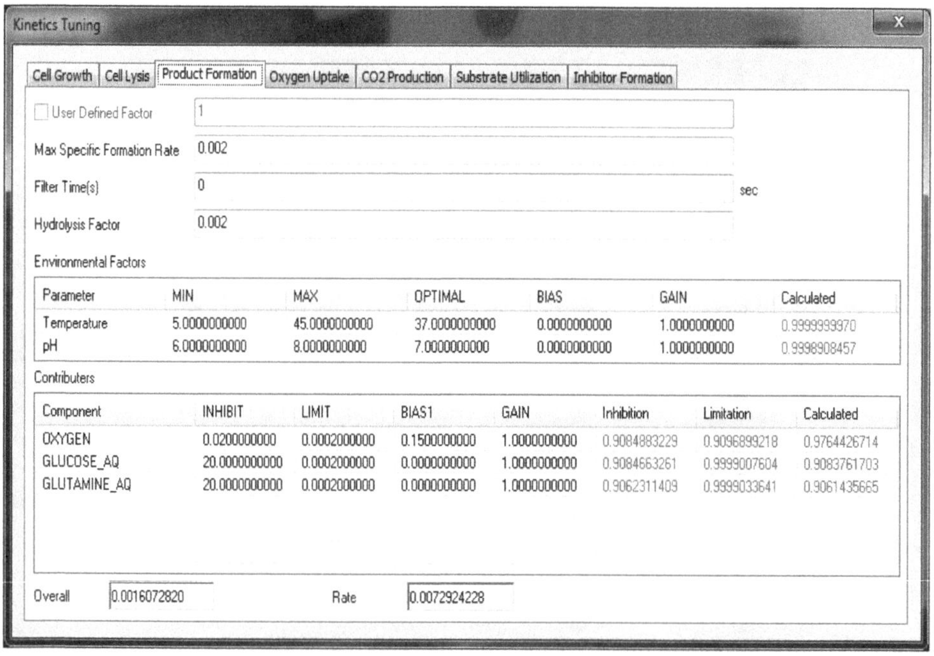

Kinetics Tuning

Cell Growth | Cell Lysis | Product Formation | Oxygen Uptake | CO2 Production | Substrate Utilization | Inhibitor Formation

User Defined Factor: 1

Max Specific Formation Rate: 0.002

Filter Time(s): 0 sec

Hydrolysis Factor: 0.002

Environmental Factors

Parameter	MIN	MAX	OPTIMAL	BIAS	GAIN	Calculated
Temperature	5.0000000000	45.0000000000	37.0000000000	0.0000000000	1.0000000000	0.9999999970
pH	6.0000000000	8.0000000000	7.0000000000	0.0000000000	1.0000000000	0.9998908457

Contributers

Component	INHIBIT	LIMIT	BIAS1	GAIN	Inhibition	Limitation	Calculated
OXYGEN	0.0200000000	0.0002000000	0.1500000000	1.0000000000	0.9084883229	0.9096899218	0.9764426714
GLUCOSE_AQ	20.0000000000	0.0002000000	0.0000000000	1.0000000000	0.9084663261	0.9999007604	0.9083761703
GLUTAMINE_AQ	20.0000000000	0.0002000000	0.0000000000	1.0000000000	0.9062311409	0.9999033641	0.9061435665

Overall: 0.0016072820 Rate: 0.0072924228

Figure M-6. Mammalian bioreactor product formation rate kinetics.

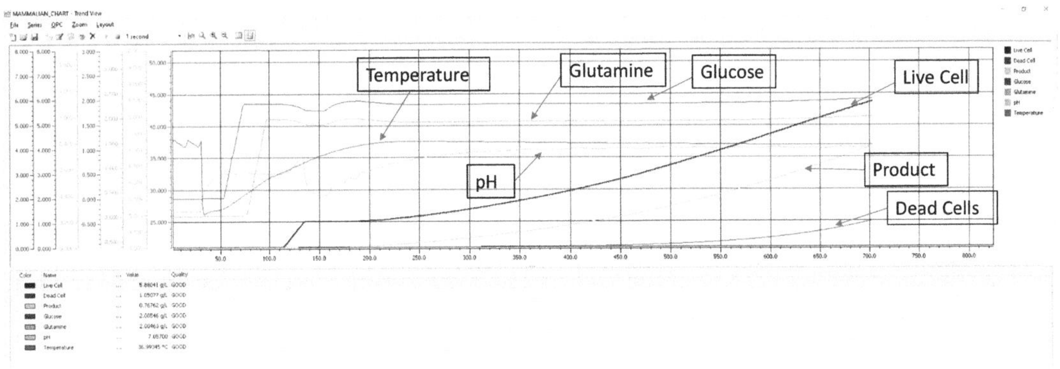

Figure M-7. Mammalian bioreactor batch trend chart.

Appendix N: Debottlenecking Using Sensitivity Analysis

There are two major challenges associated with optimizing biomanufacturing operations today. The first is variability: how to understand and improve processes when there is significant variation in process times throughout all unit operations. The second is complexity: modern biomanufacturing facilities are complex and interconnected, with piping segments, transfer panels, and valve arrays as well as utilities and other shared resource constraints. This complexity is becoming even greater with multiproduct facilities, new treatments such as cell and gene therapies, and the need to process higher titers.

In such an environment, debottlenecking is becoming increasingly important as a means of quantitative process optimization. The technique allows biomanufacturing facilities to run new products with minimal retrofits and increase the run rate of existing legacy products without significant regulatory impact. Debottlenecking thus allows a biomanufacturing facility to significantly extend its useful life by making the facility more flexible and efficient. Retrofits—when appropriately applied to the correct areas of the facility—can be orders of magnitude less expensive than building new capacity, and can be made within a year, versus multiple years for a new facility.

In this white paper, we discuss the impact of variability and complexity on the debottlenecking and optimization process. We show historical approaches to debottlenecking based on resource utilization and why they can lead to incorrect assessments of the real bottlenecks in biomanufacturing facilities. We look at alternatives to collecting data solely from subject matter expert (SME) opinion, which may be a biased estimator of facility performance.

The "gold standard" for debottlenecking methodology today is based around perturbing cycle times or resources using design of experiment (DOE) and evaluating a meaningful key performance indicator (KPI) such as throughput, labor requirements, or utility load. This involves collecting data from process historians and performing sensitivity analysis on those cycle times. Debottlenecking is important in both resource-constrained facilities (where increasing throughput has a positive return on investment) and throughput-constrained facilities (where we are seeking to make the same amount of product with increased efficiency and at lower cost). In either case, debottlenecking provides a quantitative method of accurately finding the critical processes in the facility and assessing potential improvements to overall facility performance.

N.1 Introduction to Debottlenecking

Debottlenecking is the process of improving efficiency through finding the rate-limiting steps in a facility and fixing them. Addressing these rate-limiting steps will improve performance, while changing other (non-rate-limiting) steps will have no impact on performance. Thus, while the complete manufacturing system may have several hundred key activities, only a subset of them define the run rate of the facility.

To understand this, consider a serial two-stage production process. In the first stage, a single bioreactor produces material and is cleaned, steamed, and prepped for the next batch—a process that takes 300 hours. In the second stage, a single downstream purification train processes the material produced by the production bioreactor—a process that takes 72 hours.

Because each unit operation can only process one batch at a time, the maximum processing rate of batches through the first step is one batch every 300 hours. The processing rate for the downstream operation is 72 hours; therefore, the purification skid will sit waiting for the bioreactor to complete for 228 hours. Any improvement to the purification train (e.g., by reducing its cycle time from 72 to 60 hours) will have no impact on throughput or run rate because the rate-limiting step, or bottleneck, is the fermentation area. Therefore, the only way to improve the process is to reduce the cycle time of the upstream process.

Debottlenecking is typically thought of as a two-step process:

1. **Bottleneck identification –** We identify the rate-limiting steps among the hundreds or thousands of potential resources and activities in the facility.
2. **Bottleneck alleviation –** We make the changes to those rate-limiting steps either by adding equipment, reducing cycle time, or some other means to improve the process.

The objective is to produce a *debottlenecking pathway*, or a sequence of steps that identify the bottleneck, evaluate potential solutions to the bottleneck, and quantify the impact the bottleneck has on the KPI up until the desired target KPI.

In Figure N-1, we show the variability in completion times for this production fermentation activity. The data is collected directly from the batch historian. Note that there is significant variability in this process step, which takes between 234 and 410 hours. There is also *process drift* (i.e., the amount of time that it takes to do the activity is not constant from year to year).

This data suggests that using a single number—say, the average time—for planning and optimization may not identify the correct bottleneck. Our experience in this area has been that running the same model, with and without variability, produces different bottlenecks. Such an analysis is particularly concerning because finding and fixing bottlenecks can be a capital-intensive process if retrofitting a validated facility. "Fixing" the wrong area of the facility will not improve the run rate, no matter how much improvement is made in that area.

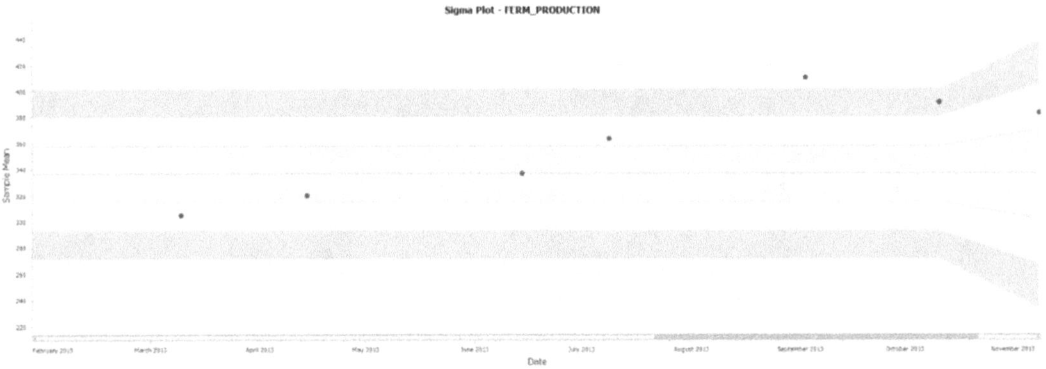

Figure N-1. Production fermentation times.

N.2 Data, Complexity, and Variability in Biomanufacturing

Another issue that must be overcome when debottlenecking and optimizing a process s *process complexity*. In Figure N-2, we show a map of the places in one facility where resources interact with processes. A delay in the availability of a person or a piece of shared equipment (e.g., a clean-in-place [CIP] skid, a transfer line, automation conflicts, and utilities) will delay the process. Such times are "hidden" because they are not normally considered in cycle time calculations. These interactions make it difficult even for highly trained operators to understand where hidden cycle times and other waiting periods will cause the facility to slow down.

Thus, it is important to use software such as Emerson's Real-Time Modeling System (RTMS) to capture both the variability and the interconnectivity of the entire process [1,2]. Discrete event simulators are ideal to implement because they reflect actual facility operations for the process level and capture automation and business constraints.

N.3 Bottleneck Identification Using Resource Utilization?

A traditional approach to identifying bottlenecks focused on the resources in the facility with the highest utilization. The theory was that those resources with the highest utilization are the "busiest," and therefore improvements would have the most impact

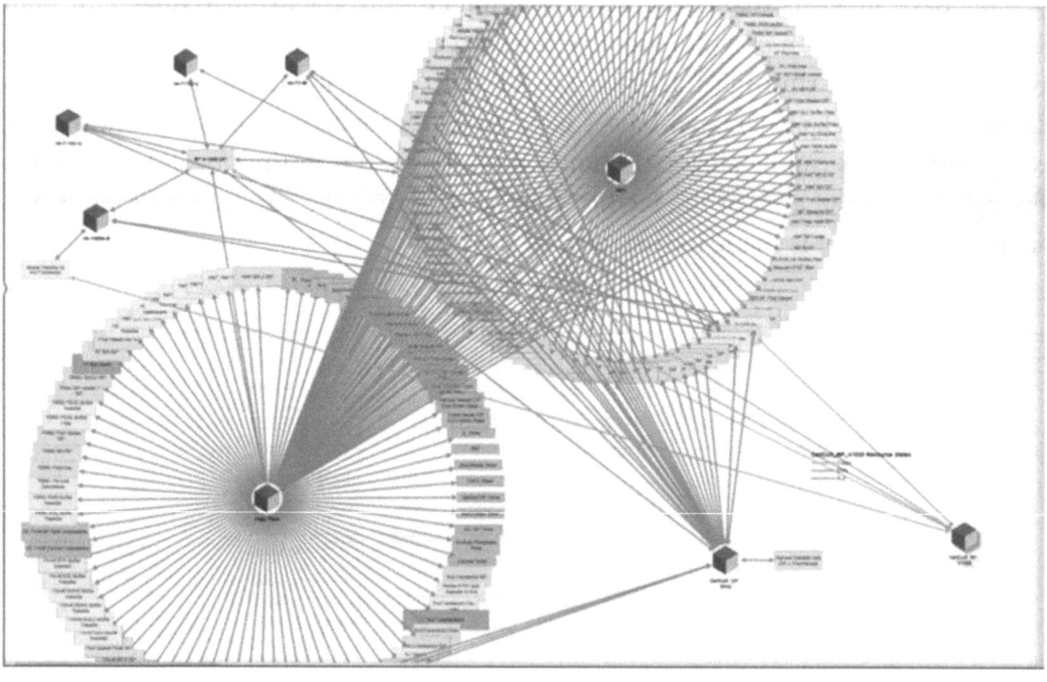

Figure N-2. Interconnectivity of facility resources and processes.

on them. However, such an approach is not guaranteed to locate the real bottlenecks in a facility. Empirical evidence suggests the approach is particularly ill-suited for biomanufacturing facilities.

Consider, for example, a single buffer tank that is only required in the first and last unit operations of a five-unit operation downstream process. The overall utilization of this tank may be low, as it is only required for two of the five steps, but because it prevents the first unit operation from starting until the last one is completed, it effectively bottlenecks the entire downstream process.

N.4 Bottleneck Detection

As the example in the previous section shows, a simplistic utilization-based approach to debottlenecking does not find the true process constraints in a biomanufacturing facility. The current "gold standard" is to use a DOE model to perturb—or make controlled changes—to a model of the facility and observe the results on some metric. A simple approach is to reduce operation times in each process step and observe the impact on throughput. By simulating a facility in which a process step (e.g., a CIP duration) takes zero time, we can examine the effect of having that activity "for free." By repeating this experiment for each activity in the facility, we can understand the impact of a potential project to reduce every activity's cycle time. Such a sensitivity analysis can correctly identify even complex bottlenecks, because it makes an actual change to a model of the system and observes the impact of that change.

In Figure N-4, we show the output from such a sensitivity analysis. In this analysis, runs per week (A_RPW) was the KPI chosen—so a higher number indicates better performance. Each point in the graph represents an activity whose cycle time has been reduced to 0 hours, and the resulting throughput was observed over a 100-day campaign. The activities, or experiments, have been sorted to show those most impactful to the cycle time on the right and least impactful on the left.

This analysis shows that of the several hundred activities in this manufacturing facility, only one improves throughput. The approach allows for very rapid bottleneck detection and is guaranteed to find real process constraints in a system because it simulates the actual impact of a change to each constraint in turn.

N.5 Bottleneck Evaluation and Alleviation

After identifying bottlenecks, bottleneck alleviation becomes the central focus for analysis. The perturbation analysis described in the previous section found that the

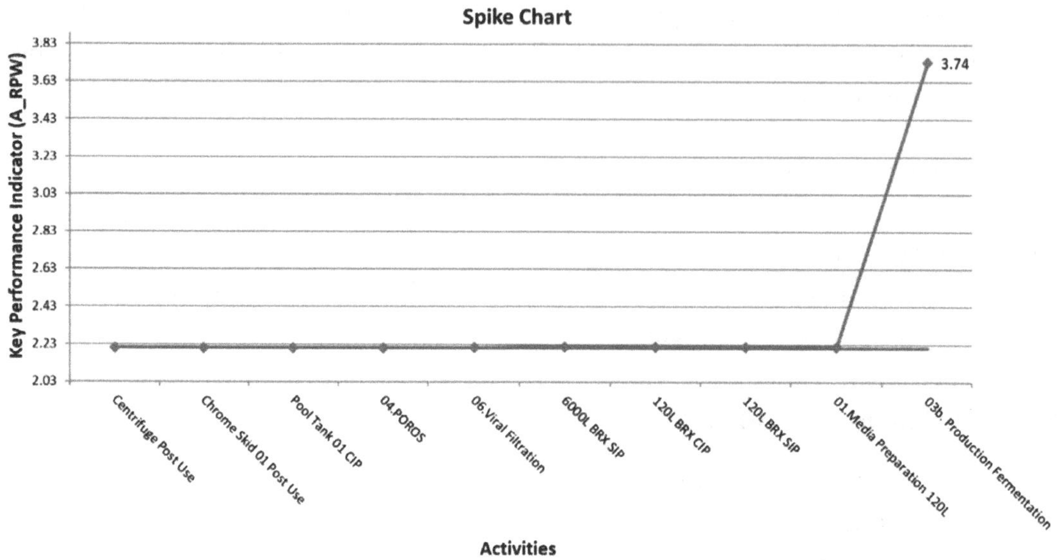

Figure N-4. Activity-based sensitivity analysis.

key constraint in the process is the cycle time of the production fermentation (bioreaction); however, the proposed change (reducing the fermentation time to 0 hours) is not a feasible engineering solution. Fortunately, it may be enough to improve the cycle time of the production bioreactor by a smaller amount to achieve the desired result. In order to quantify how much change is required before an activity or resource is no longer "on the critical path," we can use a similar form of sensitivity analysis that makes changes to one or a small number of activities or resources but does so at a higher level of granularity. For example, we could choose to reduce the bioreactor's cycle time by a specified period of time (e.g., 1, 2, or 3 hours) and observe the impact of each change on the throughput or cycle time.

Such an approach gives valuable quantitative information to engineering teams because it provides an explicit tradeoff between the KPI at various levels of change. In Figure N-5, a 46.89% reduction in cycle time improves throughput from 2.2 to 3.72 A_RPW, but any further reduction only increases the throughput marginally. Such an approach allows the user to choose the optimal solution based on cost and impact.

N.6 Debottlenecking Pathway

The approach of bottleneck detection, evaluation, and removal presented in the previous section is typically repeated iteratively in a biomanufacturing facility to provide a series of incremental improvements. As we identify and fix bottlenecks, these items move off the critical path and other activities or resources become the rate-limiting

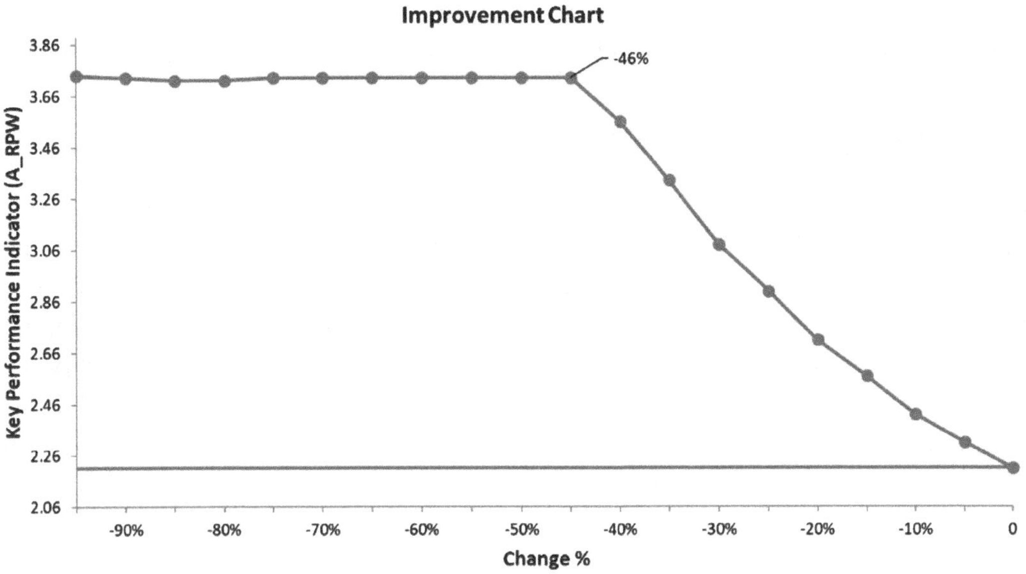

Figure N-5. Sensitivity analysis performed on a bottleneck activity.

steps. The process of incrementally debottlenecking a facility—where we find the first bottleneck and fix it, then move on to the second, and so on—produces what is known as a *debottlenecking pathway*. This pathway is a sequence of steps required to fix bottlenecks, along with changes in KPIs that are achieved by each step. We outline the results from one such analysis in Figure N-6.

In the debottlenecking pathway shown, a sequence of three engineering changes is made. Each debottlenecking iteration increases the run rate by a different amount (and in general, multiple engineering changes may be required in different areas of the facility to increase the run rate). It is important to note that the order of the pathway is important

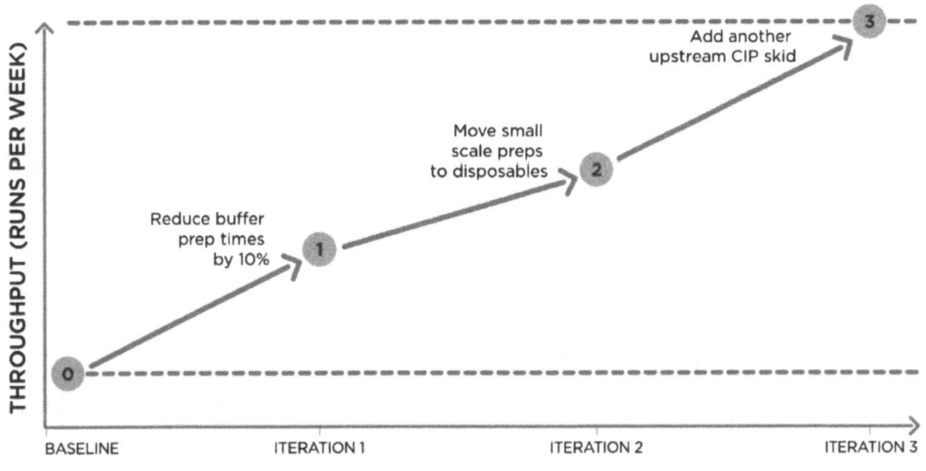

Figure N-6. Debottlenecking pathway.

(i.e., moving to disposables without reducing buffer prep times will yield no improvement). The pathway must be executed in the order shown to achieve the target run rate.

N.7 Synergy with Digital Twin Simulation

Once a debottlenecking pathway has been determined, it is a natural extension to use this RTMS with a Mimic Dynamic simulation in a digital twin to evaluate the most efficient manner of alleviating the bottleneck. The debottlenecking pathway identifies and quantifies the target cycle time reductions required to resolve the impasse, and a digital twin can provide the detailed automation modifications to be deployed. For example, the digital twin can show the benefit of more precise control valves and sensors, middle signal selection, and computations by using existing measurements to diagnose and predict abnormal operation and provide procedure automation to proactively minimize disruption.

The digital twin that uses a copy of the control system configuration with first principles modeling can detail the solution in terms that use the automation system better and more intelligently. For example, if a cleaning activity is identified as the bottleneck with a target reduction of 10 hours, multiple methods, such as adding additional cleaning skid capacity or removing phases from the cleaning recipe, can be tested in the digital twin.

Great opportunities exist to improve the upstream operations that set production capacity. The profiles of a batch operation in the digital twin are nonintrusively matched to those in the plant by first principles and experimentation. Most notably, profile slopes of key variables at key points in the batch cycle are matched. Solutions from the knowledge gained from further nonintrusive experimentation with the models are found to provide steeper, more consistent slopes. For bioreactors, the slopes would correspond to key times in the cell exponential growth and product formation phases. For high-value biologics, the advantage of generating digital twin test batches in a matter of minutes is incredible because batches may be worth $10 million or more and take 10 days or longer to complete. The digital twin can also identify conditions that lead to bad batches that are a catastrophic bottleneck.

Subsequently, the digital twin results can be passed back into an RTMS to determine how those changes would have an impact on the entire process and to ensure they would not have any unintended side effects on other areas of the facility [1, 2]. Combined, these toolsets provide the most comprehensive simulation solutions for users to make the best possible modifications to the facility to achieve the desired outcomes.

References

1. Bioproduction Group, "Biomanufacturing Debottlenecking and Process Optimization" (white paper, Emerson Real-Time Modeling System [RTMS]), https://www.bio-g.com/resources/.

2. Bioproduction Group, "The Biomanufacturing Digital Enterprise" (white paper, Emerson Real-Time Modeling System [RTMS]), https://www.bio-g.com/resources/.

Bibliography

Advanced Application Note. "MPC Implementation Methods for the Optimization of a Slow MV with Good Load Rejection by a Fast MV." Emerson (2005).

Advanced Application Note. "MPC Implementation Methods for the Optimization of the Response of Control Valves for Minimum Variability." Emerson (2005).

Aiba, Shuichi, Arthur E. Humphrey, and Nancy F. Millis. *Biochemical Engineering.* 2nd ed. Cambridge, MA: Academic Press, 1973.

Albert, Sarolta, and Robert D. Kinley. "Multivariate Statistical Monitoring of Batch Processes: An Industrial Case Study of Fermentation Supervision." *Trends in Biotechnology* 19, no. 2 (2001): 53–62.

Alford, Joseph S. "Bioprocess Control: Advances and Challenges." Presentation at CPC-7 published in *Computers and Chemical Engineering* (2007).

Alford, Joseph S. "Bioprocess Mass Spectrometry: A PAT Application." *Journal of Process Analytical Technology* (May–June 2006).

Anderson, Dave. "Watch the Water." *InTech* 52 (November–December 2005).

Annuar, M. S. M., I. K. P. Tan, S. Ibrahim, and K. B. Ramachandran. "A Kinetic Model for Growth and Biosynthesis of Medium–Chain–Length Poly–(3–Hydroxyalkonates) in Pseudomonas Putida." *Brazilian Journal of Chemical Engineering* 25, no. 2 (June 2008): 217–228.

Bailey, James E., and David F. Ollis. *Biochemical Engineering Fundamentals.* 2nd ed. New York: McGraw-Hill, 1986.

Bakshi, Bhavik R. "Multiscale PCA with Application to Multivariate Statistical Process Monitoring." *AIChE Journal* 44, no. 7 (1998): 1596–1610.

Barnett, J. A. "A History on Research on Yeasts 5: The Fermentation Pathway." *Yeast* 20, no. 6 (April 2003): 509–543.

Bézier Curves. http://www.math.ubc.ca/~cass/gfx/bezier.html.

BioNet Bioreactor Control System. Broadley-James Bulletin L0016B (2003).

Bioproduction Group, "Biomanufacturing Debottlenecking and Process Optimization." White paper, Emerson Real-Time Modeling System (RTMS). https://www.bio-g.com/resources/.

Bioproduction Group, "The Biomanufacturing Digital Enterprise." White paper, Emerson Real-Time Modeling System (RTMS). https://www.bio-g.com/resources/.

Birol, Gulnur, Cenk Undey, and Ali Cinar. "A Modular Simulation Package for Fed-Batch Fermentation: Penicillin Production." *Computers and Chemical Engineering* 26, no. 11 (2002): 1553–65.

Blevins, Terrence L., Gregory K. McMillan, Willy K. Wojsznis, and Michael W. Brown. *Advanced Control Unleashed: Plant Performance Management for Optimum Benefits.* Research Triangle Park, NC: ISA (International Society of Automation), 2003.

Bor, T. "The Static Mixer as a Chemical Reactor." *British Chemical Engineering* 16, no. 7 (1971).

Boudreau, Michael A., Gregory K. McMillan, and Grant Wilson. "Maximizing PAT Benefits from Bioprocess Modeling and Control," *Pharmaceutical Technology* (November 2006).

Boyes, Walt, Dan Hebert, Larry O'Brien, and Gregory K. McMillan. "The Light at the End of the Tunnel Is a Train (Virtual Plant Reality)." *Control* (August 2005).

Bro, Rasmus. "PARAFAC, Tutorial and Applications." *Chemometrics and Intelligent Laboratory System* 38 (1997): 149–71.

Callanan, Ken. "A Model Way to Control Fermentation." *Control* (August 2004).

"CE Refresher—Liquid Agitation." A 12-part series of articles from *Chemical Engineering* (1975–1976). Reprint, Dayton, OH: Chemineer (1985).

Champe, Pamela C., Richard A. Harvey, and Denise R. Ferrier. *Biochemistry*. 3rd ed. Philadelphia: Lippincott Williams & Wilkins, 2005.

Cherry, Gregory A. and S. Joe Qin. "Batch Synchronization and Monitoring Using Data Interpolation and Dynamic Time Warping with Principal Component Analysis." Presentation at TWMCC meeting, Austin, TX, 2005.

Chiang, Leo H., Riccardo Leardi, Randy J. Pell, and Mary Beth Seasholtz. "Industrial Experiences with Multivariate Statistical Analysis of Batch Process Data." *Chemometrics and Intelligent Laboratory Systems* 81, no. 2 (2005): 109–19.

Chiang, Leo H., Evan L. Russell, and Richard D. Braatz. *Fault Detection and Diagnosis in Industrial Systems*. New York: Springer-Verlag, 2001.

Dong, D., and Thomas J. McAvoy. "Nonlinear Principal Component Analysis—Based on Principal Curves and Neural Networks." *Computers and Chemical Engineering* 20, no. 1 (1996): 65–78.

Downey, Brandon J., Lisa J. Graham, Jeffrey F. Breit, and Nathaniel K. Glutting. "A Novel Approach for Using Dielectric Spectroscopy to Predict Viable Cell Volume (VCV) in Early Process Development." *Biotechnology Progress* (March 2014).

Emerson, David. "Production Performance Ratings Measure Effectiveness of Batches." Presented at ISA's online Technical Information and Communities, 2004.

Fersht, Alan. *Structure and Mechanism in Protein Science*. New York: W. H. Freeman, 2000.

Fickelscherer, Richard J., Douglas H. Lenz, and Daniel L. Chester. "Fuzzy Logic Clarifies Operations." *InTech* 52, no. 10 (October 2005).

Geladi, Paul, and Bruce R. Kowalski. "Partial Least Squares Regression: A Tutorial." *Analytica Chimica Acta* 185 (1986): 1–17.

Guidance for Industry: PAT—A Framework for Innovative Pharmaceutical Development, Manufacturing, and Quality Assurance. Department of Health and Human Services, US Food and Drug Administration (September 2004).

Haag, Jens, Alain Vande Wouwer, and Philippe Bogaerts. "Dynamic Modeling of Complex Biological Systems: A Link between Metabolic and Macroscopic Description." *Mathematical Biosciences* 193 (2005): 25–49.

Haag, Jens, Alain Vande Wouwer, and Marcel Remy. "A General Model for Reaction Kinetics in Biological Systems." *Bioprocess and Biosystems Engineering* 27, (2005): 303–309.

Hägglund, Tore. "A Control Loop Performance Monitor." *Control Engineering Practice* 3, no. 11 (1995): 1543–51.

Harriott, Peter. *Process Control.* New York: McGraw-Hill, 1964.

Hokuldsson, Agnar. "PLS Regression Methods." *Journal of Chemometrics* 2, no. 3 (1988): 211–28.

Jackson, J. Edward, and Govind S. Mudholkar. "Control Procedures for Residuals Associated with Principal Component Analysis." *Technometrics* 21, no. 3 (1979): 341–49.

Jackson, J. Edward. *A User's Guide to Principal Components.* New York: John Wiley, 2005: 8–21.

Jerden, Cleat, Tom Folger, and Matt Dee. "Advanced Control in Small Scale Biotechnology Development." Presentation at Emerson Exchange, Nashville, TN, October 2003.

Jolliffe, Ian T. *Principal Component Analysis.* 2nd ed. New York: Springer-Verlag, 2002.

Julian, C., and W. Whitford. "Bioreactor Control." *BioProcess International Supplement* (January 2007).

Kassidas, Athanassios, John F. MacGregor, and Paul A. Taylor. "Synchronization of Batch Trajectories Using Dynamic Time Warping." *AIChE Journal* 44, no. 4 (1998): 864–75.

Kiviharju, Kristiina, Kalle Salonen, Ulla Moilanen, and Tero Eerikäinen. "Biomass Measurement Online: The Performance of in Situ Measurements and Software Sensors." *Journal of Industrial Microbiology and Biotechnology* 35 (July 2008): 657–665.

Kourti. Theodora. "Monitor." *Chemometrics and Intelligent Laboratory Systems* 76 (2005): 215–20.

Kourti, Theodora. "Multivariate Dynamic Data Modeling for Analysis and Statistical Process Control of Batch Processes, Start-ups and Grade Transitions." *Journal of Chemometrics* 17, no. 1 (2003): 93–109.

Kourti, Theodora. "The Process Analytical Technology Initiative and Multivariate Process Analysis, Monitoring and Control." *Analytical and Bioanalytical Chemistry* 384 (2006): 1043–48.

Kourti, Theodora, Jennifer Lee, and John F. Macgregor. "Experiences with Industrial Applications of Projection Methods for Multivariate Statistical Process Control." *Computers and Chemical Engineering* 20, supplement 1 (1996): 745–50.

Kresta, J. V., J. F. MacGregor, and T. E. Marlin. "Multivariate Statistical Monitoring of Process Operating Performance." *Canadian Journal of Chemical Engineering* 69, no. 1 (1991): 35–47.

Ku, Wenfu, Robert H. Storer, and Christos Georgakis, "Disturbance Detection and Isolation by Dynamic Principal Component Analysis." *Chemometrics and Intelligent Laboratory Systems* 30, no. 1 (1995): 179–96.

Lachman-Shalem, Sivan, Nir Haimovitch, Eitan N. Shauly, and Daniel R. Lewin. "MBPCA Application for Fault Detection in NMOS Fabrication." *IEEE Transactions on Semiconductor Manufacturing* 15, no. 1 (2002): 60–70.

Lawler, D. M. "Turbidity, Turbidimetry and Nephelometry." *Encyclopedia of Analytical Science.* 3rd ed. Amsterdam: Elsevier, 2016.

Lee, Jong-Min. "Statistical Process Monitoring Based on Independent Component Analysis and Multivariate Statistical Methods." PhD diss., Department of Chemical Engineering, Pohang University of Science and Technology, Pohang, South Korea, 2004.

Lee, Jong-Min, Chang Kyoo Yoo, and In-Beum Lee. "Enhanced Process Monitoring of Fed-Batch Penicillin Cultivation Using Time-Varying and Multivariate Statistical Analysis." *Journal of Biotechnology* 110, no. 2 (2004): 119–36.

Lennox, B., G. A. Montague, H. G. Hiden, G. Kornfeld, and P. R. Goulding. "Process Monitoring of an Industrial Fed-Batch Fermentation." *Biotechnology and Bioengineering* 74, no. 2 (2001): 125–35.

Li, Weihua, H. Henry Yue, Sergio Valle-Cervantes, and S. Joe Qin. "Recursive PCA for Adaptive Process Monitoring." *Journal of Process Control* 10, no. 5 (2000): 471–86.

Lopes, J. A., J. C. Menezes, J. A. Westerhuis, and A. K. Smilde, "Multiblock PLS Analysis of an Industrial Pharmaceutical Process." *Biotechnology and Bioengineering* 80, no. 4 (2002): 419–27.

MacGregor, John F., Christiane Jaeckle, Costas Kiparissides, and M. Koutoudi. "Process Monitoring and Diagnosis by Multiblock PLS Methods." *AIChE Journal* 40, no. 5 (1994): 826–38.

MacGregor, John F., and Theodora Kourti, "Statistical Process Control of Multivariate Processes." *Control Engineering Practice* 3 (1995): 403–14.

Mansy, Michael, Gregory K. McMillan, and Mark Sowell. "Step into the Virtual Plant." *Chemical Engineering Progress* (February 2002): 56–61.

Mason, Robert L., and John C. Young, *Multivariate Statistical Process Control with Industrial Applications*. Philadelphia: ASA-SIAM, 2002.

Mattes, Robert, Denise Root, David Chang, Mike Molony, and Mahalia Ong. "In Situ Monitoring of CHO Cell Culture Medium Using Near-Infrared Spectroscopy." *BioProcess International* (January 2007).

Mazzocchi, F. "Complexity on Biology. Exceeding the Limits of Reductionism and Determinism Using Complexity Theory." *EMBO Reports* 9, no. 1 (January 2008): 10–14.

McMillan, Gregory K. *Advances in Reactor Measurement and Control*. Research Triangle Park, NC: ISA (International Society of Automation), 2015.

McMillan, Gregory K. "A Fine Time to Break Away from Old Valve Problems." *Control* (November 2005).

McMillan, Gregory K. "At the IIoT Crossroads." *Control* (February 2019).

McMillan, Gregory K. *Essentials of Modern Measurements and Final Elements in the Process Industry*. Research Triangle Park, NC: ISA (International Society of Automation), 2010.

McMillan, Gregory K. "Full Throttle Batch and Startup Response." *Control* (May 2006).

McMillan, Gregory K. *Good Tuning: A Pocket Guide*. 4th ed. Research Triangle Park, NC: ISA (International Society of Automation), 2015.

McMillan, Gregory K. "Process Modeling" and "Advanced Process Control." In *A Guide to the Automation Body of Knowledge*, 3rd ed., edited by Nicholas P. Sands and Ian Verhappen. Research Triangle Park, NC: ISA (International Society of Automation), 2018.

McMillan, Gregory K. "The Future Is Here." *Control* (November 2020).

McMillan, Gregory K. *Tuning and Control Loop Performance*. 2nd ed. Research Triangle Park, NC: ISA (International Society of Automation), 1990.

McMillan, Gregory K. "What Is Your Control Valve Telling You?" *Control Design* (May 2004): 43–48.

McMillan, Gregory K. "Wireless: Overcoming Challenges in PID Control & Analyzer Applications." *InTech* 57 (July–August 2010).

McMillan, Gregory K., Rehman Fazeem, and Christopher Stuart. "Bioreactor Control Breakthroughs: Biopharm industry turns to advanced measurements, simulation and modeling." *Control* (February 2020).

McMillan, Gregory K., and Martin E. Berutti. "Virtual Plant Virtuosity." *Control* (August 2017).

McMillan, Gregory K., and Robert Cameron. *Advanced pH Measurement and Control*. 3rd ed. Research Triangle Park, NC: ISA (International Society of Automation), 2005.

McMillan, Gregory K., and Robert Cameron. *Models Unleashed: Virtual Plant and Model Predictive Control Applications*. Research Triangle Park, NC: ISA (International Society of Automation), 2004.

McMillan, Gregory K., Mark Sowell, and Peter W. Wojsznis. "The Next Generation Adaptive Control Takes a Leap Forward." *Chemical Processing* (September 2004).

McMillan, Gregory K., and Christopher Stuart. "Simulation Breeds Innovation." *Control* (November 2019).

McMillan, Gregory K., Chris Stuart, and Thomas Hildebrand. "Compressor Surge Control: Deeper Understanding, Simulation Can Eliminate Instabilities." *Control*

(October 2018). https://www.controlglobal.com/articles/2018/compressor-surge-control-deeper-understanding-simulation-can-eliminate-instabilities/?stage=Live.

McMillan, Gregory K., and P. Hunter Vegas. *Process/Industrial Instruments and Controls Handbook*. 6th ed. New York: McGraw-Hill Education, 2019.

McMillan, Gregory K., and P. Hunter Vegas. "Best Practices for PID." *Control* (December 2018).

McMillan, Gregory K., and Stan Weiner. "The Bad Hall of Fame." *Control Talk Column. Control* (March 2006).

McMillan, Gregory K., and Stan Weiner. "Grounded in Reality." *Control Talk Column. Control* (December 2005).

McMillan, Gregory K., and Stan Weiner. "Life Is a Batch." *Control Talk Column. Control* (June 2005).

McMillan, Gregory K., and P. Wu. "How to Specify Valves and Positioners That Don't Compromise Control." *Control* (March 2016). http://www.controlglobal.com/articles/2016/how-to-specify-valves-and-positioners-that-dont-compromise-control/.

McMillan, Gregory K., and P. Wu, "Valve Response—Truth or Consequences." *Control* (March 2016).

McPherson, Lindesay, Julian Morris, and Elaine Martin. "Super Model-Based Techniques for Batch Performance Monitoring." Presented at the 12th European Symposium on Computer Aided Process Engineering, 2002.

Mehta, Ashish, Willy Wojsznis, Dirk Thiele, and Terrence L. Blevins. "Constraints Handling in Multivariable System by Managing MPC Squared Controller." Presentation at ISA Expo, Houston, TX, October 2003.

Miletic, Ivan, Shannon Quinn, Michael Dudzic, Vit Vaculik, and Marc Champagne. "An Industrial Perspective on Implementing Online Applications of Multivariate Statistics." *Journal of Process Control* 14, no. 8 (2004): 821–36.

Moo-Young, M. "Section 2: Transport Phenomena in Bioprocesses." In *Comprehensive Biotechnology*, 30–33. Oxford, England: Pergamon Press Ltd., 1985.

Myers, John A., Brandon S. Curtis, and Wayne R. Curtis. "Improving Accuracy of Cell and Chromophore Concentration Measurements Using Optical Density." *BMC Biophysics* 6, no. 4 (2013): 1–15.

Nielsen, Jens, John Villadsen, and Gunnar Lidén. *Bioreaction Engineering Principals.* 2nd ed. New York: Springer Science + Business Media, 2003.

Nienow, Alvin W., Christian Langheinreich, Neil C. Stevenson, Nicholas A. Emery, Timothy M. Clayton, and Nigel K. H. Slater. "Homogenisation and Oxygen Transfer Rates in Large Agitated and Sparged Animal Cell Bioreactors: Some Implications for Growth and Production." *Cytotechnology* 22 (1996): 87–94.

Nomikos, Paul, and John F. MacGregor. "Monitoring of Batch Processes Using Multiway Principal Component Analysis." *AIChE Journal* 40, no. 8 (1994): 1361–75.

Nomikos, Paul, and John F. MacGregor. "Multivariate SPC Charts for Monitoring Batch Processes." *Technometrics* 37, no. 1 (1995): 41–59.

Nomikos, Paul, and John F. MacGregor. "Multi-way Partial Least Squares in Monitoring Batch Processes." *Chemometrics and Intelligent Laboratory Systems* 30, no. 1 (1995): 97–108.

Pharmaceutical CGMPs for the 21st Century–A Risk-Based Approach. Final report, Department of Health and Human Services, US Food and Drug Administration (September 2004).

Pisano, Gary P. *The Development Factory: Unlocking the Potential of Process Innovation.* Boston: Harvard Business School Press, 1997.

Pramanik, K. "Use of Artificial Neural Networks for Prediction of Cell Mass and Ethanol Concentration in Batch Fermentation Using Saccharomyces Cerevisiae Yeast." *IE (I) Journal – CH* 85 (September 2004).

Pravdova, V., B. Walczak, and D. L. Massart. "A Comparison of Two Algorithms for Warping of Analytical Signals." *Analytica Chimica Acta* 456 (2002): 77–92.

Qin, S. Joe. "Statistical Process Monitoring: Basics and Beyond." *Journal of Chemometrics* 17, no. 8–9 (2003): 480–502.

Ramaker, Henk-Jan, Eric N. M. van Sprang, John A. Westerhuis, and Age K. Smilde. "Dynamic Time Warping of Spectroscopic BATCH Data." *Analytica Chimica Acta* 498, no. 1–2 (2003): 133–53.

Rosso, L., J. R. Lobry, S. Bajard, and J. P. Flandrois. "Convenient Model to Describe the Combined Effects of Temperature and pH on Microbial Growth." *Applied and Environmental Microbiology* (February 1995): 610–616.

Rotem, Y., A. Wachs, and D. R. Lewin. "Ethylene Compressor Monitoring Using Model-Based PCA." *AIChE Journal* 46, no. 9 (2000): 1825–36.

Russell, Evan L., Leo H. Chiang, and Richard D. Braatz. "Fault Detection in Industrial Processes Using Canonical Variate Analysis and Dynamic Principal Component Analysis." *Chemometrics and Intelligent Laboratory Systems* 51, no. 1 (2000): 81–93.

Sandnes, J. M., T. Ringstad, D. Wenner, P. H. Heyerdahl, T. Källqvist, and H. R. Gislerød. "Real-Time Monitoring and Automatic Density Control of Large-Scale Microalgal Cultures Using Near Infrared (NIR) Optical Density Sensors." *Journal of Biotechnology* 122 (2006): 209–215.

Shinskey, F. Greg. *Process Control Systems: Application, Design, and Tuning.* 4th ed. New York: McGraw-Hill Companies, Inc., 1996.

Shuler, Michael L., Fikret Kargi, and Matthew DeLisa. *Bioprocess Engineering: Basic Concepts.* 3rd ed. Boston: Prentice Hall, 2017.

Silva, R. G., A. J. G. Cruz, C. O. Hokka, R. L. C. Giordano, and R. C. Giordano. "A Hybrid Feedforward and Neural Network Model for the Cephalosporin C Production Process." *Brazilian Journal of Chemical Engineering* 7, no. 4–7 (December 2000).

Skoog, D. A. *Principles of Instrumental Analysis.* 2nd ed. Philadelphia: Saunders College, 1980.

Smilde, Age K. "Three-Way Analysis: Problems and Prospects." *Chemometrics and Intelligent Laboratory Systems* 15, no. 2–3 (1992): 143–57.

Smilde, Age K., and Durk A. Doornbos. "Three-Way Methods for the Calibration of Chromatographic Systems: Comparing PARAFAC and Three-Way PLS." *Journal of Chemometrics* 5, no. 4 (1991): 345–60.

Standards Library for Automation and Control: Guidelines for Quality, Safety, & Productivity. 2nd ed., CD. Research Triangle Park, NC: ISA (International Society of Automation), 2006.

Strang, Gilbert. *Linear Algebra and Its Applications.* 2nd ed. New York: Academic Press, 1980.

Stuart, Christopher, Rehman Fazeem, and Gregory K. McMillan. "Improve pH Control." *Chemical Processing* (March 2016). https://www.chemicalprocessing.com/articles/2016/improve-ph-control/.

Tinoco Jr., Ignacio, Kenneth Sauer, and James C. Wang. *Physical Chemistry Principles and Applications in Biological Sciences.* 2nd ed. Englewood Cliffs, NJ: Prentice-Hall, 1985.

Tomasi, Giorgio, Frans van den Berg, and Claus Andersson. "Correlation Optimized Warping and Dynamic Time Warping as Preprocessing Methods for Chromatographic Data." *Journal of Chemometrics* 18, no. 5 (2004): 231–41.

Ude, Christian, Jörg Schmidt-Hager, Michael Findeis, Gernot Thomas John, Thomas Scheper, and Sascha Beutel. "Application of an Online-Biomass Sensor in an Optical Multisensory Platform Prototype for Growth Monitoring of Biotechnical Relevant Microorganism and Cell Lines in Single-Use Shake Flasks." *Sensors* 14 (2014): 17390–17405.

Valle-Cervantes, Sergio. "Plant-Wide Monitoring of Processes Under Closed-Loop Control." PhD diss., Chemical Engineering Department, University of Texas at Austin, Austin, TX, 2001.

Valle, Sergio, Weihua Li, and S. Joe Qin. "Selection of the Number of Principal Components: The Variance of the Reconstruction Error Criterion with a Comparison to Other Methods." *Industrial & Engineering Chemistry Research* 38, no. 11 (1999): 4389–4401.

van Sprang, Eric N. M., Henk-Jan Ramaker, Johan A. Westerhuis, Stephen P. Gurden, and Age K. Smilde, "Critical Evaluation of Approaches for On-Line Batch Process Monitoring." *Chemical Engineering Science* 57, no. 18 (2002): 3979–91.

Wachs, A., and D. R. Lewin. "Process Monitoring Using Model-Based PCA." *Proceedings of IFAC Symposium on Dynamics and Control of Process Systems.* Corfu, Greece (1998): 86–92.

Westerhuis, Johan A., Theodora Kourti, and John F. MacGregor. "Comparing Alternative Approaches for Multivariate Statistical Analysis of Batch Process Data." *Journal of Chemometrics* 13, no. 3–4 (1999): 397–413.

Wilson, Grant, McMillan, Gregory K. and Boudreau, Michael A., "PAT Benefits from the Application of Modeling and Advanced Process Control of Bioreactors," *Emerson Global Users Exchange,* Nashville (October 2005).

Wise, Barry M., and Neal B. Gallagher. "The Process Chemometrics Approach to Process Monitoring and Fault Detection." *Journal of Process Control* 6, no. 6 (1996): 329–348.

Wold, Svante, Nouna Kettaneh, Hakan Friden, and Andrea Holmberg. "Modelling and Diagnostics of Batch Processes and Analogous Kinetic Experiments." *Chemometrics and Intelligent Laboratory Systems* 44, no. 1–2 (1998): 331–40.

Yu, Cheng-Ching, and William L. Luyben. "Analysis of Valve-Position Control for Dual Input Processes." *Industrial Engineering Chemical Fundamentals* 25, no. 3 (1986).

Yue, H. Henry, and S. Joe Qin. "Reconstruction-Based Fault Identification Using a Combined Index." *Industrial & Engineering Chemistry Research* 40, no. 20 (2001): 4403–4414.

Zhang, Yang, and Thomas F. Edgar. "Application of MB-PCA and Golden Batch to Bioreactor Monitoring and Control." Presentation at TWMCC meeting, Austin, TX, 2006.

Ziegler, John G., and Nathaniel B. Nichols. "Process Lags in Automatic Control Circuits." *ASME Transactions* (1943): 433–444.

Index

A

C

D

E

F

G

H

I

N

O

P

Q

R

T

Printed in the USA/Agawam, MA
September 17, 2026
910661.229